FOR REFERENCE ONLY

Fish Processing Technology

2nd
BPP, 1997   0751402737

Fish Processing Technology

JOIN US ON THE INTERNET VIA WWW, GOPHER, FTP OR EMAIL:

WWW: http://www.thomson.com
GOPHER: gopher.thomson.com
FTP: ftp.thomson.com
EMAIL: findit@kiosk.thomson.com

A service of I(T)P®

Fish Processing Technology

Second edition

Edited by

G.M. HALL
Lecturer
Food Engineering and Biotechnology Group
Loughborough University

BLACKIE ACADEMIC & PROFESSIONAL
An Imprint of Chapman & Hall
London · Weinheim · New York · Tokyo · Melbourne · Madras

Published by Blackie Academic and Professional, an imprint of
Chapman & Hall, 2–6 Boundary Row, London
SE1 8HN, UK

Chapman & Hall, 2–6 Boundary Row, London SE1 8HN, UK

Chapman & Hall GmbH, Pappelallee 3, 69469 Weinheim, Germany

Chapman & Hall USA, 115 Fifth Avenue, New York, NY 10003, USA

Chapman & Hall Japan, ITP-Japan, Kyowa Building, 3F, 2-2-1 Hirakawacho, Chiyoda-ku, Tokyo 102, Japan

DA Book (Aust.) Pty Ltd, 648 Whitehorse Road, Mitcham 3132, Victoria, Australia

Chapman & Hall India, R. Seshadri, 32 Second Main Road, CIT East, Madras 600 035, India

First edition 1992
Second edition 1997

© 1997 Chapman & Hall

Typeset in 10/12pt Times by Thomson Press (India) Ltd, New Delhi
Printed in Great Britain by TJ Press Ltd, Padstow, Cornwall

ISBN 0 7514 0273 7

Apart from any fair dealing for the purposes of research or private study, or criticism or review, as permitted under the UK Copyright Designs and Patents Act, 1988, this publication may not be reproduced, stored, or transmitted, in any form or by any means, without the prior permission in writing of the publishers, or in the case of reprographic reproduction only in accordance with the terms of the licences issued by the Copyright Licensing Agency in the UK, or in accordance with the terms of licences issued by the appropriate Reproduction Rights Organization outside the UK. Enquiries concerning reproduction outside the terms stated here should be sent to the publishers at the London address printed on this page.

The publisher makes no representation, express or implied, with regard to the accuracy of the information contained in this book and cannot accept any legal responsibility or liability for any errors or omissions that may be made.

A catalogue record for this book is available from the British Library

∞ Printed on acid-free text paper, manufactured in accordance with ANSI/NISO Z39.48-1992 (Permanence of Paper)

Contents

List of contributors		ix
Preface		xi

1 Biochemical dynamics and the quality of fresh and frozen fish   1
R.M. LOVE

1.1	Introduction	1
1.2	Sequential changes during the spawning cycle	1
1.3	The condition of fish	3
1.4	The role of body constituents in governing fish quality and processability	4
	1.4.1 Lipids	4
	1.4.2 Proteins	10
	1.4.3 Carbohydrates	13
	1.4.4 Pigmentation	22
	1.4.5 Flavour compounds	24
	1.4.6 Minerals	24
1.5	Summary of considerations of biological condition and quality	24
References		26

2 Preservation of fish by curing (drying, salting and smoking)   32
W.F.A. HORNER

2.1	Introduction	32
2.2	Water content, water activity (a_w) and storage stability	32
	2.2.1 Basic definitions	33
	2.2.2 Water activity and microbial spoilage	35
	2.2.3 Water activity and water relationships in fish	36
	2.2.4 Water relationships, preservation and product quality	40
2.3	Drying	42
	2.3.1 Air or contact drying	43
	2.3.2 Drying calculations	46
2.4	Salting	54
	2.4.1 Water activity and shelf-life	54
	2.4.2 The salting process	55
	2.4.3 Storage: maturing and spoilage	59
	2.4.4 Other salted fish products	61
2.5	Smoking	62
	2.5.1 Introduction: preservation, titivation or camouflage	62
	2.5.2 Smoke production	63
	2.5.3 Quality, safety and nutritive value	66
	2.5.4 Processing and equipment	68
References		72

3 Surimi and fish-mince products   74
G.M. HALL and N.H. AHMAD

3.1	Introduction	74
3.2	Fish-muscle proteins	75
	3.2.1 Nature of muscle proteins	75
	3.2.2 Properties of actin and myosin	76

		3.2.3 The action of salt	78
		3.2.4 Surimi-based products	79
	3.3	The surimi process	80
		3.3.1 Basic concepts	80
		3.3.2 Process elements	82
		3.3.3 Appropriate species for surimi production	85
		3.3.4 Quality of surimi products	86
		3.3.5 Microbial aspects of surimi	87
	3.4	Fish mince	88
		3.4.1 Sources of raw material	88
		3.4.2 Fish-mince products	89
		3.4.3 Comparison of surimi and fish-mince products	89
	References		90

4 Chilling and freezing of fish — 93
G.A. GARTHWAITE

	4.1	Introduction	93
		4.1.1 Relationship between chilling and storage life	93
		4.1.2 Relative spoilage rates	94
	4.2	Modified-atmosphere packaging (MAP)	95
		4.2.1 Introduction	95
		4.2.2 Modified-atmosphere packaging systems	95
	4.3	Freezing	98
		4.3.1 General aspects of freezing	98
		4.3.2 Prediction of freezing times by numerical methods	100
		4.3.3 Freezing systems	103
	4.4	The application of freezing systems in fish processing	108
		4.4.1 Freezing on board	108
		4.4.2 Onshore processing	111
	4.5	Changes in quality on chilled and frozen storage	113
		4.5.1 Chilled storage	113
		4.5.2 Frozen storage	114
		4.5.3 Thawing	116
	References		117

5 Canning fish and fish products — 119
W.F.A. HORNER

	5.1	Principles of canning	119
		5.1.1 Thermal destruction of fish-borne bacteria	119
		5.1.2 Thermal processing: quality criteria	129
		5.1.3 Storage of canned fish	134
		5.1.4 Choice of heat process	134
	5.2	Design of packaging for fish products	135
		5.2.1 Glass jars	135
		5.2.2 Flexible containers	135
		5.2.3 Rigid metal containers	137
		5.2.4 Rigid plastic containers	138
		5.2.5 Labelling	139
	5.3	Process operations and equipment	139
		5.3.1 Pre-processing operations	139
		5.3.2 Exhausting	143
		5.3.3 Heat processing and heat-processing equipment	146
		5.3.4 Post-process operations	152
	5.4	Cannery operations for specific canned-fish products	153
		5.4.1 Small pelagics	153
		5.4.2 Tuna and mackerel	155

		5.4.3 Crustacea and molluscs	155
		5.4.4 New canned-fish products	156
	References		158

6 Methods of identifying species of raw and processed fish   160
I.M. MACKIE

6.1	Introduction		160
6.2	Requirements for non-sensory methods of fish species identification		164
6.3	Principles of electrophoresis and isoelectric focusing		166
	6.3.1	Electrophoretic systems	167
	6.3.2	Separation systems	169
6.4	Fish flesh proteins		172
	6.4.1	Structure of muscle	172
	6.4.2	Structure of myofibrils	173
	6.4.3	Muscle proteins	173
6.5	Experimental procedures for electrophoretic methods		176
	6.5.1	Raw fish flesh	176
	6.5.2	Cooked but not autoclaved fish	184
	6.5.3	Heat-sterilised and autoclaved products	188
6.6	Alternative protein-based methods of fish species identification		190
	6.6.1	Immunoassay procedures	190
	6.6.2	Capillary electrophoresis	190
	6.6.3	HPLC	191
6.7	DNA techniques of fish species identification		191
6.8	Fish eggs		192
6.9	General conclusions		196
References			197

7 Modified-atmosphere packaging of fish and fish products   200
A.R. DAVIES

7.1	Introduction		200
7.2	Microbial flora of fresh fish		202
7.3	Pathogenic flora of fresh fish		203
	7.3.1	*Clostridium botulinum*	205
	7.3.2	Other pathogens	206
7.4	Present applications of MAP to fish and fish products		206
7.5	Experimental approach		207
	7.5.1	Introduction	207
	7.5.2	Materials and methods	208
	7.5.3	Results	210
	7.5.4	Discussion	213
7.6	Future developments		214
	7.6.1	Combination treatments	215
	7.6.2	Predictive/mathematical modelling	215
	7.6.3	Intelligent packaging	217
	7.6.4	Developments in packaging films and equipment	218
	7.6.5	Quality assurance of MAP	219
Acknowledgements			220
References			220

8 HACCP and quality assurance of seafood   224
M. DILLON and V. McEACHERN

8.1	Introduction	224
8.2	Defining HACCP	225

8.3		Application of QMP	228
	8.3.1	ISO 9002 elements not addressed by QMP	228
	8.3.2	ISO 9002 elements partially addressed by QMP	229
8.4		Practical aspects of planning and implementing HACCP systems	229
8.5		HACCP verification	238
	8.5.1	Defect definitions	240
8.6		Future developments of seafood quality systems	243
		References	246

9 Temperature modelling and relationships in fish transportation   249
C. ALASALVAR and P.C. QUANTICK

9.1		Introduction	249
9.2		Transportation of fish	250
	9.2.1	Road transportation	250
	9.2.2	Air transportation	251
	9.2.3	Sea transportation	252
9.3		Containers and cooling gels	252
	9.3.1	In developing countries	253
	9.3.2	In developed countries	255
	9.3.3	Use of cooling gels in fish transportation	255
9.4		Safety, quality and spoilage of fish during transportation	257
	9.4.1	Effect of temperature on the growth of micro-organisms during transportation	257
	9.4.2	Temperature control and legislation in fish transportation	259
	9.4.3	Application of HACCP in seafood	260
	9.4.4	Factors affecting the shelf-life of fish	262
9.5		Types of predictive modelling in fish transportation	263
	9.5.1	Time–temperature function integrators and rate of spoilage	263
	9.5.2	Heat transfer/mathematical approach	265
	9.5.3	Computer modelling of time–temperature	275
9.6		Food MicroModel	282
	9.6.1	Types of model in Food MicroModel	282
	9.6.2	Use of Food MicroModel in fish transportation	283
9.7		Conclusion	284
		References	284

Index   **289**

Contributors

N.H. Ahmad	Department of Chemical Engineering, Loughborough University, Ashby Road, Loughborough, Leicestershire LE11 3TU, UK
C. Alasalvar	Food Research Centre, University of Lincolnshire and Humberside, Humber Lodge, 61 Bargate, Grimsby DN34 5AA, UK
A.R. Davies	Department of Food Microbiology, Leatherhead Food Research Association, Randalls Road, Leatherhead, Surrey KT22 7RY, UK
M. Dillon	Midway Technology, 14 Farndon Road, Woodford Halse, Northants NN11 6TT, UK
G.A. Garthwaite	School of Applied Science & Technology, University of Lincolnshire and Humberside, Humber Lodge, 61 Bargate, Grimsby DN34 5AA, UK
G.M. Hall	Department of Chemical Engineering, Loughborough University, Ashby Road, Loughborough, Leicestershire LE11 3TU, UK
W.F.A. Horner	University of Hull International Fisheries Institute, Cottingham Road, Hull HU6 7RX, UK
R.M. Love	East Silverburn, Kingswells, Aberdeen AB1 8QL, UK
V. McEachern	Quality Management Program, Inspection Service Branch, Dept of Fisheries & Oceans, 200 Kent Street, Ottawa, Ontario, Canada K1A 0E6
I.M. Mackie	CSL Food Science Laboratory, PO Box 31, 135 Abbey Road, Aberdeen AB9 8DG, UK
P.C. Quantick	Food Research Centre, University of Lincolnshire and Humberside, Humber Lodge, 61 Bergate, Grimsby DN34 5AA, UK

Preface

As with the first edition this book includes chapters on established fish processes and new processes and allied issues. The first five chapters cover fish biochemistry affecting processing, curing, surimi and fish mince, chilling and freezing and canning. These established processes can still show innovations and improved theory although their mature status precludes major leaps in knowledge and technology.

The four chapters concerned with new areas relevant to fish processing are directed at the increasing globalisation of the fish processing industry and the demands, from legislation and the consumer, for better quality, safer products. One chapter reviews the methods available to identify fish species in raw and processed products. The increased demand for fish products and the reduced catch of commercially-important species has lead to adulteration or substitution of these species with cheaper species. The ability to detect these practices has been based on some elegant analytical techniques in electrophoresis. A second chapter describes work in modified atmosphere packaging with emphasis on pathogenic organisms including these which are just emerging into our consciousness. The following chapter describes the application of hazard analysis critical control point (HACCP) into fish processing management. As fish processing becomes more sophisticated and located nearer to the catching grounds the processors, in developing and developed countries, must be able to show compliance with the hygiene regulations of their export markets. The importance of HACCP as a management tool is increasing in the fishery sector and this chapter describes its application. Finally, reflecting again the increase transportation of fish to distant markets, there is a chapter on temperature relationships and fish quality. The chapter indicates the success of temperature monitoring schemes in predicting quality changes during transportation but also includes information on simple heat transfer calculations which can be done to estimate, for example, ice usage in less sophisticated distribution systems.

Finally, as with the previous edition we have tried to emphasise quality aspects throughout. This edition also shows that product innovation and increased trading raise new opportunities (or problems?) for the technologist to solve.

G.H.

1 Biochemical dynamics and the quality of fresh and frozen fish
R.M. LOVE

1.1 Introduction

Unlike pure chemical substances, which always have the same composition, the musculature of a fish enfolds a variety of constantly changing interactive systems. The balance between these systems can vary widely without causing the death of the fish but, after capture and killing, the variations are often found to have influenced the acceptability of the flesh as food for human consumption. They can also affect its suitability for processing.

The variations, their causes, and their significance for the food industry form the basis of this chapter. Their quantification, especially by the simultaneous measurement of two or more parameters, has great potential in assessing the biological 'condition' of fish, and as the chapter continues, the aim is to highlight parameters that might be useful in this connection.

Some changes in the biochemistry of the musculature are brought about by environmental influences, but the most radical recasting results from the spawning cycle and its attendant depletion. Since eggs and sperm are usually shed at a season when the natural food supply is optimal for the development of the larvae, rather than for the health of the parent fish (Sundararaj *et al.*, 1980), it follows that many fish perforce synthesise large amounts of germinal tissue within their bodies during periods when food is scarce. At such times, the food supply may even be insufficient to satisfy the requirements of ordinary metabolism or physical activity. The problem is solved within the fish by plundering existing stores or potential stores of energy, sometimes to an extreme degree.

The manner in which such resources are mobilised can vary quite considerably between different species, therefore it is difficult to formulate general principles. Observations made on one species cannot be extrapolated to others; nevertheless many investigations have been carried out on single species. Throughout this chapter, therefore, an incomplete, rather than a comprehensive, scene will be reviewed; gaps will be apparent, and some conclusions must be tentative.

1.2 Sequential changes during the spawning cycle

The different energy reserves are not mobilised simultaneously but in a sequence that changes with the progress of depletion. In general, lipids are mobilised first (Nagai and Ikeda, 1971; later authors listed by Love, 1980, p. 182) but the pattern varies according to species. In herrings and similar fatty fish, most of the lipid reserves are found in the flesh and begin to decrease from the outset of depletion. In contrast, cod and other non-fatty species carry most of their lipid reserves in their livers, and consequently little change occurs in the flesh for some time.

Proteins, because of their structural importance, are mobilised from the flesh late in the depletion process and are the first to be restored after the completion of spawning (Black and Love, 1986).

Detailed studies on the depletion of energy from cod experimentally starved are summarised in Figure 1.1. It can be seen that the lipid from the liver, and glycogen from both liver and white muscle, are all mobilised from the outset. Dark muscle, like heart muscle, is almost continuously active in fish. It is therefore more important than the large bulk of white muscle, which is used only intermittently for vigorous pursuit or escape (Boddeke *et al.*, 1959). Its glycogen level during depletion is preserved for a considerable time and only mobilised when the protein structures also begin to be broken down.

Figure 1.1 does not consider the lipids of cod flesh; however, the total concentration in the white muscle is only about 0.5%. Of this, only about 1% is readily mobilised (triacylglycerols), the rest being phospholipids which are essential components of the cellular structures (Ross, 1977). Consequently, appreciable breakdown of white muscle lipids occurs in cod only when the actual contractile protein structures begin to disintegrate.

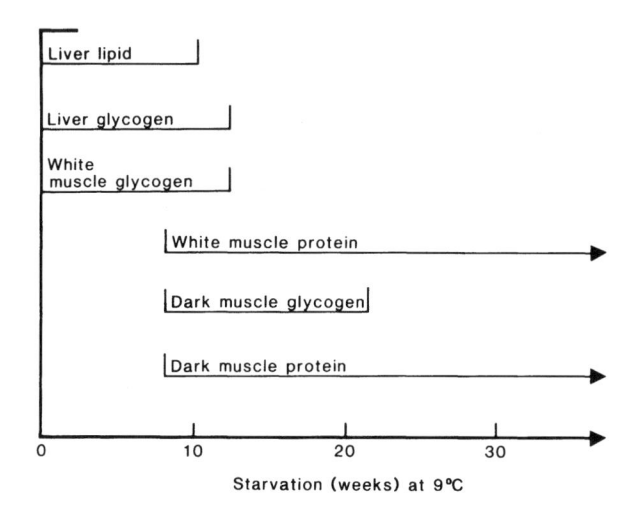

Figure 1.1 A diagrammatic representation of the beginning and end of mobilisation of the main energy reserves in cod starved at 9°C. Time values are approximate. From Love and Black (1986) by courtesy of Springer.

In cod, liver glycogen and white muscle glycogen decrease together, but in carp (Murat, 1976) and goldfish (Chavin and Young, 1970) the glycogen levels are preserved during long periods of depletion – energy being supplied by lipids and proteins. In starving eels, the protein reserves are drawn upon at a greater rate than are the lipid reserves, although the two reserves contribute energy at the same levels (Boëtius and Boëtius, 1985). Doubtless, further species differences will come to light in the future.

1.3 The condition of fish

Objective measurements have long been used by biologists to try to assess the nutritional 'condition' of fish. This concept is closely linked to the acceptability of the fish as food, so is also of interest here. The trouble is that no single measurement on its own can describe nutritional condition adequately, and can be misleading without the support of other measurements.

The 'weight/length ratio' ($W/L^3 \times 100$) gives a figure for visible emaciation, but is not realistic in non-fatty species. Firstly, Figure 1.1 showed that only minor components are removed from the muscle of cod for a considerable period, while liver lipids are steadily utilised. Secondly, even when protein is being removed from the flesh, much of the volume of the flesh is retained by a corresponding incursion of water, so that, in this species, the water and protein contents form an inverse relationship (Love, 1970, figure 85), as with water and lipids in the herring (Brandes and Dietrich, 1958). The fish gradually appear thinner in advanced depletion (Love, 1988, figure 38), but the running down of energy reserves is greatly underestimated by weight/length measurements. Such measurements may be more useful in estimating the condition of herrings and other fatty fish because mobilisation of lipids from their flesh occurs from the start of depletion.

Despite their disadvantages, measurements of the weight/length ratio are still popular. The reason is that their simplicity enables the investigator to examine large numbers of fish without using sophisticated apparatus. Bolger and Connolly (1989) reviewed many papers on the statistical evaluation of weight/length measurements, concluding that most of the trouble arose through the data being analysed regardless of the assumptions on which the method was based.

The gonadosomatic index (weight of gonads as a proportion of the whole fish weight) and hepatosomatic index (weight of liver as a proportion of the whole fish weight) both give some information – the latter being quite useful in species with much of their energy stored in the liver. The water content of cod muscle (Love, 1960) gives a good idea of its protein loss, but misses the early stages of depletion where liver energy reserves are being utilised; the lag period for cod muscle depletion is as much as 9 weeks at 9°C (Love, 1969) and would be still longer at lower temperatures.

None of these methods can tell us that a fish has been depleted but is now recovering, or that depletion is actively in progress. However, Love (1980, figure 139) showed that when cod are starving, the gall bladder is large and blue, whereas during active feeding it becomes small and yellow. A cod with intense blue bile has been starving for at least 3 days. Several constituents of fish blood (*e.g.* lipids, cortisol, glucose) have been shown to decline within the first 2 days of starvation (White and Fletcher, 1986: plaice, *Pleuronectes platessa*), while the concentration of free fatty acids increases (Black, 1983; White and Fletcher, 1986). It has also been observed by Heming and Paleczny (1987: brook trout, *Salvelinus fontinalis*) that the concentration of ketone bodies in the skin mucus is positively related to the duration of starvation.

Combined with other observations, signs such as these might, on further investigation, begin to give a fuller picture of the state of the fish and its suitability as food.

1.4 The role of body constituents in governing fish quality and processability

1.4.1 *Lipids*

Lipids are the most concentrated form of energy stored in the fish, and it is no coincidence that active species such as salmon, tuna or herring carry more lipids than less-active species such as cod or plaice.

They occur in fish as two broad groups. The first consists of triacylglycerols (triglycerides), and is the main form in which energy resources are stored. The lipids are often observable as actual globules of oil that have accumulated in the flesh, liver and, in some species, around the intestine also. The second lipid group, mostly phospholipids and cholesterol, is an essential component of cell walls, mitochondria and other sub-cellular structures. Consequently, it cannot be readily drawn on to supply energy and, in cod at least, its mobilisation coincides with the breakdown of actual contractile proteins.

The lipids in the edible part of fish are important to the food scientist in three respects. Firstly, any oily deposits noticeably influence the sensation of the cooked flesh in the mouth of the eater. Herrings, for example, when well-fed and fat-rich, taste very smooth and succulent ('juicy'), although the sensation is produced by oil, not water. After spawning, when the oil is at its lowest level, the main sensation is of dryness or fibrousness; perhaps 'rough' or 'coarse' describes it better – at any rate the taste is disappointing.

Secondly, fish lipids, as is now widely recognised, are very beneficial to the health of the consumer. In cases of myocardial infarcts, patients put on a diet of fatty fish appear to have a greatly reduced likelihood of a recurrence, and atherosclerosis is reduced (Lands, 1986). When Eskimos and Japanese used fish as the main part of their food intake, they almost never suffered from heart attacks (Dyerberg and Bang, 1979). Many other diseases, such as rheumatoid arthritis and even cancer, appear to be alleviated by eating fish oils (reviewed by Drevon, 1989).

The beneficial substances in fish oils are the polyunsaturated fatty acids, especially eicosapentaenoic acid, which has 20 carbon atoms in the chain and 5 double-bonds (written 20:5), and also the fatty acid dodecahexaenoic acid (22:6). Both acids belong to the n-3 series, that is, with the first unsaturated linkage at the third carbon atom along the carbon chain from the methyl group.

Finally, flesh lipids contribute to the flavour of the fish. The lipids themselves have a slight taste, but of greater importance is their propensity to develop an off-flavour in the frozen state. This is caused by atmospheric oxidation, especially of the unsaturated phospholipids. Each of these aspects is now considered in turn.

1.4.1.1 *Oiliness of the flesh in relation to the spawning cycle.* The oiliness of the flesh of fatty species is linked to the time of spawning and varies in a regular annual cycle. Lipids are deposited during a feeding period when the gonads are inactive, and still continue to be deposited as they start to develop. Beyond a certain stage of gonadal development, the rate at which lipids are transferred to the gametocytes exceeds the dietary intake and stocks run down steadily thereafter. There appears to be further depletion for a while after spawning is completed (Campbell and Love, 1978: haddock, *Melanogrammus aeglefinus*; Goldenberg *et al.*, 1987: hake, *Merluccius hubbsi*).

There is a difference between the sexes that modifies the oiliness of the flesh or liver. This stems, in part, from the greater size of mature female gonads compared with male, for example 18% of the body weight compared with 4.2%, respectively, in the flounder, *Pleuronectes flesus* (Ziecik and Nodzynski, 1964). The mature gonads of a male goby (*Gobius melanostomus*) contain in total only about 10% of the lipids of a corresponding female goby, so require little of the stored lipids during maturation (Chepurnov and Tkachenko, 1973). Shatunovskii and Novikov (1971) found that more lipids are removed from the muscle of female trout (*Salmo trutta*) than from that of the male during maturation, and the female mackerel (*Scomber scombrus*) has been shown to be the more depleted with regard to flesh lipids (Ackman and Eaton, 1971).

Corresponding to their greater need for lipids at the spawning time, female fish appear to accumulate more lipid reserves during the feeding season. However, to the author's knowledge, all published observations relate to species that store their lipids in the liver, rather than the flesh (reviewed by Love, 1980). It is probably not established that the flesh of fatty species is actually oilier in females than males during the run-up to spawning.

Although male fish require relatively little lipid material for their developing gonads, they are more physically active than the females, both in the sexual act and in fighting each other. This observation is well known (J.A. Lovern, personal communication), and is supported by the fact that the number of circulating red blood cells is higher in mature males than mature females or immature fish (Pottinger and Pickering, 1987: brown trout, *Salmo trutta*). This may explain why Baltic cod (*Gadus callarias*) females have been reported to withdraw lipids from their reserves while males withdraw mostly glycogen (Bogoyavlenskaya and Vel'tishcheva, 1972). However, much needs to be done to establish the relative succulence of male and female fatty fish flesh. Love (1980) summed up the literature on the subject as showing that female fish lay down stores of lipids for transfer to the ovary, while males mobilise both lipid and glycogen as fuel for physical activity.

1.4.1.2 *Fish lipids and human health.* When discussing the beneficial effects of fish lipids, it must be remembered that the proportions of the various polyunsaturated fatty acids in fish muscle are not constant: we are again dealing with a dynamic system. The lipid composition of the food eaten by the fish is probably the most important influence on the lipid composition of the fish itself (Lovern, 1935). Worthington and Lovell (1973) concluded that it accounts for 93% of the variance in the fatty acid composition of channel catfish (*Ictalurus punctatus*) – genetic and other factors accounting for the remainder.

The extent to which polyunsaturated fatty acids can be synthesised by the fish from less unsaturated fatty acids in the diet varies with the species. Chinook salmon (*Oncorhynchus tshawytscha*) grow very slowly on a fat-free diet but recover a normal growth rate completely when fed only fatty acid 18:2 (Lee and Sinnhuber, 1973). Rainbow trout (*Salmo gairdneri*) can produce substantial quantities of fatty acids 20:3, 22:5 and 22:6 when fed only 18:2 and 18:3 (Owen et al., 1975). On the other hand, turbot (*Scophthalmus maximus*) can convert only 3–15% of labelled precursors into fatty acids of longer chain length and cannot increase their unsaturation (*idem*). The same authors suggested that turbot in the ocean would receive adequate polyunsaturated fatty acids in their diet, which they therefore have no need to modify. Similarly, Ross (1977) showed that the elongation (addition of carbon atoms to the chain) and desaturation (increase in the number of double bonds) of 18:3 fatty acid administered to another marine teleost, the cod (*Gadus morhua*), were both slight. Where fish are cultured for human consumption, therefore, it is sensible to ensure that fresh marine oils are used as the basis of their dietary lipids, and that they are not admixed with vegetable oils, which are deficient in the n-3 series of polyunsaturates (Sargent, 1989). Futhermore, if the marine oils have oxidised before being fed to cultured fish, they cause pathological symptoms (Ono et al., 1960).

The annual cycle of water temperature also has an important influence on lipid unsaturation. Phospholipids are, as already stated, important constituents of cell membranes and, as their polyunsaturation increases, the melting point of the lipid mixture is lowered. This phenomenon appears to be central to the control of the flexibility and motility of cells so that they do not become rigid at lower temperatures. Farkas and Herodek (1964) observed that the unsaturation of the lipids of crustacean plankton increased in winter and decreased in summer, and that it changed to a greater extent in plankton from a small lake than from a large lake because of the wider fluctuations in temperature. The phospholipids of tropical fish are more saturated than those from cooler water (Gopakumar and Nair, 1972; Irving and Watson, 1976), and Kemp and Smith (1970) showed that raising the environmental temperature by 20°C actually halved the quantity of 20:4 and 22:6 in the lipids of goldfish (*Carassius auratus*), and doubled the quantity of the (fully saturated) 18:0 fatty acid. The changes were complete in 3 or 4 days (Smith and Kemp, 1971), so there is no doubt that they occur within the fish by enzymic activity rather than by a changed diet. Several other authors have studied this interesting phenomenon, and their studies and conclusions are reviewed by Love (1970, pp. 216, 217 and 1980, pp. 339, 340).

Finally, cultured salmon can be more beneficial to the eater than are their wild counterparts. This stems from the fact that they contain a greater quantity of lipids, and hence a greater absolute quantity of n-3 fatty acids per unit weight of muscle (Thomassen and Austreng, 1987, cited by Skjervold, 1989).

1.4.1.3 *The development of rancidity in frozen fish.* Fish that are frozen and cold-stored gradually develop an off-flavour and off-odour, which have been likened to boiled clothes, wet cardboard, cold tea, etc. In the case of very oily fish such as herring or mackerel, the eater, unless experienced in tasting cold-stored fatty fish, does not immediately think that what he or she is eating is rancid. More usually, the fish simply tastes more 'oily' than usual and the oiliness is subtly unpleasant.

In cod, the compound responsible for the off-flavour is *cis*-4-heptenal (McGill, 1974; McGill *et al.*, 1974). McGill (personal communication) regards its origin in cod (but not in fatty fish) largely as the oxidation, by atmospheric oxygen, of the polyunsaturated fatty acids in phospholipids. In cod at least, this means the oxidation of 22:6 (the fatty acid which comprises over 40% of cod white muscle lipids) and 20:5 (which comprises 16%), as other polyunsaturates are present in much smaller amounts (Ross, 1977). If fish dry out in the cold-store the oxygen reaches the susceptible fatty acids much more readily, enhancing the development of cold-store flavour (Hardy and McGill, 1990: fish of the cod family).

Cod and other non-fatty species. Although the muscle of cod (*Gadus morhua*) contains only about 0.5% of lipids, it soon develops a strong undesirable taint during frozen storage, not only because of the very large proportion of 22:6 and 20:5 but also because over 82% of the total lipids are phospholipids (Ross, 1977).

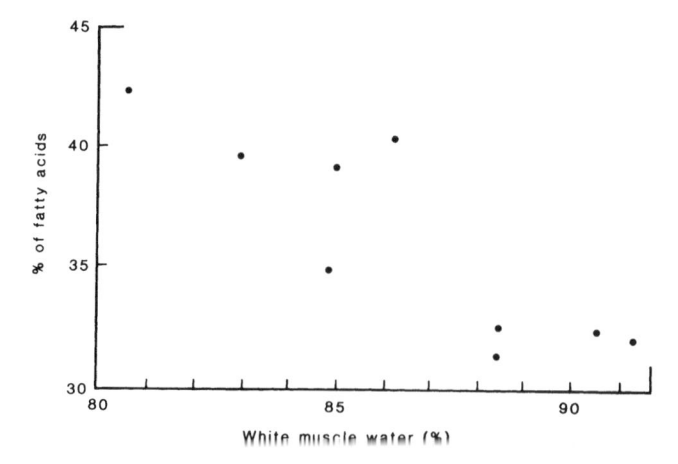

Figure 1.2 The decrease in the proportion of docosahexaenoic acid (C22:6n3) in the total white-muscle lipids of starving cod. The increase in depletion is monitored by the increase in water content of the muscle. After Ross (1977), from Love (1988) by courtesy of Farrand Press.

When cod starve, the proportion of polyunsaturated fatty acids decreases in the muscle lipids, the greatest decrease occurring in 22:6 (Figure 1.2). In this figure, the progress of starvation is monitored by the increase in muscle water content, which is approximately equivalent to the extent of removal of protein (Love, 1970, figure 85). It is possible that the preferential disappearance of 22:6 indicates only the physical breakdown and removal of sub-cellular structures incorporating it, but the polyunsaturates may also be destroyed by catabolism over the starvation period, not being replaced by dietary polyunsaturates (Ross, 1977). Be this as it may, Ross and Love (1979) starved cod for 2 months in an aquarium, then cold-stored them at $-10°C$, a treatment which is known to cause rapid oxidation of the lipids. The results (Table 1.1) show that the fed controls tasted and smelled much worse than fish subjected to moderate starvation, which can easily occur in the wild. Also, much less *cis*-4-heptenal was produced in the starved fish.

There is a geographical corollary to these observations. Cod caught on the Faroe Bank (S.E. Faroe Islands) are unusually thick-bodied and contain very large, oily livers. The total lipids in the white muscle of this race are about 16% higher than in cod from, for example, the Aberdeen Bank off the east coast of Scotland (0.78% lipids compared with 0.67%, respectively, in autumn-caught fish (Love *et al.*, 1975)). Despite the relatively slight superiority of their lipid content, cod from the Faroe Bank developed far more off-flavour and off-odour than the cod from four other grounds, even after only 3 months' storage at $-30°C$ (Table 1.2). Such conditions normally yield fish of first-class eating-quality, but those from the Faroe Bank were actually rejected by the taste panel. Thus, cold-store off-flavour generated in cod undergoes a big decrease

Table 1.1 Taste panel assessment of off-odour and off-flavour developed in the muscle of fed and starved cod, frozen and stored at $-10°C$ for 5 or 10 weeks, then thawed and cooked. The higher the panel score, the poorer the quality. *Cis*-4-heptenal was determined on pooled samples of muscle from both 5 and 10 weeks' storage (nmol/1000 g wet weight). After Ross and Love (1979) by courtesy of Blackwell Scientific Publications

	Off-odour		Off-flavour		*Cis*-4-heptenal
	5 weeks	10 weeks	5 weeks	10 weeks	(pooled)
Fed controls (5 fish)	1.5	1.55	3.43	3.55	23.0
Starved (5 fish)	0.55	0.4	1.28	1.8	3.5
Difference	0.95	1.15	2.15	1.75	
Significance level	1%	5%	0.1%	5%	

Table 1.2 Taste panel assessment of off-odour and off-flavour developed in the muscle of cod caught in the spring of 1970 on different fishing grounds, stored for 3 months at $-30°C$, then thawed and cooked. The upper limit of commercial acceptability is a score of about 3, where $n = 8$. The means of Faroe Bank results were significantly different from those from other grounds for both odour ($P < 0.05$) and flavour ($P < 0.01$). After Love (1975) by courtesy of Environment Canada, Fisheries and Marine Service

Fishing ground	Map reference	Off-odour	Off-flavour
Aberdeen Bank	57-05N 01-15W	1.32 ± 0.51	1.68 ± 0.60
Faroe Bank	60-53N 08-20W	2.29 ± 0.84	3.02 ± 0.95
Faroe Plateau	62-34N 06-24W	0.91 ± 0.51	1.45 ± 0.63
S.E. Iceland	65-27N 13-08W	1.04 ± 0.35	1.70 ± 0.49
N.W. Iceland	65-35N 25-00W	0.84 ± 0.24	1.37 ± 0.37

due to seasonal depletion, but can be increased by fishing grounds characterised by rich feeding.

In the majority of these studies, differences between the effects of simple starvation, and of synthesis of eggs or sperm have not been identified. Experimental fish were starved in aquaria and it was assumed that gonad maturation would have the same effect. The results of one publication (Takama *et al.*, 1985) seem to settle the issue. Although changes in the proportions of the different fatty acids in the flesh of cod starved while synthesising sex products could not be distinguished from those starved with gonads surgically removed, significantly more 22:6 was removed from the *livers* of the cod that were generating gonads. According to the same authors, 22:6 is the most important fatty acid in the gonads of this species. The development of rancidity during the cold storage of cod flesh should therefore be the same for a given degree of depletion of 22:6, whether caused by maturation or starving.

Salmonids. A complex situation arises in the case of salmonids, where appreciable energy reserves of triacylglycerols are stored in the flesh. Since triacylglycerols contain much smaller quantities of polyunsaturated fatty acids than do phospholipids (Fraser et al., unpublished data, cited by Sargent et al., 1990), the depletion of lipids from the muscle by starvation results in a relative increase in polyunsaturation, not an absolute decrease as in cod (Ludovico-Pelayo et al., 1984). As the 'increase' is seen only through the removal of less-unsaturated lipids it is not surprising that starvation does not result in increased rancidity during subsequent frozen storage.

The situation is not straightforward. The cold-store off-flavour of rainbow trout is uniformly low after starvation, despite a wide range of 22:6 content; in contrast, the off-flavour score can be high where the trout are re-fed after starvation, despite uniformly low relative values for 22:6.

In seasonal studies by Mochizuki and Love (unpublished data, illustrated by Love, 1988, figures 45, 46), the least cold-storage flavour was detected in fish killed in April when 22:6 was maximal, while the reverse was true in September. Mochizuki and Love observed that both rainbow trout and Atlantic salmon (*Salmo salar*) developed much less cold-storage off-flavour and off-odour than cod stored for the same period. Noting that the triacylglycerol deposits in trout muscle were concentrated in the connective tissue sheets that wrap the blocks of muscle fibres, these workers regarded the phospholipids of trout muscle as being 'protected' by the film of triacylglycerols around them. In contrast to the findings in rainbow trout, no clear seasonal variation in the 22:6 content or the cold-store off-flavour has been found in the flesh of immature farmed Atlantic salmon (Mochizuki and Love, unpublished data). Further conjecture is unprofitable at present because of the number of variables involved.

Factors that influence lipid oxidation are reported by Burlakova et al. (1988), the fish species investigated being the whitefish, *Coregonus peled*. They found that the oxidisability of fish lipids correlated with the content of polyunsaturates, the content of phospholipids and the content in the latter of phosphatidyl ethanolamine and cardiolipin. As a complication, however, the natural antioxidants, tocopherol, ubiquinone and ubichromenol, were found to increase with the proneness to oxidation of the lipid substrate. They also noted that the lipids of red muscle were more oxidisable than those of white muscle, a phenomenon first noted by Banks (1938) in Atlantic herrings (*Clupea harengus*).

1.4.2 *Proteins*

Figure 1.1 showed that the proteins of cod muscle are utilised only when depletion is fairly far advanced. Red muscle and white muscle are eroded together but, in view of the more consistent use made of red muscle in swimming (Boddeke et al., 1959), its proteins are broken down less rapidly than those of white muscle (Black and Love, 1986).

The inverse relationship between the protein and water contents observed from analytical data on starving cod (Love, 1970, figure 85) is vividly illus-

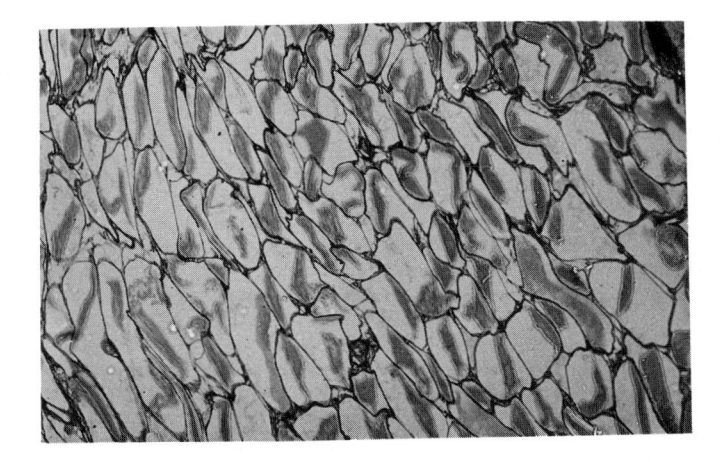

Figure 1.3 Cross-section of the muscle of cod starved to a water content of 95.3%. Black outlines are connective tissue; the remains of contractile tissue are grey shaded areas, which share with fluid the contents of former muscle cells. From Lavéty, unpublished (Crown Copyright). The bar below the photograph represents 100 μm.

trated in histological section (Figure 1.3). The water content of fully-nourished white muscle from cod appears to be 80.8% or less (Love, 1960). In such tissue, the contractile cells are packed tightly with very little extracellular fluid separating them, so that the greatest possible contractile power can be obtained (Best and Bone, 1973). Figure 1.3, however, illustrates an extreme case, in which the water content has risen to 95.3%. The outlines of connective tissue (intensely black in the picture) resemble those in nourished fish, but the contractile elements within them have been greatly reduced and replaced by fluid. A few 'cells' seem to contain no contractile material at all.

After such fish are filleted, much of the watery infill flows freely out, so the fillet rapidly shrivels and seems to be composed almost entirely of the very distinct connective tissue septa (myocommata). When cooked, the texture of such a fillet is so insubstantial that it can be sucked through the teeth without chewing. However, the cause of such repugnant texture is not solely the removal of protein. Provided that the *post mortem* pH of the flesh is constant, the progressive increase in water content, even to over 85%, affects the texture only slightly (Love *et al.*, 1974b). A more important factor is the pH, which rises at the same time. This phenomenon is dealt with in a later section.

Apart from textural considerations, however, the removal of proteins as described affects the quality, since the remaining fillet leaks and looks opaque.

How best can we assess this aspect of condition in a batch of fish? In the case of non-fatty species, the measurement of the water content is a good guide and, in some cases, it is even possible to measure the increasing opacity of the muscle itself to get a rough estimate of the extent of starvation (Love, 1962a). In fatty fish, however, an initial increase in the water content relates to a decrease in lipids and could be misleading.

What is really needed is a measure of the vigour with which the protein is being broken down; clearly as other resources are used up this will accelerate. Such a measure might help to fulfil another need – to know whether starving fish are actually getting better or still deteriorating.

Cellular lysis and tissue degeneration are closely linked with the activity of lysosomes, and the activity of acid phosphatase has often been employed as an index of lysosomal activity (De Duve, 1963). Figure 1.4 shows that there is a very close correlation between the activity of acid phosphatase and the water content of cod muscle.

In addition to this enzyme, Beardall and Johnston (1985) investigated the activities of acid proteinase, aryl sulphatase, acid ribonuclease and β-glucuronidase in saithe (*Pollachius virens*) starved for 66 days. With one exception in red muscle, all these lysosomal enzymes increased by 70–100% during starvation in red and white muscle. Another batch of fish (starved for 74 days) was re-fed, and these authors showed that the activities of acid proteinase and aryl sulphatase dropped to non-starved levels in as little as 10 days. Here, surely, is a superb new method for identifying the beginnings of recovery in severely starved fish.

Another possible marker for protein degradation is 3-methyl histidine, which has been investigated in this connection by, for example, Ward and Buttery (1978). It is said to be present in muscle in the free form only when muscle proteins are being catabolised. Ando and Hatano (1986) have shown

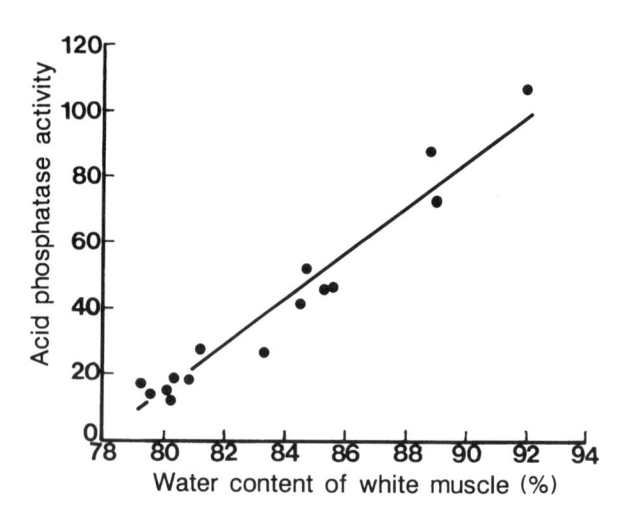

Figure 1.4 The activity of acid phosphatase in the muscle of increasingly starved cod (depletion shown by increasing water content). Activity is represented as μmol of n-nitrophenol released by the enzyme, per mg protein in 30 min, from p-nitrophenol phosphate. After Black (1983) by courtesy of Dr Darcey Black.

that the level markedly increases in chum salmon (*Oncorhynchus keta*) during spawning migration. Interestingly, the increase is especially marked in females (see p. 5). There is room for much further work here in the field of measurement of condition in fish.

In this section, the catabolism of myofibrillar protein to provide energy has been examined. There is no clear evidence of catabolism of the proteins of connective tissues, which seem to retain their integrity during starvation. The marked thickening observed in the myocommata of starving cod by Lavety and Love (1972) and Love *et al.* (1976) probably resulted from the addition of the empty collagen tubules (see Figure 1.3) to the surface of the myocommata during the isolation of the latter. Experiments by Love *et al.* (1982) with labelled proline failed to provide any positive evidence of enhanced collagen synthesis in starved cod.

1.4.3 Carbohydrates

1.4.3.1 The nature of carbohydrates. As in mammals, fish store most of their carbohydrate reserves in the liver. 'Resting' levels in muscle are much lower than in the liver, but red muscle is richer in carbohydrate than white muscle (several authors listed by Love, 1980, p. 73).

Carbohydrates are stored in the liver as glycogen, a polysaccharide built of glucose units. When required, for example to supply the energy for muscular work, the glycogen is broken down and transported by the blood stream to the appropriate site as glucose. On arrival, it may be used at once or temporarily re-converted into glycogen. Thus both glucose and glycogen are found in muscle, but only glucose is found in the blood.

The levels of reserves can be increased if fish are fed with a diet rich in carbohydrates (Tunison *et al.*, 1940: brook trout, *Salvelinus fontinalis*; Hochachka and Sinclair, 1962: rainbow trout). However, apart from eating the livers of prey and, in herbivorous species, vegetation, fish are not accustomed to consuming much carbohydrate. Metabolic disorders have been reported as a result of feeding massive amounts of carbohydrates to goldfish, *Carassius auratus* (Palmer and Ryman, 1972), and the proportion of dietary carbohydrate actually assimilated declines as its proportion in the diet increases (Cowey and Sargent, 1972). The main sources of energy in starved catfish (*Rhamdia hilarii*) are still lipids and proteins, even after adapting the fish to a high carbohydrate diet (Machado *et al.*, 1988).

The effects of carbohydrates on the growth of rainbow trout are unclear. Luquet *et al.* (1975) reported that where the diet is rich in proteins the growth is appreciably suppressed when sucrose is added as a supplement (the same amount of protein being ingested by experimental and control groups). Conversely, Kaushik *et al.* (1989) have found that high levels of various carbohydrates improve the availability of dietary energy and do not adversely affect overall growth or nutrient retention.

14.3.2 *Dynamics.* Since muscular activity uses glucose as its source of energy, active fish maintain higher levels of glucose in their blood than do sluggish fish (several authors reviewed by Love, 1970, p. 150). There is more glycogen in the red muscle of Atlantic salmon (*Salmo salar*) reared in a swimming raceway than in that of inactive salmon from a cage (Totland *et al.*, 1987). This has important consequences for the texture of cultured fish.

Carbohydrate reserves are drawn upon during maturation, since both glycogen and glucose accumulate in the growing ovaries of various species (Greene, 1926; Chang and Idler, 1960; Yanni, 1961). Maturing males, as already pointed out, also expend much carbohydrate in physical activity. Figure 1.1 showed that the glycogens of the liver and the white muscle decrease from the outset of starvation or the depletion associated with maturation. Black and Love (1986) showed that in cod, their concentrations are linked at all levels. Since an estimate of the carbohydrate reserves of a fish is another aspect of nutritional condition, it could be useful to know that the level of muscle glycogen indicates the level of the main reserve in the liver.

There is, however, a problem. Muscle glycogen is the main fuel for swimming activity, and during strenuous threshing about, as in capture, half of the reserves can be depleted in as little as 15 s (reviewed by Love, 1980, p. 423). Determinations of glycogen in the muscle of captured fish are therefore meaningless.

Nevertheless, the physical activity converts muscle glycogen into lactic acid, and the pH of the muscle falls. In mammals, such lactic acid is rapidly removed and transported to the liver for reprocessing, but for some reason fish muscle retains it whenever the muscular activity is stressful (Wardle, 1972: plaice, *Pleuronectes platessa*). After death, a proportion of any residual muscle glycogen is likewise converted to lactic acid, which lowers the pH further. The remainder is converted into glucose (Burt, 1966), which does not affect the pH. The proportions of the two end-products appear not to change under different circumstances.

Thus, the struggle of capture converts some muscle glycogen to lactic acid which remains in the muscle and, after death, a proportion of the remainder is also converted to lactic acid. Experiments by Love and Muslemuddin (1972) showed that, in a group of rested cod, it did not matter whether they were killed instantly or subjected to various periods of stress before killing: the pH of the muscle 24 h after death was always the same, varying only with the initial carbohydrate reserves of the fish. Black and Love (1988) established that the simple determination of the pH of the white muscle some 24 h after death is in fact a valid measure of the carbohydrate reserves of the fish, in this way providing us with another simple tool with which to study 'condition'. Changes in the *post mortem* pH of the muscle are also of great technological significance, and will be dealt with fully in a later section.

1.4.3.3 *Gluconeogenesis.* Carbohydrate as an energy source differs from protein and, to some extent, lipid in that it can be created from other substances within the body during starvation. In sockeye salmon (*Oncorhynchus nerka*) the quantity of liver glycogen doubles during the spawning migration upstream, although no food has been eaten (Chang and Idler, 1960). When eels (*Anguilla japonica*) starve in the summer, the concentration of glycogen in the liver falls but then rises again from gluconeogenesis as starvation continues (Inui and Yokote, 1974). Maksimovich (1988) noted that although the muscle proteins of starving Pacific salmon (*Oncorhynchus* sp.) are the major source of energy, the fish increase their secretion of insulin and their activity of glycolytic enzymes so as to utilise the glucose 'generated in the fish organism during endogenous feeding'.

This phenomenon is not universal. Fifteen per cent of the weight of the livers of male lampreys (*Petromyzon marinus*) at the beginning of spawning migration consists of glycogen and in this species it is all used up by the time the fish have reached the spawning ground (Kott, 1971).

In contrast to the increase in insulin secretion observed in Pacific salmon during starvation (Maksimovich, 1988), Ross (1977) found that the plasma insulin levels of starved cod (*Gadus morhua*) were less than half those of cod in which feeding had been resumed. In a seasonal survey the actual weight of insulin present in the Brockman Body[1] of cod was found to be high only in the months of heavy feeding, rising steeply from May to July and falling to very low values from August onwards when feeding is reduced (Brayne, 1980). There is, however, no correlation between the weight of the Brockman Body (which varies during the year) and either its insulin concentration or the total insulin resource of the fish, so the simple observation cannot be used to help assess the nutritional condition of the fish.

Black (1983) made a detailed study of the effects of starvation and the resumption of feeding on the carbohydrates of both cod and rainbow trout. He also studied variations in the activities of some of the enzymes involved. Figure 1.5 shows that starvation reduces the glycogen levels in the liver, red muscle and white muscle of cod. Re-feeding results in an overcompensation to very high levels, which spontaneously decrease on further re-feeding (not shown in Figure 1.5). The re-feeding phenomenon will be discussed later, but it is worth pointing out here that re-feeding of starving fish can also increase the liver lipids to a level higher than in fish fed continuously (Miglavs and Jobling, 1989: Arctic char, *Salvelinus alpinus*).

Rainbow trout subjected to a similar regime (Figure 1.6) behave differently. As in cod, the liver glycogen is greatly reduced, but in both red and white muscle the level of glycogen is maintained. The concentrations of glycogen in liver and muscle do not therefore go hand in hand as they do in cod. Black (1983) also showed that whereas the concentration of blood glucose in cod decreased linearly from 63 to 18 mg/100 ml of blood over 107 days, there was

---

[1] The Brockman Body is almost pure islet tissue, which forms a separate organ on the tip of the gall bladder in this species.

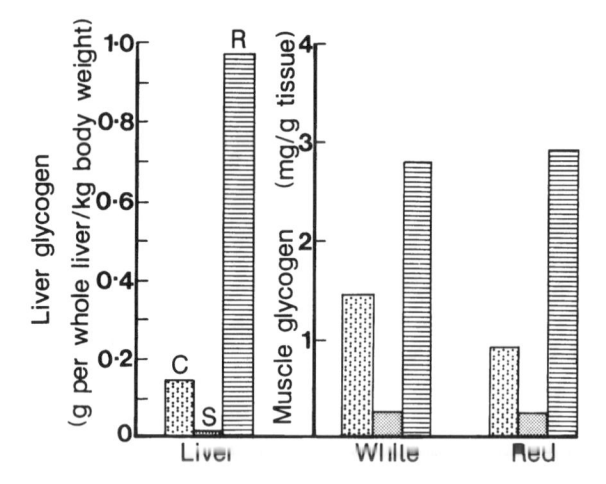

Figure 1.5 Glycogen in liver, white muscle and red muscle of cod. C = control (continuously fed); S = starved at 9°C for 77 days; R = re-fed for 97 days after starvation. The sequence is the same for each tissue. After Black (1983) by courtesy of Dr Darcey Black.

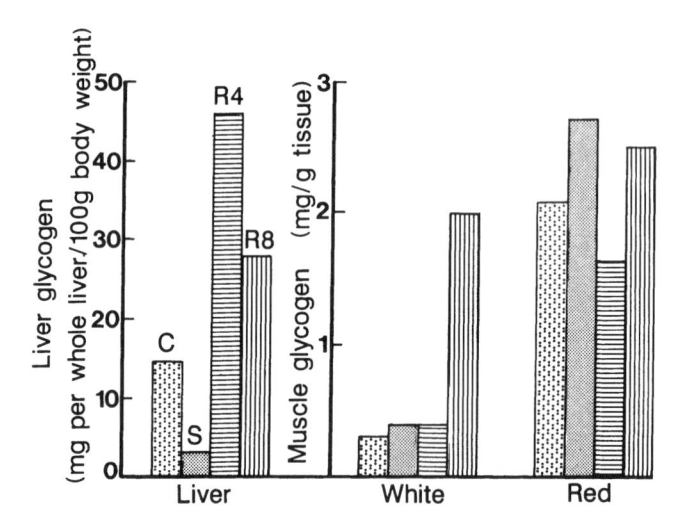

Figure 1.6 Glycogen in liver, white muscle and red muscle of rainbow trout starved for 8 weeks and re-fed for 4 weeks (R4) and 8 weeks (R8). C and S as in Figure 1.5. Note that, unlike the situation in cod, the glycogen decreases in the liver with starvation but is maintained in the two muscle tissues. After Black (1983) by courtesy of Dr Darcey Black.

a small *increase* in the same constituent in rainbow trout over the 56-day period over which they were starved.

Since the glycogen decreased in the livers, these findings alone are insufficient to demonstrate gluconeogenesis in starving rainbow trout. The changes in enzyme activity, however, tell a convincing story. Two glycolytic enzymes (enzymes that break down glycogen) were investigated. Glycogen phosphorylase activity decreased in the livers of starving cod (not measured in trout). Pyruvate kinase (PK) decreased in the livers of both species. Since the amounts of liver glycogen had decreased in both species at the chosen times of sampling, it is logical for the enzymes that mobilise it to decrease also (Black, 1983).

Three gluconeogenic enzymes were studied in cod. The specific activity of phosphoenol pyruvate carboxykinase (PEPCK) declined significantly in the liver of cod and increased significantly in the liver of trout during starvation (Black, 1983). Fructose 1-6 diphosphatase (FDPase) activity also decreased in the liver of cod (not significantly) but in trout liver the increase was significant during starvation. Alanine amino transaminase activity decreased significantly in the liver of cod during starvation but was not measured in trout.

The evidence for gluconeogenesis in trout and not in cod is therefore quite strong. Knox *et al.* (1980) used the ratio of PEPCK to PK as an indication of the relative importance of gluconeogenesis and glycolysis in rainbow trout. The convincing change in this ratio shown by Black (1983) leaves no doubt about the increase in gluconeogenesis in starved trout and its return to previous levels on re-feeding (Figure 1.7).

The low value of liver glycogen in trout (Figure 1.6) presumably shows that at the point of sampling the gluconeogenesis had been insufficient to maintain carbohydrate supplies. However, according to Lim and Ip (1989: mudskipper, *Boleophthalmus boddaerti*), any increase in the degradation of glycogen

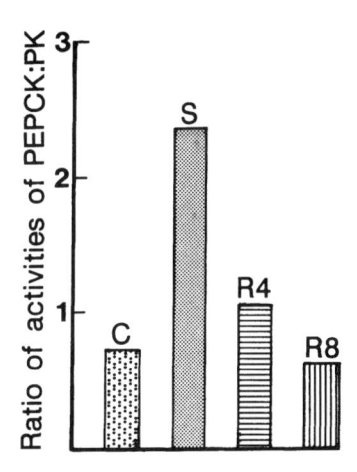

Figure 1.7 Ratio of phosphoenolpyruvate carboxykinase (PEPCK) (gluconeogenic) to pyruvate kinase (PK) (glycolytic) to show the relative importance of gluconeogenesis and glycolysis in the liver of rainbow trout. C = control (continuously fed); S = starved for 8 weeks; R4 = re-fed for 4 weeks after starvation; R8 = re-fed for 8 weeks after starvation. After Black (1983) by courtesy of Dr Darcey Black.

reserves is coupled with increasing activities of key gluconeogenic enzymes in the liver, therefore may also trigger gluconeogenesis in trout; further work should be done.

How does the cod adapt to steadily decreasing stores of glycogen in the muscle? Personal observation has shown that as starvation progresses the fish become more and more inactive, spending much time motionless on the bottom of the aquarium. Trout seem to be more active.

1.4.3.4 *Overcompensation with re-feeding.* Figure 1.5 showed the effect of starvation and re-feeding on the levels of glycogen in cod muscle. Figure 1.8 shows the corresponding change in *post mortem* pH: a rise caused by starvation and a striking fall on re-feeding for a particular period.

The period of re-feeding that gives rise to maximum glycogen values (lowest pH) appears to be about 100 days in the white muscle of cod (Love, 1979, Black and Love, 1986), but it is only 60 days in the red muscle (Black and Love, 1986). Herein lies a clue as to the purpose of the overcompensation. The RNA/DNA ratio, which indicates the vigour of protein synthesis in animals (Bülow, 1970, demonstrated it in fish), is maximal after the same two periods for the same two tissues (Black and Love, 1986). This suggests that the glycogen supplies energy for protein restoration in muscle tissue after depletion. It also reinforces the idea that red muscle is the more important to the fish, since it is restored earlier than the white muscle, having been depleted later (Figure 1.1).

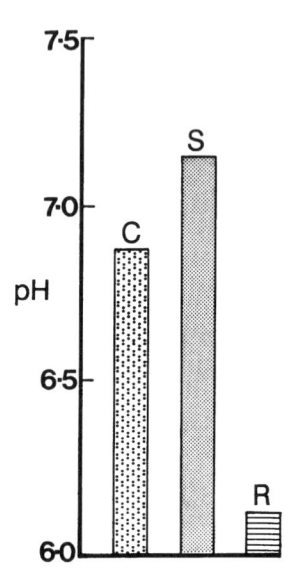

Figure 1.8 The pH of the white muscle of cod 24h after death. Values vary inversely with the glycogen content. C = controls (continuously fed); S = starved for 77 days; R = re-fed for 97 days after starvation. After Black (1983) by courtesy of Dr Darcey Black.

In a large survey of cod caught commercially over several years, Love (1979) showed that most cod in a batch exhibit a *post mortem* pH value of over 6.6 for much of the year. However, at a point in the summer there is a sudden fall to lower values, presumably the 'overcompensation' effect. Cowie and Little (1966) give the entire range of muscle pH values for cod caught commercially as lying between 5.9 and 7.0, although in the author's experience 5.9 is very unusual. In any one year, the pH values are low in fish from a particular ground for only about 2 weeks, after which they rise again. The phenomenon appears in cod from all grounds investigated and it can be seen from Figure 1.9 that the short period of low pH values can occur at any time between May and July. This suggests that restoration of cod tissues in the wild can commence at any time between February and April, depending perhaps on the abundance of food in a given year.

There is little information on this phenomenon in other species. Lavéty *et al.* (1988) illustrated a well-marked seasonal variation in the pH of farmed salmon (*Salmo salar*) with minima in June to July, but there appears to be nothing comparable in haddock (*Melanogrammus aeglefinus*). On the other hand, the range of pH of haddock muscle over the year is consistently lower than that of cod (Love, 1979). The biochemical background to this observation has not been investigated.

1.4.3.5 *The importance of pH in fish quality.* Both texture and gaping are influenced by pH.

Texture. As mentioned earlier, the removal of contractile proteins from cod muscle during starvation does, on its own, soften the texture of the cooked product, although the effect is small. There is also a comparable small effect

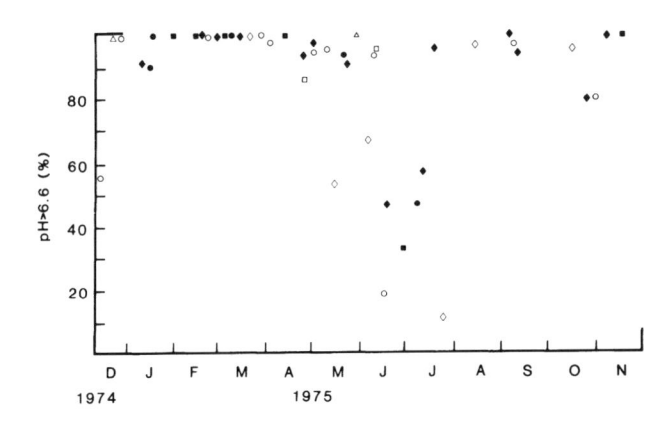

Figure 1.9 Seasonal variation in the pH of cod muscle 24 h after death. Results are expressed as the percentage of a batch of 20 fish in which the pH is greater than 6.6. Different symbols represent different grounds in the North Atlantic, all fish being caught during 1974 and 1975. After Love (1979) by courtesy of the Society of the Chemical Industry.

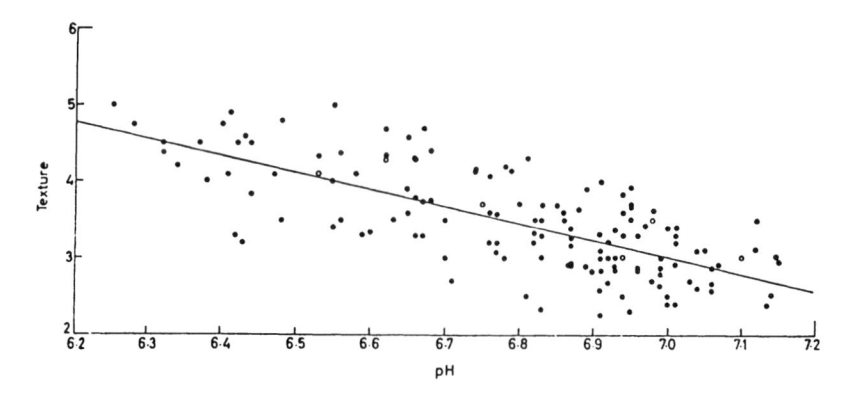

Figure 1.10 The influence of *post mortem* pH on the texture of cooked cod muscle as eaten. Texture scores above 3 represent firm or tough fish; below 3, fish are soft or sloppy. Hollow symbols represent two congruent points. From Love *et al.* (1974b) by courtesy of Kluwer Academic.

from body length, larger cod being tougher (Love *et al.*, 1974b). For the most important factor influencing texture we must look to the pH of the muscle (Figure 1.10). On the scale shown, a score of '3' represents the most acceptable 'normal' texture, a score of 2 is unacceptably sloppy, and a score of 4 and over is unacceptably firm.

When fish are stored in the frozen state, they gradually toughen (many authors, reviewed by Love, 1966) at a rate which varies widely between species (Love and Olley, 1964). It is comparatively rapid in gadoid species and slow in salmonids. The species that toughens at the slowest rate appears, at present, to be the lemon sole (*Pleuronectes microcephalus*: Kim *et al.*, 1977). Cold storage can therefore be used to good effect on cod that are unpleasantly sloppy through having a high pH: the texture can be made to firm up and improve acceptability (T.R. Kelly, 1969). In the case of cod with a low pH, the texture will already be firm, therefore it will cross the boundary of unacceptability after even a little cold storage – the act of freezing and immediate thawing alone toughens the texture appreciably (Love, 1962b).

From what we have seen, there is a short period in the summer when cod, especially large cod, are not really suitable for freezing because of their low pH. However, almost by way of compensation, such fish keep better when chilled in ice. Spoilage bacteria flourish in a neutral pH, but their growth is inhibited to some extent at lower pH values (Jay, 1970). This fact probably explains the observation of Reay (1957) that cod caught on the North Cape Bank (Norway) spoil more rapidly than those caught on the Faroe Bank. In the latter case, the pH of the muscle is often lower than that of cod from any other ground (Love *et al.*, 1974a).

It may surprise those accustomed to assessing the cooked texture of cod to learn that the effect of fish size on texture is 'small', because it is well-known that large fish can be considerably tougher than smaller fish. The principal reason for this is that the muscle of larger fish, although intrinsically tougher, also tends to equilibrate after death at a lower pH than that of smaller fish (K.O. Kelly, 1969). Consequently, two reinforcing factors are involved.

Gaping. Gaping in cod fillets is illustrated in Figure 1.11. It is a troublesome and costly defect, not only because of the damaged appearance that makes the fillets difficult to sell, but also because such fillets cannot be mechanically skinned, hung for smoking or sliced (as in smoked salmon). In brief, the factors that cause gaping in fillets are:

- freezing whole fish (filleted after thawing) (Love and Haq, 1970b);
- whole fish entering *rigor mortis* at raised temperatures (Love and Haq, 1970a: cod; Lavéty *et al.*, 1988: salmonids);
- increasing time after death before freezing (Love *et al.*, 1969);
- mechanical damage, *e.g.* bending fish in *rigor mortis* (Love, 1988);
- shorter body lengths (Love *et al.*, 1972a);
- very slow freezing (Love, 1988);
- the pH of the muscle (Love *et al.*, 1972a).

Gaping appears not to be affected by the method of thawing, rate of freezing (apart from very slow rates) or length of cold storage (Love, 1988).

There are, however, important species differences that stem from intrinsic differences in the mechanical strengths of the connective tissues involved.

Figure 1.11 Gaping in a cod fillet. From Love *et al.* (1969) by courtesy of the *International Journal of Food Science and Technology*.

Strength increases markedly in the following series: hake (*Merluccius merluccius*), cod (*Gadus morhua*) and catfish (*Anarhichas lupus*) (Yamaguchi *et al.*, 1976). The latter species can be very roughly handled without causing gaping. Haddock (*Melanogrammus aeglefinus*) gape more than cod because of their lower muscle pH after death; the intrinsic strength of haddock connective tissue is the greater of the two at the same pH (Love *et al.*, 1972b). This fact emphasises the central role of pH in the phenomenon of gaping: it is probably the most important of the seven causes listed on p 21. Love *et al.* (1972b) showed that cod connective tissue buffered at pH 7.1 is more than four times as strong as that at pH 6.2: it is the rupture of connective tissue that underlies gaping.

There is a marked seasonal variation in the pH of cod (Figure 1.9) and Atlantic salmon (Lavéty *et al.*, 1988) and a mirror image of the pattern is seen in the gaping of these two species (Love, 1980; Lavéty *et al.*, 1988, respectively) and probably others.

In summary, therefore, glycogen overcompensation (low pH) causes acute problems with both gaping and texture, whilst the subject of carbohydrate dynamics is of enormous significance for fish technology.

1.4.4 *Pigmentation*

1.4.4.1 Red muscle. Red muscle is a strip of brownish-red tissue lying immediately under the skin of most species of fish. It also occurs near the spine of very active species such as tuna (*Thunnus* spp.). A technological problem arises only when the pigmentation in an expensive white fish species, such as cod, becomes excessive. Purchasers may then conclude that they have been sold saithe (*Pollachius virens*), a more heavily pigmented species, which is cheaper and less popular with some people. The problem arises because the pigmentation is dynamic, increasing if the fish become more active and fading when they rest. The red muscle in cod is therefore much darker in the stock that migrates yearly many hundreds of miles, from Lofoten to Bear Island to Spitzbergen and back again, than in any other stocks investigated. In other stocks, it intensifies when the fish become more active in summer, fading steadily during the winter (Love *et al.*, 1977).

1.4.4.2 Skin colour. Skin colour varies according to the colour of the seabed. Norwegian coastal cod are unusually rich in red pigment (Dannevig, 1953), a feature that is completely absent from oceanic cod, which have only yellow and black pigments.

There also appears to be a genetic difference. All species of fish found on the Faroe Bank, which is composed of white shell material, are extremely pale in colour. If cod from the Faroe Bank are kept in captivity with cod from a darker ground, both races change their colour in the direction of the colour of the aquarium. However, after many months there is still a difference between them

(Love, 1974), implying either that a genetic factor also is involved, or that a *range* of potential pigment intensities is fixed early in the life of the fish. Coloured pictures of cod from different grounds are given by Love (1970). These colour differences are merely of aesthetic interest; there does not appear to be any relationship between them and quality or marketability.

1.4.4.3 *Flesh carotenoids.* Where carotenoids occur in fish, they are not just 'a playful diversion of nature' as was often thought in the past (Deufel, 1975), but a benefit to the fish in several ways. Red-coloured trout eggs hatch better than pale ones, and canthaxanthin and astaxanthin appear to be sperm activators. Deufel (1975) considered that they also acted as hormones influencing growth, fertilisation, maturation and embryonic development. Male guppies (*Poecilia reticulata*) with rich carotenoid pigmentation in their skins are preferred by ripe females and have greater mating success than siblings raised on carotenoid-free diets (Kodric-Brown, 1989). This is of interest here since customers are unlikely to buy even the most tasty salmon if the flesh is white.

The point to remember is that carotenoids are laid down in the flesh of salmonids during the feeding season prior to maturation, then transferred to the eggs in the developing ovary. There is some evidence that carotenoids enter the flesh only when maturation has at least started (Lewtas, unpublished, cited by Love, 1980), but if the fish are slaughtered when maturation is well-advanced, the flesh will have become pale again, – indeed, Reid *et al.* (1993) concluded that the colour of the musculature was a better indicator of the state of maturity of chum salmon (*Oncorhynchus keta*) than were the contents of muscle lipid, protein or water. Mochizuki and Love (unpublished, illustrated in Love, 1988, figure 49) showed that the flesh colour of rainbow trout from one farm decreased steadily from September to March during maturation, then increased to maximum values in June, July and August, the 'recovery' period.

As a consequence of the close link between pigmentation and maturation, the concentration of pigment in the flesh of rainbow trout increases progressively with age from 2 to 4 years (Sivtsieva and Dubrovin, 1981). In addition, taking just 1-year-old salmon, Torrissen and Naevdal (1988) showed that the level of pigmentation was influenced by genetic factors as well as the stage of maturation and weight.

Increasing the level of lipids in the diet appears to increase the deposition of dietary canthaxanthin in salmonids (Malak *et al.*, 1975), and unpublished work by Robertson and Love (cited by Love, 1988) seems to suggest that some dietary free fatty acids are better carriers of carotenoids across the intestinal wall than others. However, dietary canthaxanthin is always absorbed with difficulty, and only a small percentage ends up in the flesh (Choubert *et al.*, 1987). Some appears to be oxidised in the alimentary canal, and a dietary supplement of α-tocopherol increases the amount reaching the flesh (Pozo *et al.*, 1988).

On the other hand, astaxanthin, the principal carotenoid in wild salmonids, is deposited more efficiently, and a combination of astaxanthin and canthaxanthin in the diet results in a higher deposition of total muscle pigment than is possible when either compound is fed individually (Torrissen, 1989). In the wild, most of the flesh colour is contributed by astaxanthin (48.5–99.8% of total colour), while canthaxanthin contributes only 1–15.4% (Bird and Savage, 1990: chinook salmon, *Oncorhynchus tshawytscha*).

1.4.5 Flavour compounds

This subject has been reviewed earlier (Love, 1988). The most salient points are that different species vary in cooked flavour, but that sometimes experts are required to detect differences; to most people the different species of non-fatty fish taste more-or-less the same. The pleasant sweetness of cooked newly-killed fish is almost all derived from the amino acid glycine, with a small contribution from free glucose.

In chilled non-fatty fish, the maximum sweetness is developed some 2–4 days after death. After 7–10 days, the fish become rather tasteless; spoilage flavours then develop. The flavours of spoiled fish differ between freshwater and marine species, since trimethylamine (the 'fishy' taste) develops only in the latter.

Flavours can be acquired by fish from a restricted body of water, so cultured fish tend to taste different from, although not necessarily inferior to, wild fish (Hume *et al.*, 1972). In oceanic species off-flavours such as 'seaweedy', 'iodine-like', 'egg-like' or 'blackberry' appear to be invariably derived from dietary organisms, especially algae. These flavours are often strongly seasonal in their occurrence.

1.4.6 Minerals

A diet low in minerals is recommended for patients with heart trouble, therefore the minerals present in the edible parts of fish are of interest. The total mineral contents of the flesh of marine and freshwater fish are not markedly different, although trace elements such as boron, bromine and lithium, are more plentiful in the former (Vinogradov, 1953). Since connective tissue is much richer in sodium than is contractile muscle, the concentration of sodium at the tail end of cod is twice that of the head end, containing as it does many more of the connective tissue septa per unit length (Love *et al.*, 1968). Love (1961) and Love *et al.* (1968) found that gutted cod kept in melting ice rapidly lose sodium from the flesh, and that potassium loss, negligible at first, steadily increases as time passes.

These 'snippets' of information may be useful to hospital dietitians.

1.5 Summary of considerations of biological condition and quality

'Biological condition' is a general term referring to the state of the energy reserves of a fish, either replete from a period of feeding or impoverished from starvation, spawning, or a combination of the two. Reserves of lipids, carbohydrates and proteins are mobilised and later replenished in a definite order, which differs among species.

Removal of lipid reserves from the flesh results in a product that tastes 'dry' or fibrous. On the other hand, moderate starvation of some fatty or non-fatty species yields a product that develops less off-flavour and off-odour during cold storage. Further depletion, which initiates the mobilisation of tissue proteins, results in very watery flesh of reduced nutritional value.

A decrease in flesh lipid reserves results in an increase in the water content. A subsequent decrease in protein reserves results in a further increase in the water content, also seen when protein is removed from the muscle of non-fatty species. The two sorts of water augmentation cannot readily be distinguished, but an increase in lysosomal enzyme activity in the flesh unequivocally identifies the mobilisation of proteins.

A high water content in non-fatty fish flesh, accompanied by low values of lysosomal enzymes and pale-coloured bile in the gall bladder, indicates a poor biological condition, which is currently improving. High lysosomal enzyme activity and voluminous dark-green bile indicate that the condition, already poor, is getting worse.

Carbohydrate reserves usually start to decrease at the outset of depletion, but in some species they can be maintained by synthesis within the fish from protein or lipid precursors. In cod, and probably other species, the level of muscle glycogen reflects the level of the main reserve in the liver. An approximate simple assessment can be made by measuring the pH of the flesh about 1 day after death.

The pH values rise as carbohydrate reserves diminish and vice versa. Starvation results in a muscle pH around neutral. Fish after death exhibit pH values lower than neutral for much of the year, and at a particular time, coinciding with a late stage in the recovery from winter starvation, the pH is lower still.

The pH of the flesh is of great importance to food technology. It is the most important factor governing the texture of the cooked flesh. Cod having muscle of relatively low pH are tough and so are unsuitable for cold storage, which toughens them further; conversely, they keep well when chilled in ice because spoilage bacteria are inhibited at low pH values.

If the fish have spawned recently the carbohydrate reserves are very low, the *post mortem* pH is 7 or more, and the texture is unacceptably sloppy. In this case, it can be improved by a period of cold storage, which firms the product.

Soft (raw) flesh is a common failing in cultured salmon. It is now suspected that such fish have probably been reared in relatively still water. Raceways or the natural environment of wild fish seem to confer a more elastic texture to the flesh. The observations again appear to relate to variations in the pH of the flesh, originating in variations in the glycogen content. Totland *et al.* (1987)

showed that the red muscle of Atlantic salmon (*Salmo salar*) reared in a raceway contains more glycogen than that of the same species reared in a cage. Subsequently, Tachibana et al. (1988) have reported that the flesh texture of red sea-bream (*Pagrus major*) is less soft where the fish have been habitually forced to swim.

The pH of the flesh also exerts a big influence on the strength of the connective tissue that holds the fillets together. Connective tissue is strong at neutral pH but greatly weakened at more acid values, such that fillets gape or fall to pieces. Consequently, they cannot be mechanically skinned, hung on a tenter for smoking, or sliced.

Off-flavours in fresh fish originate either from dietary organisms or from keeping the fish in a limited volume of water. They are often seasonal and disappear when the live fish are placed in 'pure' running water for a week or so.

Variations in the colour of the flesh or skin of fish may affect saleability, but do not reflect any real quality defects.

References

Ackman, R.G. and Eaton, C.A. (1971), Mackerel lipids and fatty acids. *J. Can. Inst. Food Technol.* **4**, 169–174.

Ando, A. and Hatano, M. (1986), Myofibrillar protein degradation in spawning-migrating chum salmon, as evaluated by extractive N-methyl histidine level. *Bull. Japan. Soc. Sci. Fish.* **52**, 1237–1241.

Banks, A. (1938), The storage of fish with special reference to the onset of rancidity: 1. The cold-storage of herring. *J. Soc. Chem. Ind.* **57**, 124–128.

Beardall, C.H. and Johnston, I.A. (1985), Lysosomal enzyme activities in muscle following starvation and refeeding in the saith *Pollachius virens* L. *Europ. J. Cell Biol.* **39**, 112–117.

Best, A.C.G. and Bone, Q. (1973), The terminal neuro-muscular junctions of lower chordates. *Z. Zellforsch. Mikrosk. Anat.* **143**, 495–504.

Bird, J.N. and Savage, G.P. (1990), The use of CIE (1976) L*a*b* colorimetric values for the determination of fillet colour of eviscerated farmed chinook salmon (*Oncorhynchus tschawytscha*). *Chilling and freezing of new fish products, I.I.F. Commission C2, Aberdeen, UK 1990*, 219–228.

Black, D. (1983), The metabolic response to starvation and refeeding in fish. PhD Thesis, University of Aberdeen.

Black, D. and Love, R.M. (1986), The sequential mobilisation and restoration of energy reserves in tissues of Atlantic cod during starvation and refeeding. *J. Comp. Physiol.* **156**, 467–479.

Black, D. and Love, R.M. (1988) Estimating the carbohydrate reserves in fish. *J. Fish Biol.* **32**, 335–340.

Boddeke, R., Slijper, E.J. and van der Stelt, A. (1959), Histological characteristics of the body musculature of fishes in connection with their mode of life. *Koninklijke Ned. Akad. van Wetenschappen* (Series C), **62**, 576–588.

Boëtius, I. and Boëtius, J. (1985), Lipid and protein content in *Anguilla anguilla* during growth and starvation, *DANA* **4**, 1–17.

Bogoyavlenskaya M.P. and Vel'tischeva, I.F. (1972) Some data on the age changes in the fat and carbohydrate metabolism of Baltic Sea cod. *Trudy Vses. Nauchno-issled. Inst. Morsk. Ryb. Khoz. Okeanogr.* **85**, 56–62.

Bolger, T. and Connolly, P.L. (1989), The selection of suitable indices for the measurement and analysis of fish condition. *J. Fish Biol.* **34**, 171–182.

Brandes, C.H. and Dietrich, R. (1958), Observations on the correlations between fat and water content and the fat distribution in commonly eaten fish. *Veröff. Inst. Meeresforsch. Bremerh.* **5**, 299–305.

Brayne, L.J. (1980) Aspects of the physiology of North Sea cod (*Gadus morhua* L.). *PhD Thesis*, University of Aberdeen.
Bulow, F.J. (1970), RNA-DNA ratios as indicators of recent growth rates of a fish. *J. Fish. Res. Bd Can.* 27, 2343–2349.
Burlakova, Ye B., Storozhuk, N.M. and Khrapova, N.G. (1988), Relationship between the activity of antioxidants and substrate oxidisability in lipids of natural origin. *Biophysics* 33, 840–846.
Burt, J.R. (1966), Glycogenolytic enzymes of cod (*Gadus callarias*) muscle. *J. Fish. Res. Bd Can.* 23, 527–538.
Campbell, S. and Love, R.M. (1978), Energy reserves of male and female haddock (*Melanogrammus aeglefinus* L.) from the Moray Firth. *J. Cons. Int. Explor. Mer.* 38, 120–121.
Chang, V.M. and Idler, D.R. (1960), Biochemical studies on sockeye salmon during spawning migration: xii. Liver glycogen. *Can. J. Biochem. Physiol.* 38, 553–558.
Chavin, W. and Young, J.E. (1970), Factors in the determination of normal serum glucose levels of goldfish. *Carassius auratus* L. *Comp. Biochem. Physiol.* 33, 629–653.
Chepurnov, A.V. and Tkachenko, N.K. (1973), Changes in the lipid composition of females and males of the Black Sea round goby (*Neogobius melanostomus*) in the spawning period and in early ontogenesis. *Communications of the All-Union Symposium on the Study of the Black and Mediterranean Seas and the Utilisation and Preservation of Their Resources: Part 1. Biological and Ecological-physiological Studies of Fishes and Invertebrates.* Sevastopol, *Izdat. Naukova Dumka, Kiev*, pp. 212–216.
Choubert, G., Guillou, A. and Fauconneau, B. (1987), Absorption and fate of labelled canthaxathin 15, 15′-^{3}H$_2$ in rainbow trout (*Salmo gairdneri* Rich.). *Comp. Biochem. Physiol.* 87A, 171–720.
Cowey, C.B. and Sargent, J.R. (1972), Fish nutrition. *Adv. Mar. Biol.* 10, 383–492.
Cowie, W.P. and Little, W.T. (1966), The relationship between the toughness of cod stored at $-29°C$ and its muscle protein solubility and pH. *J. Food Technol.* 1, 335–343.
Dannevig, A. (1953), The littoral cod of the Norwegian Skagerak coast. *Rapp. P.-V. Réun. Cons. Perm. Int. Explor. Mer.* 136, 7–14.
De Duve, C. (1963), General properties of lysosomes: The lysosome concept. In *Ciba Foundation Symposium: Lysosomes*, De Reuck, A.V.S. and Cameron, M.P. (Eds), Little, Brown and Co., Boston, Mass. pp. 1–35.
Deufel, J. (1975), Physiological effect of carotenoids on Salmonidae. *Hydrologie* 37, 244–248.
Drevon, C.A. (1989), ω-3 fatty acids in health and disease. *Fish, Fats and Your Health. Proceedings of the Intern. Conference on Fish Lipids and Their Influence on Human Health.* Svanøy Foundation, 6965 Svanøybukt, Norway, pp. 19–25.
Dyerberg, J. and Bang, H.O. (1979), Haemostatic function and platelet polyunsaturated fatty acids in Eskimos. *Lancet* 2, 433–435.
Farkas, T. and Herodek, S. (1964), The effect of environmental temperature on the fatty acid composition of crustacean plankton. *J. Lipid Res.* 5, 369–373.
Goldenberg, A.L., Paron, L. and Crupkin, M. (1987), Acid phosphatase activity in pre- and post-spawning hake. *Comp. Biochem. Physiol.* 87A, 845–849.
Gopakumar, K. and Nair, M.R. (1972), Fatty acid composition of eight species of Indian Marine fish. *J. Sci. Food Agric.* 23, 493–496.
Greene, C.W. (1926), The physiology of the spawning salmon. *Physiol. Rev.* 6, 201–241.
Hardy, R. and McGill, A.S. (1990), The influence of cold-storage dehydration on the oxidation of white fish. *Chilling and freezing of new fish products, I.I.F. Commission C2, Aberdeen, UK 1990*, 289–295.
Heming, T.A. and Paleczny, E.J. (1987), Compositional changes in skin mucus and blood serum during starvation of trout. *Aquaculture* 66, 265–273.
Hochachka. P.W. and Sinclair, A.C. (1962), Glycogen stores in trout tissues before and after stream planting. *J. Fish. Res. Bd Can.* 19, 127–136.
Hume, A., Farmer, J.W. and Burt, J.R. (1972), A comparison of the flavours of farmed and trawled plaice. *J. Food Technol.* 7, 27–33.
Inui, Y. and Yokote, M. (1974), Gluconeogenesis in the eel – 1. Gluconeogenesis in the fasted eel. *Bull. Freshwat. Fish. Res. Lab., Tokyo* 24, 33–45.
Irving, D.O. and Watson, K. (1976), Mitochondrial enzymes of tropical fish: a comparison with fish from cold-waters. *Comp. Biochem. Physiol.* 54B, 81–92.
Jay, J.M. (1970) *Modern Food Microbiology.* Van Nostrand Reinhold, New York, pp. 26–28.

Kaushik, S.J., Medale, F., Fauconneau, B. and Blanc, D. (1989), Effect of digestible carbohydrates on protein/energy utilisation and on glucose metabolism in rainbow trout (*Salmo gairdneri* R.). *Aquaculture* **79**, 63–74.

Kelly, K.O. (1969), Factors affecting the texture of frozen fish. In *Freezing and Irradiation of Fish*, Kreuzer, R. (Ed.) Fishing News Books, Farnham, pp. 339–342.

Kelly, T.R. (1969), Quality in frozen cod and limiting factors on its shelf life. *J. Food Technol.* **4**, 95–103.

Kemp, P. and Smith, M.W. (1970), Effect of temperature acclimatisation on the fatty acid composition of goldfish intestinal lipids. *Biochem. J.* **117**, 9–15.

Kim, H.-K., Robertson, I. and Love, R.M. (1977), Changes in the muscle of lemon sole (*Pleuronectes microcephalus*) after very long cold storage. *J. Sci. Food Agric.* **28**, 699–700.

Knox, D., Walton, M.J. and Cowey, C.B. (1980), Distribution of enzymes of glycolysis and gluconeogenesis in fish tissues. *Mar. Biol.* **56**, 7–10.

Kodric-Brown, A. (1989), Dietary carotenoids and male success in the guppy: an environmental component to female choice. *Behav. Ecol. Sociobiol.* **25**, 393–401.

Kott E. (1971), Liver and muscle composition of mature lampreys. *Can. J. Zool.* **49**, 801–805.

Lands, W.E.M. (1986), *Fish and Human Health*. Academic Press, Orlando, pp. 1–170.

Lavéty, J. and Love, R.M. (1972). The strengthening of cod connective tissue during starvation. *Comp. Biochem. Physiol.* **41A**, 39–42.

Lavéty, J., Afolabi, O.A. and Love, R.M. (1988), The connective tissues of fish. IX. Gaping in farmed species. *Int. J. Food Sci. Technol.* **23**, 23–40.

Lee, D.J. and Sinnhuber, R.O. (1973), Lipid requirements. In *Fish Nutrition*, Halver, J.E. (Ed.), Academic Press, London, pp. 145–180.

Lim, A.L.L. and Ip, Y.K. (1989), Effect of fasting on glycogen metabolism and activities of glycolytic and gluconeogenic enzymes in the mud skipper, *Boleophthalmus boddaerti*. *J. Fish. Biol.* **34**, 349–367.

Love, R.M. (1960), Water content of cod (*Gadus callarias* L.) muscle. *Nature* **185**, p. 692.

Love, R.M. (1961), The expressible fluid of fish fillets. X. Sodium and potassium content in frozen and iced fish. *J. Sci. Food Agric.* **12**, 439–442.

Love, R.M. (1962a), The measurement of 'condition' in North Sea cod. *J. Cons. Perm. Int. Explor. Mer.* **27**, 34–42.

Love, R.M. (1962b), Protein denaturation in frozen fish. VI. Cold-storage studies on cod using the cell fragility method. *J. Sci. Food Agric.* **13**, 269–278.

Love, R.M. (1966), The freezing of animal tissue. In *Cryobiology*, Meryman, H.T., (Ed.), Academic Press, New York, pp. 317–405.

Love, R.M. (1969), Condition of fish and its influence on the quality of the frozen product. In *Freezing and Irradiation of Fish*, Kreuzer, R. (Ed.), Fishing News Books, Farnham, pp. 40–45.

Love, R.M. (1970), *The Chemical Biology of Fishes*. Academic Press, London.

Love, R.M. (1974), Colour stability in cod (*Gadus Morhua* L.) from different grounds. *J. Cons. Perm. Int. Explor. Mer.* **35**, 207–209.

Love, R.M. (1975), Variability in Atlantic cod (*Gadus morhua*) from the Northeast Atlantic: a review of seasonal and environmental influences on various attributes of the flesh. *J. Fish. Res. Bd Can.* **32**, 2333–2342.

Love, R.M. (1979), The post-mortem pH of cod and haddock muscle and its seasonal variation. *J. Sci. Food Agric.* **30**, 433–438.

Love, R.M. (1980), *The Chemical Biology of Fishes* (Vol. 2). Academic Press, London.

Love R.M. (1988), *The Food Fishes: Their Intrinsic Variation and Practical Implications*. Farrand Press, London.

Love, R.M. and Haq, M.A. (1970a), The connective tissues of fish. III. The effect of pH on gaping in cod entering *rigor mortis* at different temperatures. *J. Food Technol.* **5**, 241–248.

Love, R.M. and Haq, M.A. (1970b), The connective tissues of fish. IV. Gaping of cod muscle under various conditions of freezing, cold-storage and thawing. *J. Food Technol.* **5**, 249–260.

Love, R.M. and Muslemuddin, M. (1972), Protein denaturation in frozen fish. XII. The pH effect and cell fragility determinations. *J. Sci. Food Agric.* **23**, 1229–1238.

Love, R.M. and Olley, J. (1964), Cold-storage deterioration in several species of fish, as measured by two methods. In *The Technology of Fish Utilisation*, Kreuzer, R. (Ed.) Fishing News Books, London, pp. 116–118.

Love, R.M., Lavéty, J. and Steel, P.J. (1969), The connective tissues of fish. II. Gaping in commercial species of frozen fish in relation to *rigor mortis*. *J. Food Technol.* **4**, 39–44.

Love, R.M., Haq, M.A. and Smith, G.L. (1972a), The connective tissues of fish. V. Gaping in cod of different sizes as influenced by a seasonal variation in the ultimate pH. *J. Food Technol.* **7**, 281–290.

Love, R.M., Lavéty, J. and Garcia, N.G. (1972b), The connective tissues of fish. VI. Mechanical studies on isolated myocommata. *J. Food Technol.* **7**, 291–301.

Love, R.M., Robertson, I., Lavéty, J. and Smith, G.L. (1974a), Some biochemical characteristics of cod (*Gadus morhua* L.) from the Faroe Bank compared with those from other fishing grounds. *Comp. Biochem. Physiol.* **47B**, 149–161.

Love, R.M., Robertson, I., Smith, G.L. and Whittle, K.J. (1974b), The texture of cod muscle. *J. Texture Studies* **5**, 201–212.

Love, R.M., Hardy, R. and Nishimoto, J. (1975), Lipids in the flesh of cod (*Gadus morhua* L.) from Faroe Bank and Aberdeen Bank in early summer and autumn. *Mem. Fac. Fish., Kagoshima Univ.* **24**, 123–126.

Love, R.M., Yamaguchi, K., Créac'h, Y. and Lavéty, J. (1976), The connective tissues and collagens of cod during starvation. *Comp. Biochem. Physiol.* **55B**, 487–492.

Love, R.M., Munro, L.J. and Robertson, I. (1977), Adaptation of the dark muscle of cod to swimming activity. *J. Fish Biol.* **11**, 431–436.

Love, R.M., Lavéty, J. and Vellas, F. (1982), Unusual properties of the connective tissues of cod (*Gadus morhua* L.). In *Chemistry and Biochemistry of Marine Food Products*, Martin, R.E., Flick, C.J., Hebard, C.E. and Ward, D.R. (Eds). AVI, Westport, Connecticut, pp. 67–73.

Lovern, J.A. (1935), Fat metabolism in fishes. VII. The depot fats of certain fish fed on known diets. *Biochem. J.*, **29**, 1894–1897.

Ludovico-Pelayo, L., Hume, A. and Love, R.M. (1984), Seasonal variations in flavour change of cold-stored rainbow trout. In *Thermal Processing and Quality of Foods*, Zeuthen, P., Cheftel, J.C., Eriksson, C., Jul, M., Leninger, H., Linko, P., Varela, G. and Vos. G. (Eds), Elsevier Applied Science, London, pp. 659–663.

Luquet, P., Léger, C. and Bergot, F. (1975), Effects of carbohydrate suppression in diets of rainbow trout kept at a temperature of 10°C 1. – Growth in relation to level of protein ingestion. *Ann. Hydrobiol.* **6**, 61–70.

McGill, A.S. (1974). An investigation into the chemical composition of the cold storage flavour components of cod. *IFST mini-symposium on freezing*, Institute of Food Science and Technology, UK, pp. 24–26.

McGill, A.S., Hardy, R., Burt, J.R. and Gunstone, F.D. (1974), Hept-cis-4-enal and its contribution to the off-flavour in cold-stored cod. *J. Sci. Food Agric.* **25**, 1477–1489.

Machado, C.R., Garofalo, M.A.R., Roselino, J.E.S., Kettlehut, I.C. and Migliorini, R.H. (1988). Effects of starvation, refeeding and insulin on energy-linked metabolic processes in catfish (*Rhamdia hilarii*) adapted to a carbohydrate-rich diet. *Gen. Comp. Endocrinol.* **71**, 429–437.

Maksimovich, A.A. (1988), The pattern of carbohydrate metabolism in Pacific salmon during starvation. *Izv. AN SSSR (Biol.)* **4**, 500–508.

Malak, N.A., Zwingelstein, G., Jouanneteau, J. and Koenig, J. (1975), Influence of certain nutritional factors on the pigmentation of rainbow trout by canthaxanthin. *Ann. Nutr. Alim.* **29**, 459–475.

Miglavs, I. and Jobling, M. (1989), The effects of feeding regime on proximate body composition and patterns of energy deposition in juvenile Arctic char, *Salvelinus alpinus*. *J. Fish Biol.* **35**, 1–11.

Murat, J.C. (1976), Studies on the mobilisation of tissular carbohydrates in the carp. *Thesis, Docteur d'Etat Mention Sciences*, University Paul-Sabatier, Toulouse.

Nagai, M. and Ikeda. S. (1971), Carbohydrate metabolism in fish – I. Effects of starvation and dietary composition on the blood-glucose level and the hepatopancreatic glycogen and lipid contents in carp. *Bull. Jafan. Soc. Sci. Fish.* **37**, 404–409.

Ono, T., Nagayama, F. and Masuda, T. (1960), Studies on the fat metabolism of fish muscles: 4. Effects of the components in foods on the culture of rainbow trout. *J. Tokyo Univ. Fish.* **46**, 97–106.

Owen, J.M., Adron, J.W., Middleton, C. and Cowey, C.B. (1975), Elongation and desaturation of dietary fatty acids in turbot *Scophthalmus maximus* L., and rainbow trout, *Salmo gairdnerii* Rich. *Lipids* **10**, 528–531.

Palmer, T.N. and Ryman, B.E. (1972), Studies on oral glucose intolerance in fish. *J. Fish Biol.* **4**, 311–319.

Pottinger, T.G. and Pickering, A.D. (1987), Androgen levels and erythrocytosis in maturing brown trout, *Salmo trutta. Fish Physiol. Biochem.* **3**, 121–126.

Pozo, R., Lavéty, J. and Love, R.M. (1988), The role of dietary α-tocopherol (vitamin E) in stabilising the canthaxanthin and lipids of rainbow trout muscle. *Aquaculture* **73**, 165–175.

Reay, G.A. (1957), Factors affecting initial and keeping quality. *Food Invest. Bd DSIR Annual Report* p. 3.

Reid, R.A., Durance, T.D., Walker, D.C. and Reid, P.E. (1993), Structural and chemical changes in the muscle of chum salmon (*Oncorhynchus keta*) during spawning migration. *Food Res. Internat.* **26**, 1–9.

Ross, D.A. (1977), Lipid metabolism of the cod *Gadus morhua* L. PhD Thesis, University of Aberdeen.

Ross, D.A. and Love, R.M. (1979), Decrease in the cold store flavour developed by frozen fillets of starved cod (*Gadus morhua* L.) *J. Food Technol.* **14**, 115–122.

Sargent, J.R. (1989), Wax esters, long chain monoenoic fatty acids and polyunsaturated fatty acids in marine oils. Resource and nutritional implication. *Fish, Fats and Your Health, Svanøy Foundation Report No. 4*, Svanøy Foundation, 6965 Svanøybukt, Norway, pp. 43–49.

Sargent, J.R., Bell, M.V., Henderson, R.J. and Tocher, D.R. (1990), Polyunsaturated fatty acids in marine and terrestrial food webs. In *Animal Nutrition and Transport Processes: 1. Nutrition in Wild and Domestic Animals*, Mellinger, J. (Ed.), Karger, Basel, pp. 11–23.

Shatunovskii, M.I. and Novikov, G.G. (1971), Changes in some biochemical indices of the muscles and blood of the sea trout during ripening of the sex products. In *Zakonomernosti Rosta i Sozrevaniya Ryb*, Nikol'skii, G.V. (Ed.), Moscow, pp. 78–79.

Sivtsieva. L.V. and Dubrovin, B.N. (1981), Some aspects of the quantitative distribution of carotenoid pigment in the body of rainbow trout *Salmo gairdneri* Richardson. *Vopr. Ikhtiol.* **21**, 748–751.

Skjervold, H. (1989), Genetic selection for fatty acid composition in farmed salmonides. In *Fish, Fats and your Health, Svanøy Foundation Report No. 4*. Svanøy Foundation, 6965 Svanøybukt, Norway, pp. 63–73.

Smith, M.W. and Kemp, P. (1971), Parallel temperature-induced changes in membrane fatty acids and in the transport of amino acids by the intestine of goldfish (*Carassius auratus* L.) *Comp. Biochem. Physiol.* **39B**, 357–365.

Sundararaj, B.I., Lamba, V. and Goswami, S.V. (1980), Seasonal reproduction in fish: Steroid profiles in annually breeding fish. *Endocrinol. Proc. Int. Congr. Endocrinol.* **6**, 267–272.

Tachibana, K., Doi, T., Tsuchimoto, N., Misima, T., Ogura, M., Matsukiyo, K. and Yasuda, M. (1988), The effect of swimming exercise on flesh texture of cultured red sea-bream. *Bull. Japan. Soc. Sci. Fish.* **54**, 677–681.

Takama, K., Love, R.M. and Smith, G.L. (1985), Selectivity in mobilisation of stored fatty acids by maturing cod, *Gadus morrhua* L. *Comp. Biochem. Physiol.* **80B**, 713–718.

Torrissen, O.J. (1989), Pigmentation of salmonids: interactions of astaxanthin and canthaxanthin on pigment deposition in rainbow trout. *Aquaculture* **79**, 363–374.

Torrissen, O.J. and Naevdal, G. (1988), Pigmentation of salmonids – variation in flesh carotenoids of Atlantic salmon. *Aquaculture* **68**, 305–310.

Totland, G.K., Kryvi, H., Joedestoel, K.A., Christiansen, E.N., Tangeraas, A. and Slinde, E. (1987), Growth and composition of the swimming muscle of adult Atlantic salmon (*Salmo salar* L.) during long-term sustained swimming. *Aquaculture* **66**, 299–313.

Tunison, A.V., Phillips, A.M., McCay, C.M., Mitchell. C.R. and Rodgers, E.O. (1940), The nutrition of trout. *Fish. Res. Bull. N.Y.* p. 30.

Vinogradov, A.P. (1953), The elementary chemical composition of marine organisms. Efron, J. and Setlow, J.K. (Trans.), Sears Foundation, New Haven.

Ward, L.C. and Buttery, P.J. (1978), N^t methyl histidine: an index of true rate of myofibrillar degradation? An appraisal. *Life Sci.* **23**, 1103–1116.

Wardle. C.S. (1972), The changes in blood glucose in *Pleuronectes platessa* following capture from the wild: a stress reaction. *J. Mar. Biol Ass. UK* **52**, 635–651.

White, A. and Fletcher, T.C. (1986), Serum cortisol, glucose and lipids in plaice (*Pleuronectes platessa* L.) exposed to starvation and aquarium stress. *Comp. Biochem. Physiol.* **84A**, 649–653.

Worthington, R.E. and Lovell, R.T. (1973), Fatty acids of channel catfish (*Ictalurus punctatus*): variance components related to diet, replications within diets, and variability among fish. *J. Fish. Res. Bd Can.* **30**, 1604–1608.

Yamaguchi, K., Lavéty, J. and Love, R.M. (1976), The connective tissues of fish. Comparative studies on hake, cod and catfish collagens. *J. Food Technol.* **11**, 389–399.

Yanni, M. (1961), Studies on carbohydrate content of the tissues of *Clarias lazera*. *Z. Vergl. Physiol.* **45**, 56–60.

Ziecik, M. and Nodzynski, J. (1964), Chemical and weight composition of the comestible parts, waste and gonads of flounder (*Pleuronectes flesus*) during an annual cycle. *Zesz. Nauk. Wyzsz. Szk. Roln. Olsztyn.* **18**, 263–280.

2 Preservation of fish by curing (drying, salting and smoking)

W.F.A. HORNER

2.1 Introduction

Curing, as a means of preserving fish, has been practised perhaps longer than any other food preservation technique. Marine fish bones found in cave dwellings, inhabited 20 000 years ago and situated many days' walk from the coast of Spain, indicate some form of curing, probably by drying in the open air. Salting, smoking and drying have all continued as preservation techniques virtually unaltered from prehistory to the present day. Modern developments have centred around understanding and controlling the processes to achieve the standardised product demanded by today's market. A major exception has been exploitation of the sublimation of ice to dry food so that it resembled the starting material in volume and shape. This only became possible with the development of pumps which could create, and valve seals which could maintain, high vacuum.

None the less, for all the developments in cure-processing accommodating continuous production lines, the time required to achieve a long shelf-life product purely by water removal is much greater than for any other commonly used preservation method. This is because the process relies upon the diffusion rate of either water from the centre of the food to its surface, or the diffusion rate of salt (or other solute) in the opposite direction, or a combination of both.

2.2 Water content, water activity (a_w) and storage stability

Unlike canning, which engenders the destruction of micro-organisms and their spores, curing preserves by rendering the medium an unsuitable environment for microbial propagation. Increasing the concentration of soluble substances in the medium either by abstracting water or by causing soluble substances to diffuse in (salting, brining or sugar curing) are the principal means of accomplishing this. In addition to concentrating the soluble substances by brining and dehydration, smoking preserves by depositing bacteriostatic chemicals like formaldehyde and phenols in the system.

The addition of salt is more effective weight for weight than the addition of sugar because salt ionises to a sodium cation and a chloride anion each of which attracts a sheath of water molecules. These ionically associated water molecules are unavailable for use by micro-organisms and there is a tendency for the ionic forces to pull water molecules from the microbial cells dehydrating them to the point where they die or sporulate and lie dormant. Sucrose also withdraws water molecules from the system and holds them by hydrogen bonding. However, far fewer molecules become bound or unavailable in this way than is the case for an equal mass of sodium chloride.

This availability of water in the system for use by micro-organisms directly relates to the effectiveness of preservation and can be represented physically by the water activity (a_w).

2.2.1 Basic definitions

The water content of fresh white fish is about 80%. When this is reduced below approximately 25%, bacterial spoilage stops, and below approximately 15%, moulds cease to grow. These figures are calculated on a wet mass basis, where water content is defined as

$$M_w = \frac{\text{mass of water in the wet solid}}{\text{total mass of solid}} \times 100\%$$

Occasionally water content is quoted on a dry mass basis defined as

$$M_d = \frac{\text{mass of water in the wet solid}}{\text{total mass of dry solid}} \times 100\%$$

The relationship between the two modes of expression is

$$M_d = \frac{100\, M_w}{100 - M_w} \quad \text{and} \quad M_w = \frac{100\, M_d}{100 - M_d}$$

If 10 kg of such fish is to be dried to 25% water content, wet mass basis, the amount of water to be removed is calculated as follows:

At 80% water content the composition of the fish is

10 kg = 8 kg water + 2 kg dry solids

At 25% water content, the 2 kg dry solids represent $100 - 25 = 75\%$, of the mass. Therefore, the total mass of the fish at 25% water content is

$2 \times \frac{100}{75} = 2.67$ kg composed of 0.67 kg water + 2 kg dry solids

Thus the amount of water to be removed is

$8 - 0.67 = 7.33$ kg

Clearly, the removal of quite a large proportion of the water, say 7 out of the 8 kg of water contained in the fish, does not prevent bacterial growth. Hence water content is not the most useful indicator of the ability of the medium to support microbial growth. Water activity, however, directly relates to the concentration of solutes within the system, which, as has already been said, relates to the availability of that water for microbes to use in their growth and reproduction. Water activity is frequently represented as the quotient of the vapour pressure exerted by a solution (P) and the vapour pressure exerted by the pure solvent, normally water, (P_0) at the same temperature.

$$a_w = P/P_0$$

More correctly, this is the *relative equilibrium vapour pressure* which, expressed as a percentage, is the quantity measured by 'water activity' meters such as the 'Novasina' (Figure 2.1). In such equipment, the percentage vapour pressure of the closed atmosphere surrounding the sample, after allowing time for the sample and atmosphere to come into equilibrium at a standard temperature, is measured by hygrometry. The reading is termed the percentage relative humidity (%RVP) of the sample.

The term 'activity' was first used by Lewis (1907) in accounting for the difference between the thermodynamic free energy of a component in a system and that of the same component isolated from the system. This difference is related to a function Lewis termed 'fugacity', which is a measure of the excess, rather than total, free energy available for the work of either the system or the component within the system. Reid (1973) defined fugacity as 'a measure of the escaping tendency... having the form of a vapour pressure which has been corrected for non-ideality of the vapour'. Water activity, then, implies water's 'fagacity ratio', that is

$$\frac{\text{the escaping tendency of water from a system}}{\text{the escaping tendency of pure water at the same temperature}}$$

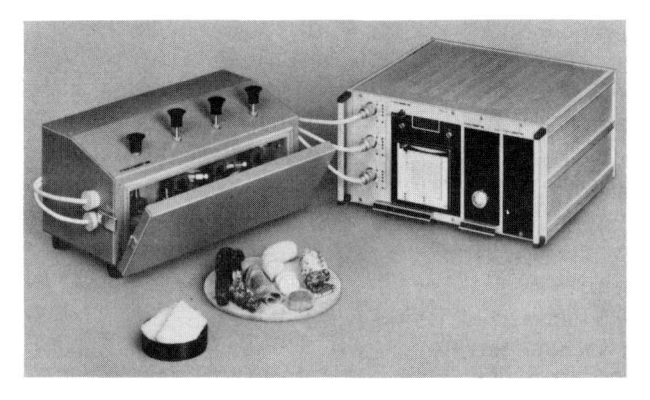

Figure 2.1 Novasina water activity meter. By courtesy of Humitec Ltd, Unit 20, Southwater Industrial Estate, Southwater, Horsham, West Sussex RH13 7UD.

It had been shown by Gal (1972), however, that there was only about 0.2% difference between a_w defined on fugacities and RVP. Thus water activity has come to be widely accepted (Gilbert, 1986) as the effective concentration of water in a substance which controls that substance's susceptibility to biological and chemical spoilage.

As long as the vapour pressure of water in the food remains the same as free, pure water at the same temperature and pressure, the $a_w = 1$. If, for example, the vapour pressure is 50% of that exerted by free water, the $a_w = 0.5$, if it is 25%, $a_w = 0.25$. The water activity falls below 1 once all the free water within the food has been removed or bound to some extent. In the case of fish being cured, the water activity would be the vapour pressure exerted by the complex solutions of the fish cells divided by the vapour pressure of pure water at the same temperature.

2.2.2 Water activity and microbial spoilage

Scott (1957) put forward the idea that, since micro-organisms compete with solute molecules for the water they require for growth over the entire range of a_w for which they are viable, knowledge of a food product's water activity, amongst other factors, can give an indication of its preservation status. Table 2.1 gives an idea of the limiting water activities for the growth of various specific micro-organisms with examples of foodstuffs in which one would expect to find this a_w.

This, of course, is an oversimplified view in that it assumes that equilibration of a_w has taken place throughout the food, when, in actual fact, this situation is probably never reached, particularly in a system as complicated as curing fish muscle. Most samples would exhibit an a_w gradient from inside, out, or vice versa, depending upon whether the food is being dehydrated or rehydrated, salted or de-salted. Mossel (1975) questioned the choice of the term 'water activity', in the context of microbial spoilage prevention, on the grounds that deterioration was always controlled by a number of factors, of which a_w was only one. Furthermore, it only had a direct influence on spoilage rate when water content was so low that the mobility of reactants was severely reduced.

Broughall et al. (1983) have made predictions of lag and generation times for *Staphylococcus aureus* and *Salmonella typhimurium* based on their observations of the growth kinetics of these organisms at different water activities and temperatures (Table 2.2).

It can be seen from Table 2.2 that in order to reduce the possibility of fish going rotten at the centre before it is sufficiently cured, the water activity must either be reduced rapidly by, for example, making sure that the diffusion distance between centre and surface is small, as in sliced or minced fish curing, or by ensuring the curing takes place at low ambient temperatures.

Table 2.1 Limiting water activities and examples of appropriate foodstuffs for the growth of various specific micro-organisms

a_w	Micro-organisms inhibited	Food examples
1.00	None	Most fresh, high water content food
0.95	Gram negative rods like *E. coli* and spores of Bacillaceae	40% sucrose or 7.5% salt (NaCl) solution; breadcrumb, cooked sausage
0.91	Most cocci and lactobacilli. Vegetative cells of Bacillaceae	55% sucrose or 12% salt solution; Parma ham
0.88	Most yeasts	65% sucrose or 15% salt solution; salami; sausage; fishmeal
0.80	Most moulds. *Staphylococcus aureus*	Wheat flour; dry cereal grains and pulses; fruitcake; dry sausage
0.75	Most halophilic bacteria	26% salt solution; jams; fondants; kench-cured cod prior to drying
0.65	Xerophilic moulds	Marzipan; marshmallow; fishmeal dried to 5% moisture; 'stockfish'
0.60	Osmophilic yeasts	Liquorice; fruit gums; kench-cured cod after drying

2.2.3 *Water activity and water relationships in fish*

Duckworth and Kelly (1973) concluded that in water/solute/polymer systems the minimum a_w at which solvent action (and, by implication Scott's 'water competition' between microbes and solutes) becomes apparent depends upon the solute not the polymer. The amount of water *unavailable* as a solvent in such a system depends upon the nature of the polymers present. The greater the solute concentration of the surface cells of fish, the more water is held by interaction with ions and polar groups such as (–CO–, –NH$_2$, –OH) on other food components and, therefore, the lower is the vapour pressure compared with that of a free water surface at the same temperature. In addition to these, water may be bound into crystals or held by surface tension effects in capillaries both of which would tend to further reduce a_w. The effect of capillary radius on a_w has, in fact, been shown by Karel (1973) and is reproduced in Table 2.3.

It would be expected that a definite capillary effect on a_w would occur in cured fish with water activities around 0.9 since capillaries with a radius of 10^{-6} cm are likely to be quite common.

Table 2.2 Predicted lag and generation times for *Staphylococcus aureus* and *Salmonella typhimurium* (days)

Temperature (°C)	Growth phase/ organism	Water activity					
		0.88	0.90	0.92	0.94	0.95	0.96
10	L/Sa			>35	21.6	4.9	1.1
	GT/Sa				2.2	1.1	0.6
	L/St					>5	3.4
	GT/St						1.0
14	L/Sa	>35	33.0	12.8	1.9	0.83	0.55
	GT/Sa		2.6	1.1	0.48	0.32	0.21
	L/St				>5	4.9	1.8
	GT/St					0.53	0.32
18	L/Sa	24.5	11.0	2.9	0.72	0.47	0.33
	GT/Sa	1.2	0.61	0.32	0.17	0.13	0.09
	L/St				>5	3.2	1.0
	GT/St					0.35	0.20
22	L/Sa	12.1	4.3	1.1	0.43	0.29	0.20
	GT/Sa	0.58	0.31	0.16	0.09	0.07	0.05
	L/St				>5	2.1	0.65
	GT/St					0.24	0.13
26	L/Sa	6.3	2.0	0.63	0.27	0.18	0.12
	GT/Sa	0.30	0.17	0.10	0.05	0.04	0.03
	L/St			>5	4.1	1.4	0.43
	GT/St				0.17	0.17	0.09
30	L/Sa	3.6	1.1	0.40	0.17	0.11	0.08
	GT/Sa	0.15	0.09	0.06	0.04	0.03	0.02
	L/St			>5	2.8	1.0	0.30
	GT/St				0.12	0.12	0.06

L, lag; GT, generation time; Sa, *Staphylococcus aureus*; St, *Salmonella typhimurium*.

Table 2.3 The effect of capillary radius (assuming water meniscus contacts the capillary sides at an angle of 0°) on water activity

Capillary radius (cm)	a_w
1	0.9999999
10^{-2}	0.999895
10^{-4}	0.99895
10^{-6}	0.90
10^{-7}	0.20

2.2.3.1 *Water sorption in fish.* Searching for a scientific explanation for the capacity of fish, or any food, to retain water requires some examination of molecular adsorption theory. Most gases adsorb onto solids by weak intermolecular forces known as 'van der Waals' forces. Figure 2.2 shows that the surface takes up gas molecules until it becomes saturated. At this point the gas

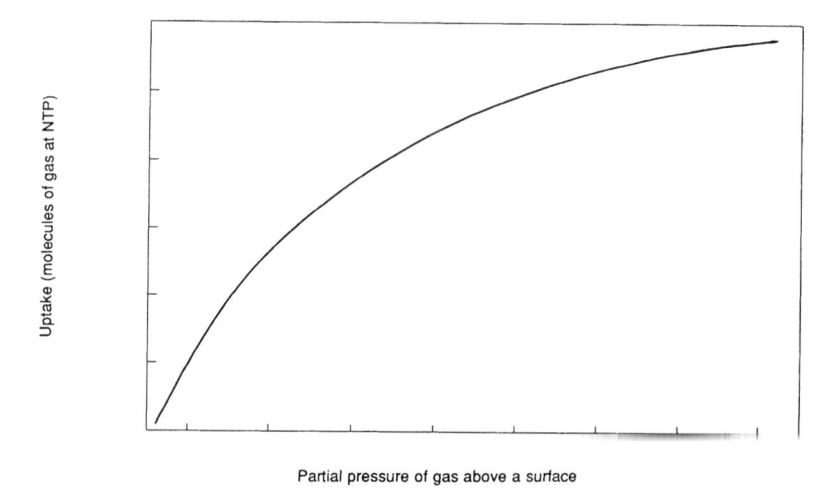

Figure 2.2 Uptake of gas against its partial pressure above the surface: Langmuir (1918) adsorption isotherm.

molecules act like a monomolecular layer of liquid on the surface. The adsorption of water molecules on to surfaces, however, was found by Brunauer et al. (1938) to be somewhat different. It is probable that the polar nature of water molecules causes them to attract each other and build up multilayers before the surface monolayer is complete (Figure 2.3). The amount of water which would be required to complete a monolayer over the surfaces of the main structural components of a food corresponds to (a_w) equivalent monolayer or a_{wm}. For starchy foods like potatoes, pulses, cereals and farina, a_{wm} is approached when the water content is approximately 6% of the dry weight, while for protein foods, like fish, where water is not held so firmly, Salwin (1959) found that a_{wm} is approached when the water content is about 4.5% of the dry weight. The importance of a_{wm} is that it corresponds to the optimum water activity for long-term preservation. At water activities greater than a_{wm}, microbiological spoilage may occur, whereas below a_{wm} the bare patches on the food surfaces are more rapidly attacked by oxygen leading to rancidification of fatty material, discoloration and lowering of nutritive value.

Generally a_{wm} does not correspond to what is commonly known as the 'bound' water content of food. The latter is defined as that water which cannot be frozen out of the food. Duckworth (1963) found 'bound' water to range between 16 and 50% of the dry solids weight, far above the water content at a_{wm}. It seems, then, that a_{wm} is the minimum amount of water associated with the food in protecting the active sites. If this equivalent monolayer is removed, these sites will react. When the water activity of foods is compared with their water content, graphs similar to that shown in Figure 2.4 are obtained.

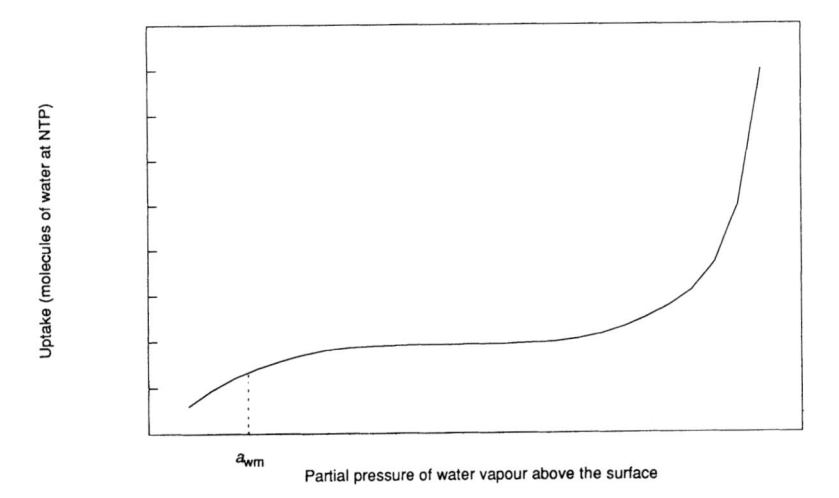

Figure 2.3 Uptake of water vapour against its partial pressure above the surface: BET type adsorption.

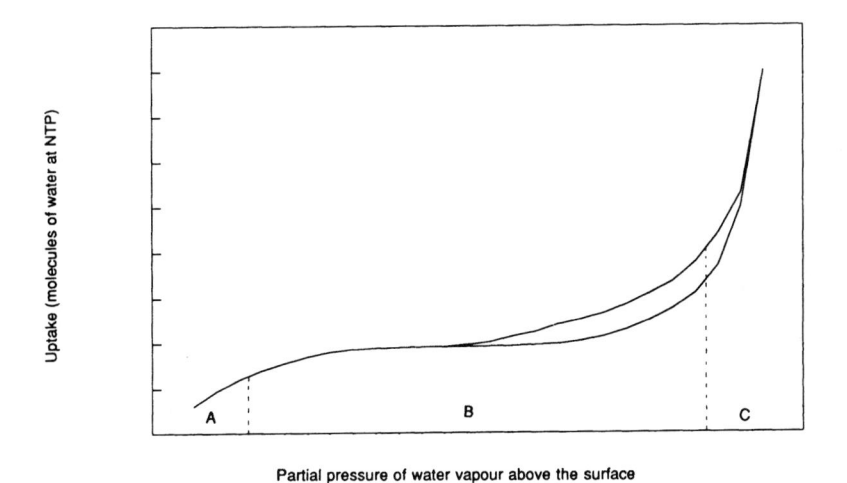

Figure 2.4 Typical food sorption isotherm showing hysteresis.

For a wide range of moisture contents, including those associated with fresh fish and many fish products, the water activity is close to 1. Dry foods, like cereals and stockfish, may have $a_w = 0.6$ and high solute content foods, like sugar preserves and salted fish, may have water activities between 0.6 and 0.8. However, Figure 2.4 shows that a great proportion of the water has to be removed from any food before there is a significant effect on water activity and, therefore, long-term storage potential.

In zone A of Figure 2.4 water is firmly bound by the structural components of the food, or is otherwise unavailable for reaction. It is not necessary to lower water activity to this level since the growth of micro-organisms is inhibited long before this zone is approached. In zone C water is said to exist in solution with food solutes and in capillaries formed by the food structure. This water is relatively free for reaction, evaporation and use by micro-organisms. In this zone the water activity is supportive of microbe growth and therefore water content must be reduced beyond this for long-term preservation if dehydration is to be the sole preservation mechanism. This leaves zone B, in which the water is loosely bound to the food structures and where changes in water content have a profound effect on a_w and, therefore, preservation.

2.2.4 Water relationships, preservation and product quality

In the curing of fish, we know that in order to achieve a state of long-term preservation, water activity in the centre of the fish must be lowered below some critical level, probably corresponding to a point on the curve within zone B in Figure 2.4, before microbial growth causes significant decomposition or, more importantly, a health hazard.

2.2.4.1 *Sorption hysteresis and quality.* If a foodstuff is completely dried and then rehydrated, the sorption isotherm does not retrace the sorption of the drying food. Figure 2.4 shows that the rehydrating dried food achieves higher water activity at lower water contents in the mid-portion of the graph. This behaviour is called 'hysteresis' and in the drying of fish is due to the fact that proteins aggregate and denature during the process and do not become associated with the same amount of water as the native proteins of fresh fish to which the desorption isotherm, obtained by Horner (1991) and given in Figure 2.5, pertains.

Cutting *et al.* (1956) showed that the temperature used for drying cooked fish samples affected the shape of the sorption isotherms obtained. The results of Curran and Poulter (1983), however, showed that neither fish species, nor salt concentration, nor variation of drying temperature within commercial norms had any significant effect on the sorption characteristics of cured tropical fish. Further investigation is, however, needed on this aspect. Horner (1991) noted that frequent over-drying of fish frequently leads to the formation of opaque yellow patches on the surface which rehydrate poorly, yielding tough, fibrous parts with a dry mouthfeel. This quality change, which may also have been due to irreversible myofibrillar aggregation, coincided with a widening of the hysteresis loop when compared to fish samples which had been less severely dehydrated.

Wolf *et al.* (1972) suggested that the width of the hysteresis loop might be used as an index of dried food quality deterioration through storage.

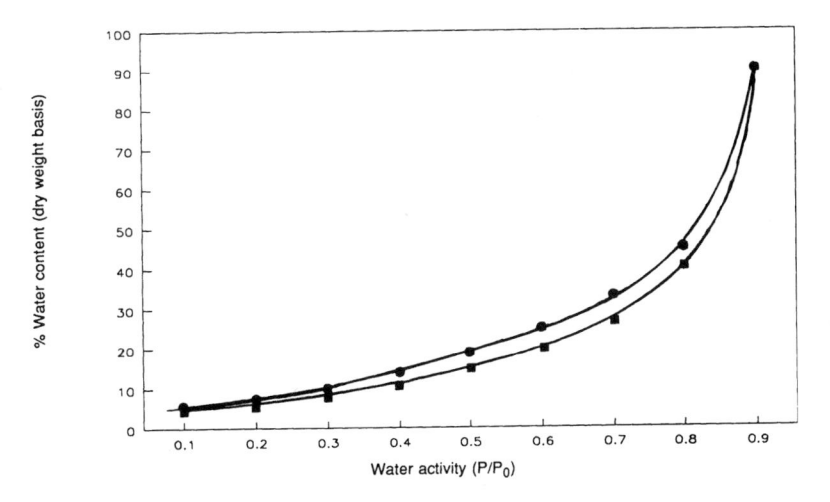

Figure 2.5 Sorption isotherm for cod (*Gadus morrhua*) muscle: (●) desorption; (■) resorption. After Horner (1991).

Haddock, for example, showed a 3% difference in water content, for a given a_w value, between freshly dried and stored samples. This increase in hysteresis during storage was accompanied by a drastic decrease in the samples' capacity to reabsorb water and, furthermore, upon sensory examination, showed a decrease in the quality attributes of colour, taste and rehydratability. It has been suggested by Labuza (1970) that protein–protein aggregation in dried foods would lead to poor solubility and toughening, engendering a gradual loss of quality during the storage period.

2.2.4.2 *Chemical changes and quality.* Water contents between 30 and 80% (dry basis) were found by Labuza *et al.* (1970) to coincide with the maximum rate of oxidation as shown in Figure 2.6. The products of oxidising lipids can interact with proteins to produce protein radicals which further react to form scission products of sufficiently low molecular weight to become mobile in the limited free water present in dried foods. If the product water content varies, as is the case in the varying humidities during cured fish storage, the increased reactivity and mobility of the protein and protein radicals would increase their tendency to aggregate. This, like over-drying, would have its widening effect on the hysteresis loop and upon the textural quality of the rehydrated product.

The quality status of the fish at the commencement of drying was shown by Williams (1976) to be capable of affecting the sorption characteristics of the product and the way in which these might change during storage. High levels of nucleic acids, as might be encountered in spoiling fish, signified high levels of ribose which readily reacted with proteins in the Maillard browning reaction. This, in turn, promoted the cross linking of proteins, causing the product

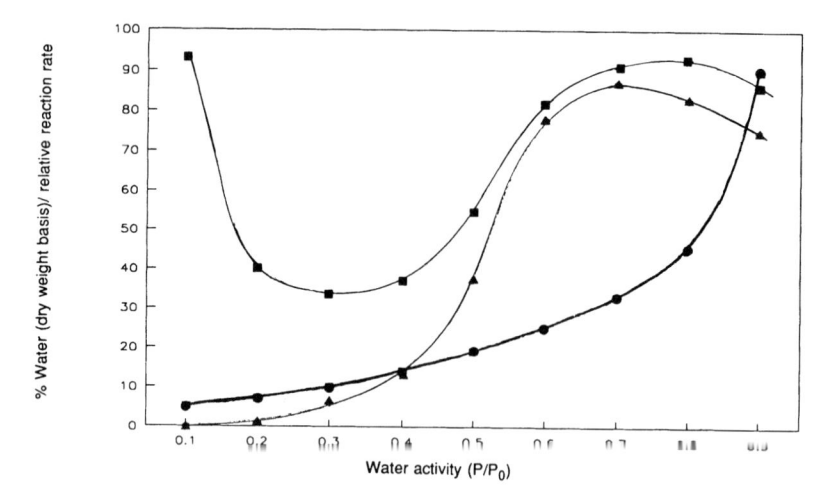

Figure 2.6 Dried product storage stability (susceptibility to lipid oxidation and Maillard browning) related to its water activity; (●) sorption isotherm; (■) lipid oxidation; (▲) Maillard browning. After Labuza et al. (1970).

to be tougher and less digestible. The maximum rate of Maillard browning was shown by Labuza (1970) to occur between a_w 0.6 and 0.9. Such a_w levels would coincide with the later stages of curing and with storage of cured fish.

All these factors, which appear responsible for both changes in sorption characteristics and product eating quality, such as levels of reducing sugars and lipids in the flesh, are subject also to variations which are associated with physiology, season and the method by which the fish is captured. Thus, although it might be a basis for a quality index, the extent to which dried fish exhibit hysteresis is not indicative of the causative agents of quality deterioration.

2.3 Drying

Three types of process can be employed in the drying of fish:

- Air or contact drying, where heat is transferred to the fish from heated air or a heated surface, utilises the air movement above the fish to carry away the moisture.
- Vacuum drying, where advantage is taken of the greater evaporation rate of water from the fish at reduced pressure, utilises conduction by contact with heated surfaces or radiation to evaporate the water which is removed by the vacuum pump.
- Freeze drying relies upon the attainment of very low pressures by highly efficient vacuum pumps in a sealed chamber containing the fish. The

PRESERVATION OF FISH BY CURING

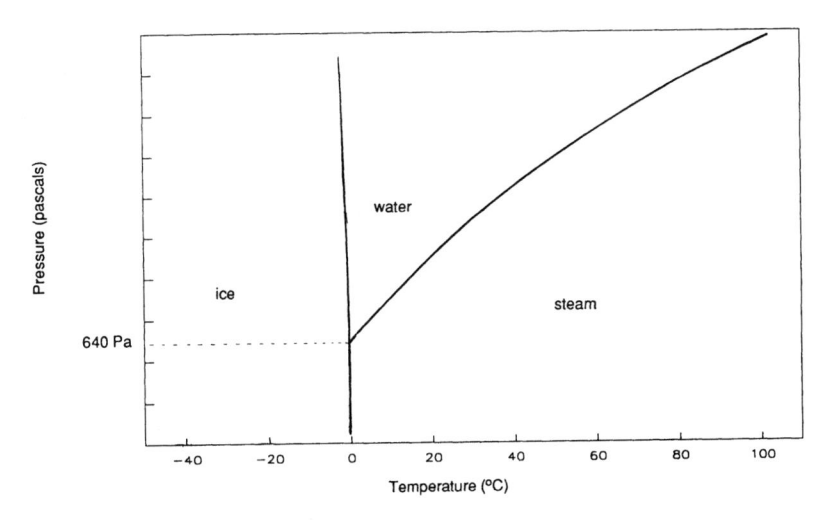

Figure 2.7 Phase diagram for water.

latter, in contact with refrigerated plates, freezes. At pressures below 0.64 kPa ice sublimes (see Figure 2.7) and the vapour is removed from the fish by the vacuum pump.

2.3.1 *Air or contact drying*

In most parts of the world, fish drying is still largely carried out in the open air, using the energy of the sun (solar drying) to evaporate the water and air currents to carry away the vapour. Modern refinements designed to accelerate this slow process have centred upon using higher temperatures to increase the evaporation rate, faster air speeds to carry away the water vapour and increase the surface heat transfer rate, and size reduction (atomisation) in the case of solids suspended in liquids to reduce the distance through which water vapour must diffuse. Such refinements and their possible applications to fish and fish products are discussed later. However, before proceeding further, it is necessary to define certain parameters which are properties of the drying medium, in this case, air.

2.3.1.1 *Psychrometrics.*
Absolute humidity (H) is the mass of water vapour contained by unit mass of air.

$$H = \frac{\text{water vapour (kg)}}{\text{dry air (kg)}}$$

Saturated air is air which, at a given temperature and pressure, cannot hold more water vapour without condensation taking place.

Relative humidity (%RH) is the absolute humidity divided by the humidity of saturated air under the same conditions of temperature and pressure assuming unit mass of dry air in both cases.

$$\%RH = \frac{\text{water vapour (kg)/dry air (kg)}}{\text{water vapour (kg)/dry air (kg) which is saturated at the same temperature and pressure}}$$

Specific heat (c) is the heat required to raise the temperature of 1 kg of dry air plus the water vapour it contains by 1°C.

Specific volume is the total volume of 1 kg of dry air plus whatever water vapour it contains at a specific temperature and pressure.

Dew point is the temperature to which an air/water vapour mixture must be reduced before the contained water vapour will condense.

Wet bulb temperature is very important in the initial stages of drying because water freely evaporates from the fish surface when that surface is at the wet bulb temperature. When water evaporates from a surface, the surface, in supplying the latent heat energy required to evaporate the water, cools to a temperature below ambient. The temperature gradient thus created causes a flow of heat energy from the air to the surface which in turn supplies more latent heat for further evaporation of water. Eventually an equilibrium is reached at a specific temperature, called the wet bulb temperature, where the loss of heat from the surface due to the evaporation of water is balanced by the heat gain by the surface from the air flowing over it.

Considering further the application of wet bulb temperature to fish drying, cod, for example, begins to cook at temperatures greater than 30°C. Practically, this could mean that the fish muscle could fall apart or drop from speats on which it was being hung to dry during the early part of the process. This would be the result of increased heat denaturation of the proteins and would also lead to a product which, for long-term storage, needed to be maintained at lower moisture levels than corresponding low temperature dried products at the same a_w level.

High wet bulb temperatures during the initial stages of fish drying and consequent protein denaturing also adversely affect the sensory properties of the fish. Denatured protein structures do not become associated with water in the same way, or in as large a quantity, as undenatured protein. Thus, when reconstituted, the mouthfeel of such fish will approach that of inert, wetted fibres, like cottonwool, departing from the succulence associated with fish and meat protein.

Wet bulb depression is the difference between temperature readings of two thermometers in an air current, one with its bulb uncovered and dry, the other with its bulb kept damp by enclosing it in a wetted muslin sleeve. In the initial stages of a fish drying operation, this corresponds to the difference between the temperature of the wet fish surface (wet bulb) and that of the drying air (dry

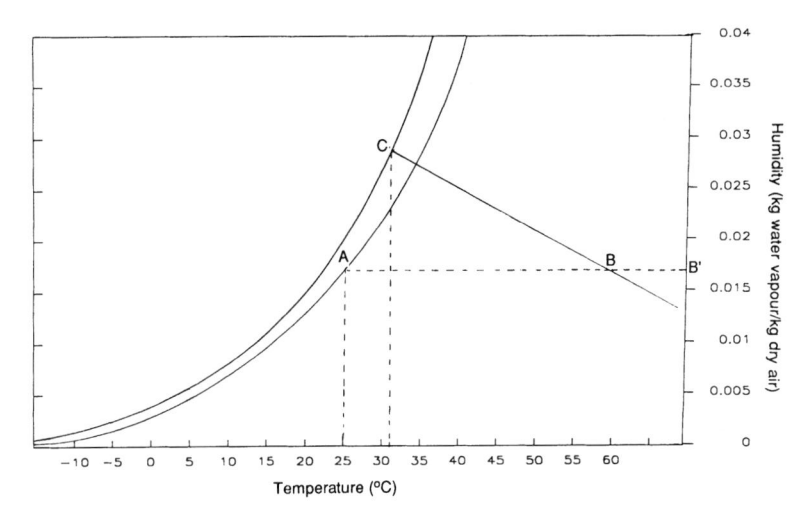

Figure 2.8 Psychrometric chart.

bulb). If the air were saturated with water vapour, the wet and dry bulb temperatures would coincide and no drying would take place. Thus in humid conditions, where the wet bulb depression was small, the rate of drying would be slow and it would be necessary to heat the air before passing it over the fish surfaces. The effect of such a heating process, applied to the drying air, on the wet bulb depression can be computed using a psychrometric chart (Figure 2.8).

Example

Air at 25°C and 80%, RH is heated to 60°C before passing over fish fillets. What would be the surface temperature of the fish fillets during the initial stages of drying whilst their surfaces remained wet?

Starting at 25°C on the ordinate, point A at the intersection of the 25°C line of the 80% RH sling represents the initial condition of the air. Moving horizontally parallel with the ordinate to the 60°C intersection, point B represents the condition of the heated air. Tracing back from point B, parallel to the closest adiabatic cooling line, to point C on the 100%, RH sling, the wet bulb temperature is given by the *x*-co-ordinate of point C and this will be the temperature of the wet fish surface.

Using this method it is possible to calculate the maximum temperature to which the drying air can be heated in order not to cause the fish surface temperature to rise above 30°C or whatever temperature has been decided as the upper limit for maintaining product quality.

2.3.2 Drying calculations

In terms of the rate at which moisture is lost, drying, as a process, has been separated into two distinct periods, the constant rate period and the falling rate period.

Constant rate period. While the surface of the fish being dried remains moist, water is considered to evaporate as from a free water surface. The evaporation rate is thus controlled by the conditions of the drying air, namely speed over the surface, its temperature and its humidity.

Falling rate period. As evaporation takes place at the fish surface, water travels to the surface by diffusion and via the capillaries formed by structural components. When the movement of this water through the fish can no longer keep pace with the evaporation rate, the surface becomes dry and the point in the drying sequence is referred to as the 'critical moisture content'. Thereafter the drying rate depends on the rate at which water vapour, evaporating from the ever-receding wet interface within the fish, travels to the surface to be carried away in the air current. Hence, the rate of drying falls progressively and, given the requirement that there must be air movement over the fish surface for this air to be capable of taking up moisture, is dependent upon the permeability of the already dry part of the fish. Temperature is the only aspect of process control which, because it affects diffusion rate, can purposefully be adjusted during the falling rate period.

To what extent the process of dehydration can be so conveniently divided as suggested in Figure 2.9, in the case of modern or traditional fish drying

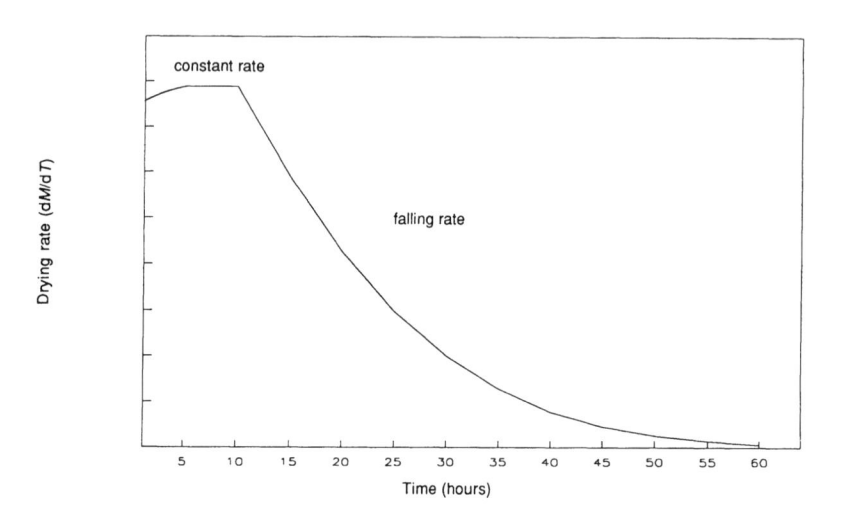

Figure 2.9 Idealised interpretation of the drying process.

PRESERVATION OF FISH BY CURING

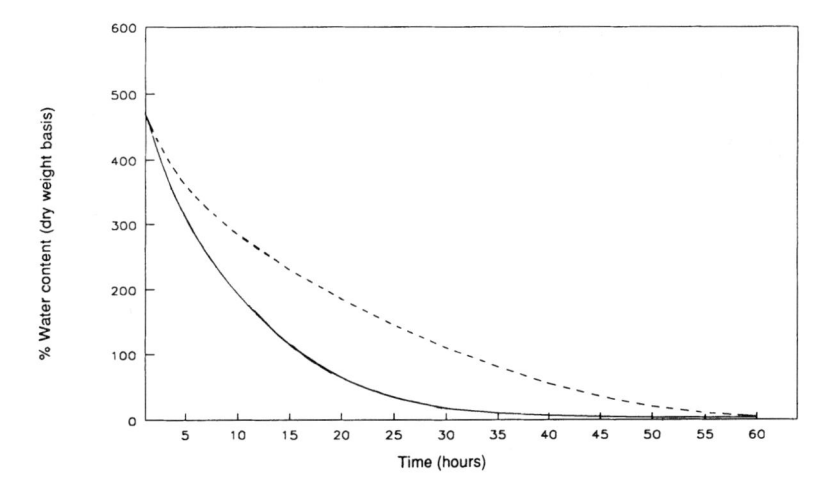

Figure 2.10 Effect of drying temperature on water loss from cod (*Gadus morhua*) fillets in a 1 ms^{-1} air current: (——) 60°C dry bulb temperature; (- - -) 30°C dry bulb temperature.

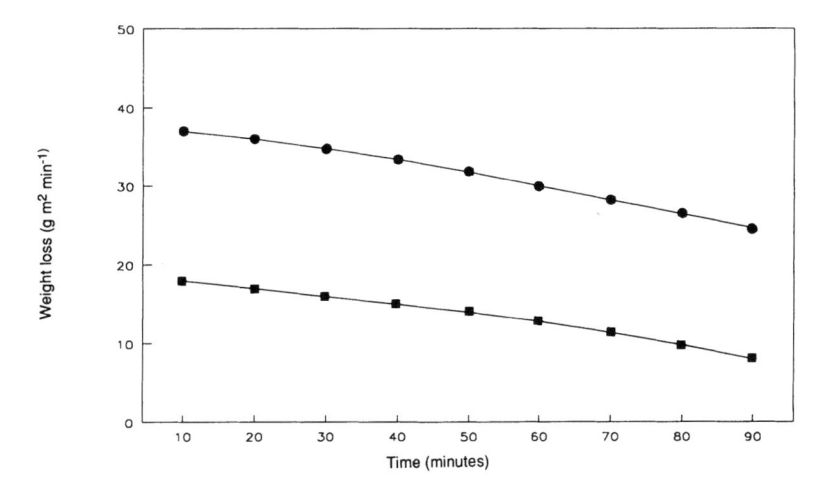

Figure 2.11 Effect of direction of water flow (parallel or perpendicular to the myofibrils) in cod (*Gadus morhua*) muscle blocks on drying in 1 ms^{-1} air current: (●) parallel: (■) perpendicular. After Horner (1991).

techniques has been seriously questioned in the light of drying curves obtained in practice with little or no identifiable constant rate period. Figure 2.10 shows water content against time curves for cod fillets drying under different conditions. Figure 2.11 shows drying rate against time curves when water movement is restricted to moving parallel or perpendicular to the myofibrils.

2.3.2.1 *Calculation of drying times in air.* During the period where the surface of fish behaves as a free water surface, a period which, as has been seen in Figures 2.10 and 2.11, may be brief or even non-existent, the vapour pressure exerted by water at the surface is greater than the vapour pressure of the air passing over it. Thus evaporation to the air stream takes place, cooling the surface and causing a temperature gradient between air and surface. Heat is transferred from the air to the surface and eventually a state of equilibrium is reached where the latent heat supplied in evaporating the water from the surface is balanced by the heat flow to the surface from the air. These are the conditions of constant rate drying and the surface remains at wet bulb temperature. Under these conditions, the rate of drying, $(dM/dt)_c$ is directly proportional to the wet bulb depression and to the air velocity raised to the power 0.8.

$$\left(\frac{dM}{dt}\right)_c \propto \theta_a - \theta_s$$

and

$$\left(\frac{dM}{dt}\right)_c \propto V^{0.8}$$

It is possible to relate mass transfer to heat transfer during the constant rate period

$$\text{Rate of water removal} \left(\frac{dM}{dt}\right)_c = -KA(p_s - p_a)$$

where K is the mass transfer coefficient, A is the area of the drying surface and $(p_s - p_a)$ is the vapour pressure difference between surface and air.

The heat transfer rate during the constant rate period dQ/dt is given by

$$\left(\frac{dQ}{dt}\right) = h_c A(\theta_a - \theta_s)$$

where h_c is the surface heat transfer coefficient. However, the dynamic equilibrium established in the constant rate period means that

$$Q = M h_{fg}$$

where h_{fg} is the latent heat of vaporisation of water at the surface temperature, θ_s. Thus

$$\left(\frac{dM}{dt}\right)_c \cdot h_{fg} = -\left(\frac{dQ}{dt}\right)_c$$

so

$$\left(\frac{dM}{dt}\right)_c = -\frac{h_c A}{h_{fg}}(\theta_a - \theta_s) \qquad (2.1)$$

For a tray of wet material of depth d

$$Ad = \text{volume} = \frac{\text{mass}}{\text{density}}$$

Thus, for a material containing unit mass of dry solids, density, ρ_s, becomes

$$Ad = \frac{1}{\rho_s} \quad \text{or} \quad A = \frac{1}{\rho_s d}$$

Assuming that no shrinkage takes place during constant rate drying, equation 2.1 becomes

$$\left(\frac{dM}{dt}\right)_c = \frac{-h_c}{h_{fg}\rho_s d}(\theta_a - \theta_s) \tag{2.2}$$

Integrating equation 2.2 over the constant rate period starting at the moisture content at time zero, M_0, and finishing at the critical moisture content, M_c, an expression for drying time, t, for a wet body containing unit mass of dry solids is

$$t = \frac{(M_0 - M_c)\rho_s h_{fg} d}{h_c(\theta_a - \theta_s)}$$

The surface of fish exposed to a stream of drying air dries out very early in the drying process and the rate of drying becomes dependent on the rate at which water moves from the ever receding 'wet band' inside the fish to the surface. This is referred to as *falling rate drying*. The point in the drying process at which constant rate drying ends and falling rate drying begins is called the *critical moisture content*. Figure 2.12 shows the drying rate falling off and gradually becoming very low as the equilibrium moisture content for the dried fish is approached. The rate-controlling factors in this falling rate period are difficult to quantify with respect to one another but involve the diffusion of water through the food, the moisture gradients and the changing energy-bonding pattern of the water molecules. Nevertheless, summation through the falling rate period, which is treated as a series of small constant rate periods, multiplying each by an approximate factor which corresponds to the mean rate in that small period divided by the rate during the true constant rate period, can be used to estimate drying time

$$t = \int_{tc}^{tf} \left\{ \frac{(M_0 - M_c)\rho_s h_{fs} d}{h_c(\theta_a - \theta_s)} \cdot f \right\} \cdot dt$$

Where t_c to t_f is the time period from the time of the critical moisture content to some time in the falling rate period at which it has been decided that the final moisture content has been reached (f is the appropriate factor denoting the actual mean drying rate divided by the constant drying rate, and dt is the time period over which this drying rate applies).

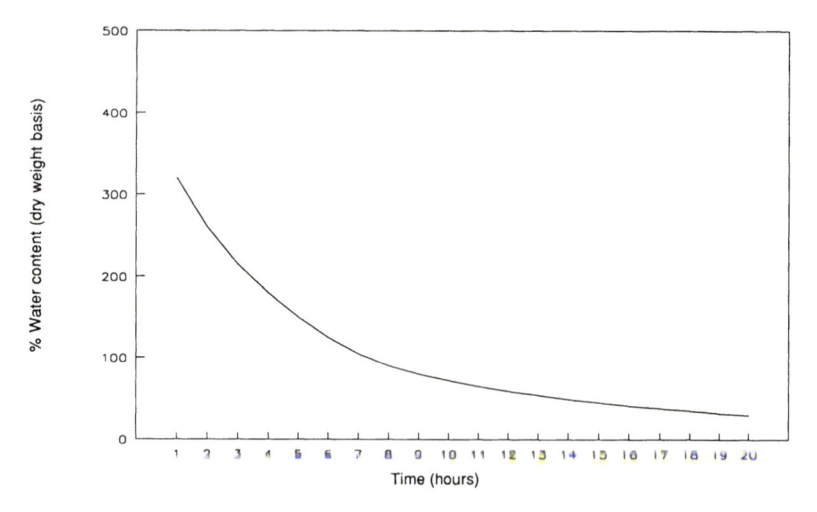

Figure 2.12 Falling drying rate.

2.3.2.2 *Inaccuracies in drying calculations.* The assumptions made in applying the constant rate drying equation, with respect to non-shrinkage of the fish during the drying process and the heat being supplied purely by convective means and being totally dissipated in the evaporation of water, are not strictly valid. In fact, shrinkage invariably occurs when fish is dried in air. Not all the heat transfer taking place is convective: there is some conduction and radiation. Also some of the heat transferred is expended in heating up the surroundings rather than supplying merely the latent heat for evaporation.

When applying the constant rate equation to the falling rate period, the values of factor f can only be discovered experimentally though they might be applied for different quantities of similar fish, *i.e.* fish of different species or sizes, or drying equipment. Constant rate drying is controlled by air velocity over the fish (since h_c is proportional to (air velocity)$^{0.8}$), the wet bulb depression $(\theta_a - \theta_s)$ and the surface area of the fish. The use of factor f in calculating the drying time in the falling rate period obscures the fact that drying rate is controlled virtually exclusively by the diffusion coefficient for water vapour through the dried fish to the surface. Since diffusion rate is proportional to temperature, the falling drying rate can be shored up by increasing the temperature of the air current. The dried out surface comes to the air temperature but, without water, can no longer suffer the denaturation and cooking associated with high temperatures during the constant rate period.

The latter part of any air-drying operation is very slow due to the inevitable formation of layers with low permeability as the fish shrinks. If this shrinkage and its associated lowering of permeability were to occur in the initial stages of drying, because of unsuitably high temperatures or air speeds over the fish surfaces, *case hardening* would occur. This formation of a relatively impermeable outer layer, so desirable in cooking by roasting, for example, effectively halts the drying operation leaving the moist interior to spoil rapidly. It may be noted that in a roasting or grilling operation the highest temperatures are used initially to seal the surface against excessive water loss as heat penetrates the centre of the meat.

If the drying process has identifiable constant and falling rate periods the transition point is called the *critical water content*. When the material is dried slowly, the water content at the surface is only slightly lower than the average water content but, when dried rapidly, the surface has a much lower water content than the average water content. Thus, under rapid initial drying conditions the critical water content is approached at higher average water content than under slow initial drying conditions. It is also true that thicker specimens of the same material exhibit higher critical water contents.

It has already been suggested in this chapter that in normal fish drying operations there appears to be no constant rate period, in other words the critical water content exceeds the initial water content so that all drying is falling rate drying.

2.3.2.3 *The falling rate period and its mathematical characteristics.* In the falling rate period, water vapour is transported from the 'dry-front' to the surface across an ever-increasing thickness of 'dry-layer'. The surface, initially at the wet bulb temperature, corresponding to the conditions of the drying air, warms gradually as heat is transferred to the dry front by conduction. In these circumstances the drying rate may be calculated if the following assumptions are made:

- the dry front temperature is constant and equal to the wet bulb temperature;
- the surface temperature is constant and equal to the temperature of the drying air (dry bulb temperature);
- the specific heat of the dry-layer is negligible;
- there is no evaporation taking place directly from the dry layer.

The depth (*d*) of the dry-front from the surface after time *t* is given by:

$$d = \sqrt{\frac{2k(\theta_a - \theta_s)t}{h_{fg}\rho_s E}}$$

where *k* is the coefficient of thermal conductivity of the dry layer and *E* is a dimensionless group representing its porosity.

An expression for calculating the drying rate, alternative to the summation of a series of diminishing constant rate periods as an approximation to the falling rate period, is determined by the relationship between water content at the surface and average water content of the product. It is a calculation of water distribution in the drying material as a function of time involving internal and external resistance to transport

$$\frac{\overline{\Delta C}}{\Delta C_0} = f\left(\frac{Dt}{d^2}\right), \left(\frac{Kd}{D}\right)$$

where $\overline{\Delta C_0}$ is the difference between the mean concentration of water in the material at $t=0$ and its mean concentration after an infinitely long drying period. $\overline{\Delta C}$ is the difference between the mean concentration of water in the material at time t and the mean equilibrium concentration of water which can be ultimately achieved, and D is the diffusion coefficient of water vapour through the dry layer. Diffusion, however, is not the only mechanism by which moisture is transported through drying material.

2.3.2.4 *Moisture transport mechanisms during drying.* The nature and structure of the food material (e.g. fibrous, gel-like, crystalline or amorphous) in its wet and dry states determines which mode of moisture transport is dominant. Also, in any given material, the dominant mode may change as drying proceeds. Four modes of mass transfer have significance in drying:

- capillary flow;
- liquid diffusion due to concentration gradients;
- vapour diffusion due to partial pressure gradients;
- diffusion in liquid layers absorbed at the solid interface.

Where mass transfer is mainly controlled by *capillary flow* in the falling rate period, the rate of drying, is expressed by

$$\left(\frac{dM}{dt}\right)_f = -K(M - M_e)$$

where M is the moisture content after a period, t, of falling rate drying and M_e is the material's equilibrium moisture content (i.e. when its a_w equals the relative humidity of the drying air). Since

$$K = -\left(\frac{dM}{dt}\right)_c \Big/ (M - M_e)$$

$$\left(\frac{dM}{dt}\right)_f = \frac{h_c(\theta_a - \theta_s)}{\rho_s h_{fg} d} \cdot \frac{(M - M_e)}{(M_c - M_e)}$$

Where mass transfer is mainly controlled by diffusion

$$\frac{(M - M_e)}{(M_c - M_e)} = \frac{8}{\pi^2}\left\{\exp[Dt(\pi/2d)^2] + \frac{1}{9}\exp[9Dt(\pi/2d)^2] + \cdots\right\}$$

but for large values of t (in fact, always in drying) this equation can be reduced to

$$\frac{(M - M_e)}{(M_c - M_e)} \approx \frac{8}{\pi^2} \{\exp[-Dt(\pi/2D)^2]\}$$

or rearranging this equation, to find the time of drying, t, in this period

$$t \approx \frac{4d^2}{\pi^2 D} \ln \frac{(M - M_e)}{(M_c - M_e)} - \ln \frac{8}{\pi^2}$$

This holds for a falling rate drying period where $(M - M_e)/(M_c - M_e) < 0.6$, and by differentiating this equation with respect to t, a rate equation similar in form to that for capillary transport may be obtained

$$\left(\frac{dM}{dt}\right)_f = \frac{D\pi^2}{4d^2}(M - M_e)$$

If $(M - M_e)/(M_c - M_e)$ is plotted against t, on a semi-logarithmic grid, it should produce either:

- a *straight line* indicating *capillarity* controlled falling rate drying; or
- an *asymptote* indicating *diffusion* controlled falling rate drying.

D may be calculated from the slope of the asymptote at any given point. The critical moisture content (M_c) must be found empirically since it varies with drying rate and the dimensions and structure of the food material.

When Jason (1958) plotted the 'free' moisture content $(M - M_e)$ of fish against time of drying on semi-logarithmic scale, he found the falling rate drying period essentially divided into two phases (Figure 2.13). The diffusion

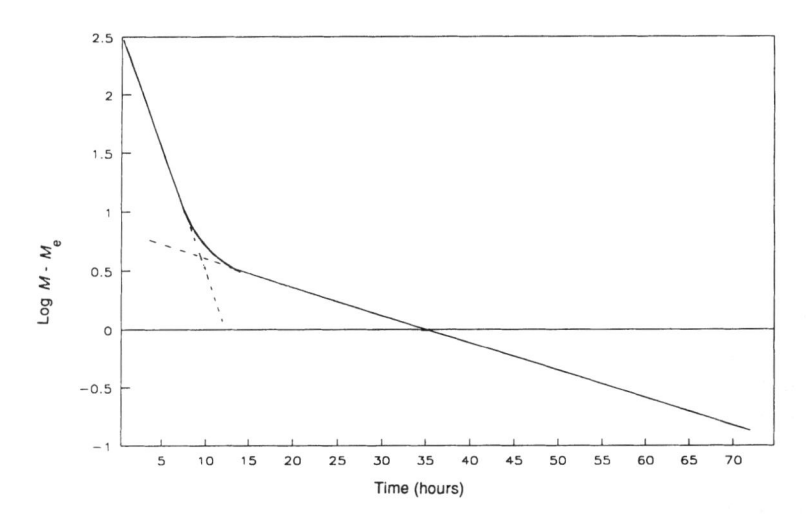

Figure 2.13 Semi-logarithmic plot of free moisture content (g) of fish against time of drying. After Jason (1958).

coefficient obtained from the straight line portion of the first phase is considerably greater than that of the second phase indicating the operation of different mass transfer mechanisms.

Kent (1985) found that the activation energies for these different mass transfer modes differed by an amount equal to the BET energy of adsorption into the monolayer. He therefore hypothesised that in the early stages of drying diffusion takes place through the bulk of the water, while in the later stages water molecules must move from one sorption site on the protein molecules to another by a kind of 'hopping' mechanism. This bimodal pattern of diffusion is also shown in the air drying of salted fish. The transitional point, at which diffusion from one mode to the other occurs is displaced towards higher water contents as the salt concentration is increased.

2.4 Salting

An alternative to lowering the water activity of fish flesh by merely extracting the water, as in simple dehydration, is to increase the concentration of solutes in lowering water activity in the flesh. Common salt, as shown in Table 2.1, is more effective than other safe, common and cheap food solutes, like sugar, even when present in relatively small concentrations. However, long-term cured product stability is only approached when the concentration of salt in the flesh reaches saturation concentrations.

2.4.1 *Water activity and shelf-life*

Where the objective of dehydration is to remove water from the deepest part of the flesh quickly enough to reduce water activity below the minimum for microbial growth, before significant spoilage takes place, the objective of salting is to ensure that the salt penetration is rapid enough to similarly lower the water activity in the deepest parts of the flesh. On completion of the process, a saline equilibrium between the muscle and the surrounding salt solutions is achieved. The maximum concentration attainable is that which corresponds to a saturated brine solution (*i.e.* around 26%) under normal temperature conditions. Practically, concentrations would be lower than this due to the presence of other solutes in the fish cells. Salt fish, then, at least theoretically, would have the water activity of a saturated common salt solution, 0.75, notwithstanding the extent to which it has been dried during and after the salting process. However, salted fish reduced to 'biscuit cure' dryness in drying kilns, or in the open air at low humidities, can absorb a considerable amount of moisture before the 0.75 water activity is exceeded, thereby initiating microbial growth leading to spoilage. Doe *et al.* (1983) used the Ross (1975) equation to predict the water activity of cured fish from their salt (M_s), water (M_w) and dry solids (M_b) contents. By observing how long it took for colonies of the dun mould, *Wallemia sebi*, growing on the cured fish, to become visible at different a_w values, Doe *et al.* were able to make a prediction of product shelf-life.

2.4.2 *The salting process*

Depending on the fish composition and size, salting may be 'dry', where the fish are stacked in salt and the brine formed is allowed to run away, or 'wet', where they are immersed in a strong brine, or 'pickle'. Perhaps the most commonly employed method is a hybrid of the dry and wet methods; the fish are placed in dry salt and eventually become immersed in the liquid pickle formed by solution of the salt in the liquid extracted from the fish. This is sometimes called 'blood pickle'. Size determines whether fish are salted whole and uneviscerated, eviscerated and split open, or in smaller pieces ranging from fillets to mince. The barrier presented by fish skins to salt penetration means that only small species, such as anchovies and small herring, can be salted whole without gutting. Larger fish salted in this manner would deteriorate at the centre before salt could exert its effects there.

Sodium chloride diffuses through the fish flesh by a dialysis mechanism and water will diffuse to the outside due to the osmotic pressure between the brine and fish muscle solution. This process does not continue indefinitely; sodium and chlorine ions form a water-binding complex with protein which itself exerts an osmotic pressure, eventually balancing that due to the surrounding brine, and an equilibrium is reached.

2.4.2.1 *Salting methods.* Four types of salting methods, brining, pickling, kench curing and Gaspé curing, will be considered:

- Where the ultimate concentration of salt in the fish is required to be sufficient only for flavouring purposes and preservation is effected by other techniques such as smoking, fish are treated for several minutes only in less than saturated brines. A secondary effect of such *brining* is the elution of soluble proteins, which form an attractive, glossy pellicle on the surface as moisture is allowed to evaporate on standing before subsequent processing continues.
- The immersion in concentrated brines for long periods, *pickling*, is generally used for longer term preservation, mainly of fatty fish. The restriction of oxygen access by immersion retards rancidity reactions, although some rancidity is desirable in the development of characteristic flavour.
- Where fish are split, opened out flat and placed in layers interspersed with layers of salt (Figure 2.14), and the liquor which exudes is allowed to drain away, a long shelf-life dry product is produced. This is called *kench curing* and is used for white, non-fatty fish.

Figure 2.14 Kench (dry salt) curing of fish.

- If, instead of allowing the exudant liquor to run away, the dry salting is carried out in tubs, the split fish float in the brine formed. Weights are used to keep the fish immersed for 2–3 days, after which they are taken out and dried in the sun or in kilns. This is called the 'Light' or *Gaspé* cure after the peninsula in Eastern Canada where it originated.

2.4.2.2 *Types of salt.* Four types of salt, solar, brine-evaporated, rock and manufactured salt, may be used:

- *Solar* salts are prepared by the evaporation of sea or salt-lake water by sun and wind. Lagoons constructed on the shore line are flooded, sealed-off and allowed to evaporate to dryness. The salt obtained is very impure due to the multiplicity of salts other than sodium chloride, and, when it is dug out, may be further contaminated by sand from the bed of the lagoon.
- *Brine-evaporated* salts are prepared by the application of heat to evaporate strong brines pumped from deep mines. The purity of such salts depends on the nature of the underground deposit but they are less likely to be contaminated with sand than solar salts.
- *Rock* salt is mined from underground deposits of varying purity from 80 to 99% sodium chloride.
- Purified *manufactured* salt may contain 99.9% sodium chloride and be derived from any one of the three types of salt mentioned above, which may contain up to one fifth of their weight of impurities.

The major impurities are sand and water followed by the chlorides and sulphates of calcium and magnesium, sodium sulphate and carbonate, together with traces of heavy metals, such as copper and iron.

Very pure sodium chloride would be the most desirable for curing purposes except that it tends to yield a slightly yellow product and many markets associate highest quality with the whitest cure possible. Nevertheless, this is by no means a universal preference since in many areas a darker fleshed product, for example, saithe, is preferred to white fleshed cod. A whiter cure is, however, promoted by the presence of up to approximately 0.5% calcium and magnesium impurities in the salt, although greater levels of such impurities impart bitter tastes and hydroscopicity to the product. Since many of the major markets for dry salted fish are in the tropics, where high relative humidity leads to very rapid moisture uptake by hydroscopic products, it may well be worth risking a slightly more yellow product rather than a white one which will spoil rapidly when exposed for sale. For such markets, even a product that acts like a pure solution of sodium chloride, which would come into dynamic equilibrium with a surrounding atmosphere of approximately 75% relative humidity at ambient temperatures, will take up moisture for much of the year when the ambient relative humidity exceeds this. The occurrence of pink patches, indicating outgrowth of halophilic bacteria, in dry-salted fish displayed for sale in tropical markets is common. Thus, fish cured with very pure (sodium chloride) salt and dried to 'biscuit cure' will maintain its quality for the greatest length of time under such conditions of display. Figure 2.15 shows the speed with which differently cured cod fillets gain and lose moisture in cycling humidity conditions. At constant high humidity (90% RH) even biscuit cure dry-salted fish could take up sufficient water to exceed the a_w growth minimum for halophiles within 24 h. Heavy-metal salt impurities (*e.g.* copper at > 0.5

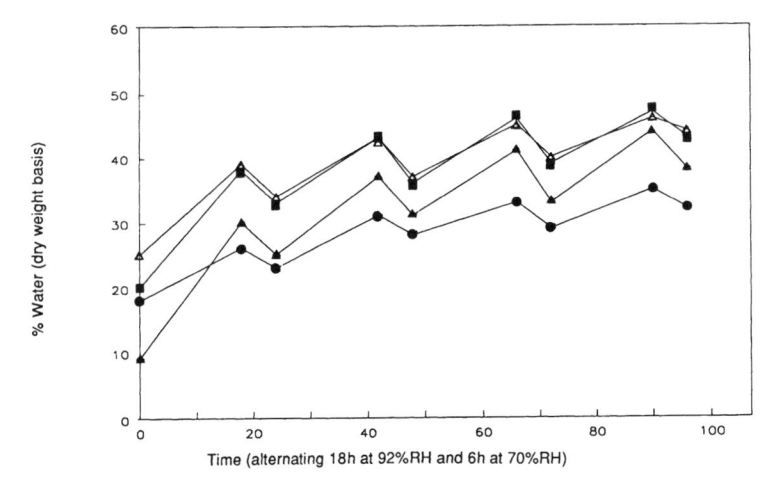

Figure 2.15 Water content changes in differently cured cod fillets under conditions of cycling humidity: (●) dried only; (■) dried/dry salt; (▲) dried/brined; (△) dried/hot smoked. After Horner (1991).

p.p.m. or iron at > 30 p.p.m.) cause undesirable yellow or brown stains in the product. It should be noted that very pure salts, as well as impure salts, have been implicated in this particular product fault.

The salt grain size is also of significance. The smaller the grain size the more readily the salt dissolves in water withdrawn from the fish and the more rapidly, therefore, will the salt diffuse through the tissues of the fish. Unfortunately, the initial speed at which water is withdrawn from the fish in wet stack causes a great deal of the salt grains to be washed away down the drains. The use of a coarse-grained salt is, consequently, less wasteful and adds only very marginally to the process time.

2.4.2.3 *Factors affecting the salting process.* The rate of diffusion of salt into, and the exosmosis of water out of, the fish tissues is proportional to the concentration gradient between the salting medium itself at the surface, and at the point in the fish most remote from the salting medium. Hence the stronger the brine, the faster will be the salt uptake and the consequent attainment of an a_w low enough for preservation. The contrary experimental observation that brine salting is initially faster than dry salting is a result of the better contact between fish surfaces and salting medium. Moreover, a saturated brine solution is, in some circumstances, unsuitable because the fish dries with a powdery film of salt on the surface where a glossy pellicle would be more desirable, for example, with smoked products.

The greatest barriers to salt penetration are the skin and scales. Where fish are salted in their whole form, anchovies for example, it appears that water activities much less than 0.90 are rarely achieved so that the risk of deterioration or contamination is high. Food poisoning organisms, like *Staphylococcus aureus*, could survive and grow at such water activities, although toxin production would be unlikely. In the case of anchovies, the oiliness of the flesh must also have a restraining effect on water and salt movement. Additionally, the high protein content (18–19%) appears to engender the attainment of osmotic equilibrium long before the internal salt concentration approaches the external salt concentration. The ultimate pH of fish muscle is usually above the isoelectric point of its constituent proteins such that the negative chlorine ion, Cl^-, is preferentially attracted to the more numerous positively charged sites on the protein. This, in turn, increases the water-holding capacity of the proteins and makes water removal more difficult. Calcium, Ca^{2+}, and magnesium, Mg^{2+}, ions from impure salts bind more readily to the negative sites on the protein than sodium, Na^+, ions so there exists a possibility that such undesirable impurities will be concentrated in the cured flesh.

Except where fish are too small and delicate for individual handling, as with anchovies, they are split and eviscerated prior to dry or brine salting. With larger species, such as shark and dogfish, and larger specimens of other species, muscles may also be scored to assist salt penetration. In dry-salting cod at 10°C, salt concentration reaches 10%, at 25 mm in 24 h, whereas, it takes

3 days to reach this concentration at 50 mm depth. At tropical ambient temperatures, such a delay would allow decomposition at the centre of thick muscles before the salt concentration had reached levels high enough to prevent bacterial spoilage. The condition is referred to as 'putty fish' and is also often associated with the Ca^{2+} and Mg^{2+} ions of impure salts binding with protein and forming a barrier to the passage of Na^+ ions to the thicker part of the flesh.

In dry curing, the salt extracts water from the fish causing a 25% weight loss in wet-stacked 1 kg split cod in the first 4–5 days. The salt dissolves in the extracted water and forms a highly concentrated solution at the surface of the fish. This gradually diffuses in, achieving an aggregate concentration of about 18% in 8 days, and ultimately, at the end of 15 days in wet stack, about 20%. At about 9–10% salt concentration in the fish, the muscle proteins denature and become fairly easy to separate from the skin, although this is rarely done because the unsupported muscle would fragment and presence of the skin is important to the consumer for product identification.

Typical changes in percentage water and percentage salt content during dry salt curing of white fish are as shown in Table 2.4.

Table 2.4 Typical changes in percentage water and percentage salt content during dry salt curing of white fish

	Water (%)	Salt (%)
Fresh fish	80	0.1
Fish after 15 days wet stack	50	20
'Biscuit' cure fish after drying subsequent to wet stack	25	26

2.4.3 Storage: maturing and spoilage

When no more water can be removed from the fish by re-piling in wet stack they can be held for months, or even years, maturing or 'pining', provided the temperature of the pining room is held below 10°C.

2.4.3.1 *Maturing or ripening process.* Maturing is an even more significant part of the flavour development in brine-pickled fatty fish. Having lost up to 20% of their weight through exosmosis of water to the brine, herring fillets regain their original weight through salt uptake within 10 days. The enzymes responsible for maturing or 'ripening' may be derived from the digestive system of the fish (sometimes the pyloric caeca are left *in situ* to encourage this kind of ripening), the fish muscles, and bacteria growing on the fish and in the brine. Although the nature of the ripening reactions is exceedingly complex, it is thought that the products of proteolysis and lipolysis are predominant in the ripened product. Lipolysis and oxidative rancidity play an important role in the flavour of even low-fat, cured white fish products. The products of Maillard browning reactions also make a significant contribution to the flavour. The sugar, ribose, released during the degradation of ATP, is particularly active in Maillard type browning. In dry, salt-cured fish, however, any browning is undesirable and can render the product unfit for sale. High levels of ribose and certain amines from protein decarboxylation, such as would occur in fish which was less than fresh, encourage Maillard browning.

2.4.3.2 *Microbiological spoilage.* Most of the micro-organisms normally associated with fish spoilage, for example, *Pseudomonas* spp., are halophobic and will not grow in salt concentrations exceeding 5%. There are, however, certain organisms that may be both common and pathogenic, and which are halotolerant, growing in a 10, or even 20% salt environment. *Staphlyococcus aureus* is a highly significant example.

The most important spoilage micro-organisms are the halophiles which actually require salt for growth and will not grow unless 10% salt is present. These bacteria, which are responsible for *pink* spoilage, so-called because of the colour of their colonies and consequent appearance of the cured fish, include *Halobacterium salinaria, H. cutirubum, Sarcina morrhuae* and *S. litoralis.* They are aerobic and usually not found in pickled fish where only limited oxygen access is possible through the brine. They are also thermophilic with an optimum growth temperature of about 42°C and minimum growth temperature of 5°C.

The first sign of pink spoilage is a delicate pink sheen on the surface of the fish in wet stack or during pining. This can be easily rubbed off without damaging the fish. Treatment with formaldehyde or sulphur dioxide vapours, or dipping the fish in a solution of sodium metabisulphite prevents recurrence, although maintenance of the ambient temperature below 10°C is likely to prevent initial germination and growth. Cases of food poisoning said to have been caused by the consumption of pink-spoiled fish have probably, in fact, been due to the growth of exotoxin-producing *Staphylococcus aureus*. The latter will commence growth at water activities slightly higher than those required for growth of pink bacteria. Pink bacteria themselves have been proved to be non-toxic and non-pathogenic.

The water activity of salt fish after drying is too low to support bacterial growth but, should temperature and humidity conditions become suitable, certain osmophilic moulds can grow. *Dun* spoilage derives its name from the brown surface discoloration caused by the growth of moulds of the *Wallemia* genus. They are able to grow in salt concentrations between 5 and 26% although they are not specific to sodium chloride and can grow on osmotic equivalent concentrations of potassium chloride, ammonium chloride, glycerol or glucose. Hence they are obligate osmophiles rather than halophiles. Other conditions for growth are: (i) temperature 10–37°C (optimum 25°C); (ii) pH 4.0–8.0 (optimum 6.0 to 7.0); and (iii) optimum relative humidity 75%.

Unlike the pink bacteria, dun moulds do not decompose the flesh but make the surface unsightly and, consequently, the product less saleable. They can be brushed off the surfaces but growth will rapidly recur if dry, cool conditions are not maintained. Old and rotting wood harbour such moulds, therefore wood should be avoided in dried fish stores or kept well covered. The causative agents of both pink and dun spoilage abound in solar curing salts, so the maintenance of low temperatures and humidity are the essential means of combating such spoilage during production and storage. Unfortunately, much of the market for dry salted fish is in areas where a hot, humid climate predominates, therefore the prevention of moisture ingress is an essential feature of any packaging used. The use of plastic bags is unsuitable because any temperature fall in the surrounding ambient would cause condensation. Dipping cured fish in vegetable oil approximately halves the rate of moisture uptake from a humid environment. This might be sufficient to delay microbiological spoilage beyond the required storage life.

2.4.4 *Other salted fish products*

Fish salting is practised world-wide, both to preserve and to extend the variety of products available. The following sections represent only a sample of fish products found.

2.4.4.1 Salt-boiled fish. The open-pan boiling of fish in brine is a much exploited method of preservation often linked to some other preservation technique, such as canning. In S.E. Asia, however, it is a commercially significant process in its own right. The products from different processes have shelf-lives at ambient temperatures of 1 or 2 days to years depending on the extent to which water activity is reduced. Boiling inactivates enzymes and kills virtually all the non-spore forming micro-organisms, thus delaying spoilage for a day or two and enabling the vending of fish in markets remote from the point of landing. This is useful in hot, humid climates where solar drying may not be practicable. However, even if the product were to be aseptically and hermetically sealed, it would spoil very quickly due to the outgrowth of bacterial spores on cooling after processing. Indeed, such products, having been recontaminated by handling during packaging, may be an even more likely source of food poisoning.

Longer shelf-life products require the addition of salt, in varying quantities, prior to or during the boiling process. Storage lives of up to 3 months in ambient tropical conditions may be achieved if the fish are cooked in salt until no free water remains at the bottom of the container, and the fish surfaces are covered with a thick layer of salt at the top of the cooking vessel and then sealed with paper. More efficient sealing, in glass jars that have lids including a rubber sealing ring, can extend storage life for up to 9 months.

In Indonesia, such salt-boiled fish products, called 'pindang', are popular and nutritious. However, occasional implication of pindang in cases of sickness or even death, emphasises the need for greater control. The use of lead-glazed cooking pots and galvanised vessels, resulting in lead and zinc poisoning, respectively, or the employment of too short a cooking time, or too little salt, resulting in microbial intoxification of the food, are serious malpractices which must be discontinued by strict adherence to codes of good manufacturing practice.

2.4.4.2 *Fish wood.* Another Indonesian product, 'ikan kaju' or 'fish wood' closely resembles its description and has a considerably longer shelf-life, perhaps several years. It is produced by repeated salt-boiling, drying, smoking, pressing and reboiling the fish to a horny translucent block that can be stored without packaging at ambient temperature, and is used as a condiment to be grated over bland, starchy foods in the cooking process.

2.4.4.3 *Dried shrimp.* Prepared as a soup condiment in many parts of the world, shrimp considered to be too small for commercialisation may be boiled, often in seawater, sun-dried or smoked to yield a long storage-life flavoursome addition to basic soup or cereal dishes.

2.5 Smoking

2.5.1 *Introduction: preservation, titivation or camouflage*

Originally, smoking of fish was an incidental occurrence when, in periods of wet or humid weather, fishermen had to resort to the use of open fires, rather than sun and wind, to dry their surplus catch. It was much later that the microbicidal and antioxidant effects of smoke processing came to be appreciated. Long before this realisation, consumers acquired a taste for smoked fish as a pleasurable alternative to the consumption of fresh fish and, by reducing the 'severity' of the process (*i.e.* the degree to which the fish is salted, dried and smoked), such products have won even wider appeal. In the UK, this reduction in process severity has increasingly led to the view that the smoking process, as practised on fish in this country, is a cosmetic rather than a preservative process to titivate the appetites of the (only slightly) adventurous consumer.

Another function of smoking, though frowned upon, has been its very widespread use as a means of eking out the shelf-life of fish remaining unsold after retail display. Thus, the process has been used to camouflage the fact that fish might be less fresh than it should be. There are great dangers in using the process for this motive, the least of which is that of alienating the purchaser against repeat fish purchases. Perhaps a more ethical way in which smoking has been used as a process of camouflage is the copying of expensive luxury fish products using inexpensive raw materials. An example of this is the German product 'seelachs', which consists of dyed, lightly smoked slices of coley fillet used as a substitute for smoked salmon.

Preservation, nevertheless, is still the prime objective of fish smoking in most parts of the world. This may be preservation for very long periods as with the Indonesian product, ikan kaju, or the addition of an extra day or two on the expected shelf-life of fresh fish to allow it to be distributed to remote markets, or to keep slightly longer on retail display. Even when the process has been greatly reduced, with the aim of attracting new consumers to a subtle taste difference, or increasing the yield of finished product from raw material, extended shelf-life still appears to be expected by manufacturers, retailers and consumer alike. This can be readily appreciated from the differences between what the Torry Advisory Notes (TAN No. 14) recommend as anticipated shelf-life and what producers and retailers seem to expect.

Smoked fish is regarded as a delicatessen food item, to be consumed on special occasions, or presented as a tempting alternative in our diet, which is nevertheless inexpensive. As such, it shows great potential as a technology which could be carefully managed to ensure the wider consumption of *safe, high quality* fishery produce.

2.5.1.1 *Preservation objectives.* The preservative effect of smoking on fishery produce is said to be due to a combination of the following four factors:

- surface drying, which provides a physical barrier to the passage of micro-organisms and a hostile environment for any aerobic microbial proliferation;
- salting, which reduces a_w and inhibits the growth of many spoilage organisms and pathogens (although a reduction of a_w below 0.95 is required before this effect becomes significant and at this level the saltiness may be too high, about 5%, for consumer tastes);
- deposition of phenolic antioxidant substances, which delays autoxidation (and rancidity) of the generally highly unsaturated fish lipids (as shown in Figure 2.16);
- deposition of antimicrobial substances such as phenols, formaldehyde and nitrites.

2.5.2 *Smoke production*

In many parts of the world, wood is used in preference to sawdust. This makes a hotter fire with less smoke and the fish is charred rather than smoked. In

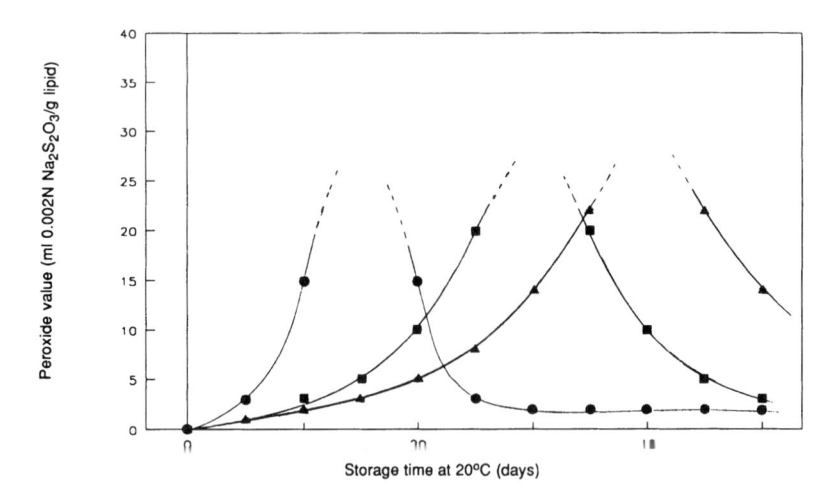

Figure 2.16 Rancidity development in differently cured samples of rainbow trout (*Salmo gairdnerii*) dried to the same water activity; (●) air dried; (■) smoked (30 °C) and dried; (▲) hot smoked (60°C). After Masette (1990).

a sawdust fire, unless there is a forced rapid flow, the air cannot easily get to the fire, so the sawdust smoulders rather than burns. Lower temperature and less oxygen give a smoke with more flavouring and preserving substances. Higher temperature and more oxygen waste these substances by oxidising them to carbon dioxide and water. Modern smoke producers feed sawdust slowly onto a very hot surface, but, worldwide, most smoke for smoking is still produced from a simple fire (Figure 2.17). The fire is usually contained in a fire box. Wood shavings, when lit in a good draught, burn fiercely and start the smouldering process in the sawdust. Both shavings and dust should be dry and free from wood preservatives. Damp material harbours moulds and the smoke will carry these onto the fish. Wood preservatives may produce harmful smoke which might make the smoked fish dangerous to eat.

2.5.2.1 *Smoke components.* The species source of the sawdust affects end-product flavour. Hard woods, such as oak, hickory, cherry, apple and beech, burn to give a smoke with more phenols, which both preserve and give a characteristic, 'medicated' flavour to the product. However, whether a hickory-smoked product can be distinguished from a similarly beech-smoked product, especially after the product has been cooled and stored, is doubtful, although the former appears to have the greater sales appeal.

Smoke is an emulsion of droplets in a continuous phase of air and vapours stabilised by electrostatic charges on the droplets. For flavouring, colouring and microbistatic purposes, the vapours are of greatest importance in

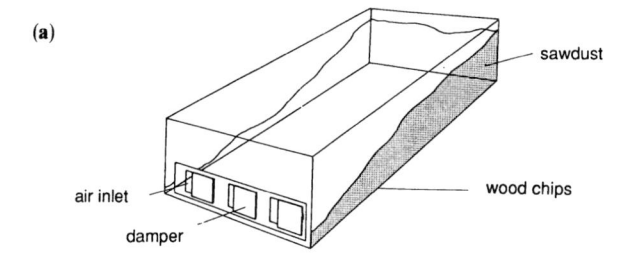

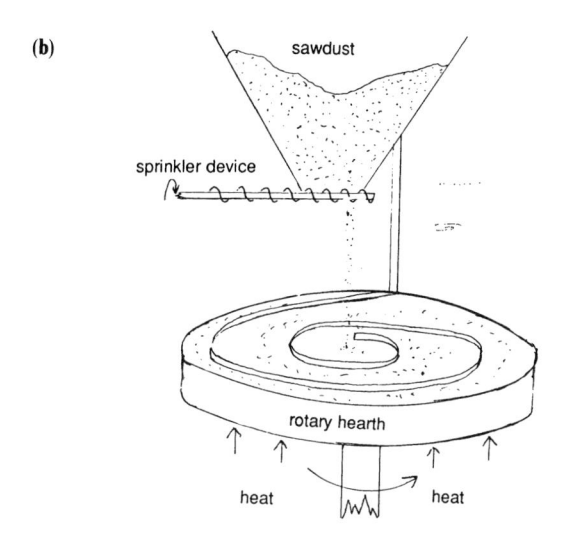

Figure 2.17 Smoke production: (a) simple; (b) continuous.

smoking. Over 200 components have been identified in the vapours, and some of the major components are listed in Table 2.5.

A dynamic equilibrium exists between droplet and vapour phase of the smoke, which, according to Tilgner *et al.* (1962), is continually changing with fluctuations in temperature, ratio of air to smoke, smoke–air velocity, and due to uptake of smoke components by the fish surfaces. The droplet phase, in fact, acts as a reservoir of volatile and non-volatile smoke components, releasing more volatiles as they are absorbed from the vapour phase.

The studies of Foster *et al.* (1961) have shown that when wood smoke is diluted by air, the concentration of the more volatile phenols falls to a greater extent than that of the non-volatile phenols. Within the temperature range 30–80°C, which covers all of the cold smoking process and most of the hot smoking process, adsorption of smoke vapours was independent of tempera-

Table 2.5 Some major components identified in smoke vapours

Acids	Phenols	Carbonyls	Alcohols	Hydrocarbons
Formic	Syringols	Formaldehyde	Ethanol	Benzpyrene
Acetic	Guaiacols	Propionaldehyde	Methanol	Benzanthracene
Butyric	Cresols	Furfuraldehydes		Indene
Caprylic	Xylenols	Octyl aldehyde		Naphthalene
Oxalic		Acrolein		Stilbene
Vanillic		Methyl ethyl ketone		Fluorene
Syringic		Methyl glyoxal		Phenanthrene
Phthalic				

ture, while deposition or absorption of the non-volatile phenols increased with temperature.

2.5.3 Quality, safety and nutritive value

Control of smoke production parameters must be complemented by control of raw materials if a standard quality product is to be achieved. Since most of the smoke components found in a smoked fish product are absorbed by the surface and interstitial water of the fish muscle, it is important that the fish remains wet, at least for part of the smoking process. Foster *et al.* (1961) have shown that the rate of absorption of smoke phenolic compounds by pre-dried fish is only 5% of that absorbed by wet fish. The objectives of modern smoking procedures should be to impart the desired sensory characteristics to the product uniformly, without undue variation from batch to batch, and to extend product shelf-life whilst avoiding the deposition of known carcinogens.

Colour imparted to the fish by the smoking process is due to carbonylamino reactions of the Maillard type and has been correlated with a quantitative decrease in carbonyl groups in the smoke. Brown pigments forming in the surface tissues, however, were said by Ziemba (1969) to inhibit further penetration of carbonylic groups and other smoke components to underlying tissues. Ribose, from the degradation of nucleotide and ribonucleic acids and free amino compounds, such as anserine and taurine, in the fish muscle extractives also contribute to browning at the surface of drying fish. Jones (1962) found that, as the fish muscle spoils, β-alanine-1-methyl histidine and lysine become increasingly important in the extent of browning. Thus for a standard smoking operation, the condition and extent of spoilage of the raw material can vary the extent of brown colour formation.

Appearance and texture of the smoked product are largely affected by the control of raw material quality and process parameters such as brining treatment, and time, temperature and air speed in the smoking kiln. However, there is evidence to suggest that some of the smoke components, for example formaldehyde, have a toughening effect on the muscle proteins.

The achievement of a consistent product in terms of aroma and flavour is considerably more complex. It is generally considered that phenols play an important role in the desirable characteristic flavour of smoked fish. Of these, eugenol, syringaldehyde, acetosyringone and acetovanillone seem to be more important in the typical flavour of hot-smoked products, and guaiacol, maltol, phenol and *m*-cresol are important in cold-smoked products. Nevertheless, eugenol is more important than guaiacol in what is considered a typically smoked flavour. Furan derivatives in the lower boiling point fraction of liquid wood smoke condensate have a sweet fragrant aroma which is thought to tone down the heavy, smoky aromas of the phenolics. Although the low molecular weight fractions of wood smoke are apparently most responsible for the palatable smoked fish flavour notes at low concentrations, higher concentrations are perceived as less desirable 'burnt' or 'phenolic' notes. Perhaps surprisingly, the middle boiling point fractions show the most uniformly good response on a hedonic scale over the widest range of concentrations in the product.

Shelf-life extension of smoked compared with fresh fish is due to a combination of lowered water activity and the uptake by the product of bactericidal and antioxidant components of wood smoke. It has been shown, for example, that traditional smoking can increase the induction period (*i.e.* up to a level of 20 milliequivalents of peroxide per kg) of autoxidation from 4 days in unsmoked controls to approximately 50 days. This antioxidant effect, however, is mainly associated with the particle phase of wood smoke (the vapour phase showing little or no antioxidant activity) and increases with increasing temperature of the oxidation stage of smoke production, but is unaffected by the pyrolysis temperature or air supply.

Concerning the avoidance of the deposition of carcinogenic substances on fish during smoking, the polynuclear aromatic hydrocarbon (PAH) group contains many highly carcinogenic compounds, 27 of which have been identified in wood smoke. One of these, 3,4-benzopyrene, has been detected at levels between 0.05 and 0.62 $\mu g\,kg^{-1}$ in a survey of German commercially smoked meat products. No correlation has been found between the uptake of phenols and the uptake of 3,4 benzopyrene in hot-smoked herring. On the contrary, the least phenols and most 3,4 benzopryene are taken up during the final, hottest (70–80°C) stage of smoking. Eight to nine times more 3,4 benzopyrene are taken up during hot smoking than during cold smoking, and PAH levels, as a whole, are highest in hot-smoked and extensively cold-smoked produce. Fish dried by hot air heated directly over gas or oil flames may also contain these substances.

Nitroso- (NO_x) substances in smoke give rise to a slight pink colour in the flesh of smoked fish, synonymous with the pink colour imparted by the nitrite/nitrate-containing curing brines for meat. It is likely, therefore, that the

NO_x substances are also capable of forming the carcinogenic N-nitrosamines by reaction with the amines in fish flesh.

It would seem that considerable scope exists for the improvement of traditional smoking processes with respect to product quality, consistency, and safety. For example, increasing smoke intensity but removing the particle phase between smoke generator and smoking chamber will remove much of the PAHs and increase the smoke flavour intensity. Great improvements could be made in this direction merely by extending the distance between fire box and chamber. However, there are two counter-arguments to this. The first is that getting rid of the particle phase greatly lowers the antioxidant effect of smoking, and the second is that heat generated from the smouldering sawdust is wasted instead of drying or cooking the fish. Notwithstanding this, safety considerations should be paramount. Smouldering of wood is a simultaneous process of pyrolysis and oxidation which is less than efficient because the oxidation, which takes place at the pyrolysis temperature, converts much of the flavour components to carbon dioxide and water. Suggested optimum parameters for smoke generation are pyrolysis of sawdust at 400°C with nitrogen flowing at $1500 \, lh^{-1}$, followed by oxidation at 200°C with air flowing at $1500 \, lh^{-1}$ per 50 g sawdust. Antioxidants and high temperatures for hot-smoked fish could be applied much more uniformly and safely than by traditional smouldering, smoking processes, such that generated smoke would be passed through charged fields to electrostatically precipitate the particle phase, and the hot part of the process would be achieved electrically.

The process of smoking affects the nutritional value of fish mainly by reducing the biological availability of proteins. Overheating, as might occur in some of the more severe procedures, such as 'pit' and 'banda' smoking, significantly reduces the availability of methionine, tryptophan and lysine. Most modern smoking processes, however, do not involve temperatures sufficiently high to reduce biological value or net protein utilisation simply due to heat-induced, cross-linking and browning reactions. It is known, nevertheless, that many smoke components react with amino acids and protein in the food. Carbonyls and phenols, in particular, react with lysine, arginine, methionine and other sulphur-containing amino acids. Even so, these losses are relatively minor. Tang (1978) reported lysine availability reduction of up to 25% in the outermost 5 mm layer, 14% in the layer 5–10 mm below the surface, and graduating down to insignificance at the centre, when coley (*Polachius vireus*) fillets were hot smoked to 115°C for 3 h.

2.5.4 *Processing and equipment*

Good smoked products cannot be made from stale fish. The quality of the product reaching the purchaser depends on the freshness of the fish before it is smoked and the care in handling it, the smoking process itself and subsequent storage history. Most poor quality smoked fish is due to bad handling before or after smoking or use of poor quality fish. Adequate smoking is nevertheless essential for a first class product that will keep well.

2.5.4.1 *Pre-smoking processes.* These can be divided into splitting and cleaning, salting and hanging.

Splitting and cleaning. Precise treatment depends on the product; care must always be taken not to bruise or tear the fish. All pieces of gut, gill and kidney must be removed, because these go bad quickly and may spoil otherwise good fish.

Salting. Fish may be soaked in a strong brine. Permitted colours may be added to the brine to intensify the colour imparted by the smoke. The contact time depends on the size and fat content of the fish, the presence of skin and the product requirement (2–3% salt is the maximum required if the product is to be eaten as a main dish rather than a condiment).

A 70–80% saturated brine (*i.e.* 80°) is the commonest strength used. One hundred per cent saturated brines leave unattractive, powdery salt crystals on the surface of the finished product, whereas a 50% saturated brine would cause the fish to swell such that excess water would have to be evaporated off during drying. Brines must be kept to strength and should be changed at least once per day. Frequent removal of scales and other debris may be necessary. Fish can be contaminated by stale brine.

Hanging. Fish is hung to drip either on racks or in the kiln. Protein dissolves in brine to give a sticky solution. During the hanging period this dries on the cut surface and produces the familiar glossy skin or 'pellicle'. The best gloss is obtained when a 70–80° brine is used. Smoked fish which has not been salted looks dull and rough.

2.5.4.2 *Smoking process control.* It is best to keep the temperature in the smoke chamber below 30°C at the start of the smoking process. In fact, for cold smoking, the temperature is kept at this level throughout the process. This is because the process must dry the fish to a certain extent, as well as deposit smoke on the fish surface. To achieve this, the air vents into the fire box should be almost closed so that the sawdust smoulders rather than burns. If the air flow is too rapid and the temperature is too high the surface becomes sealed off. No more water can be drawn out of the fish and the condition is called 'case-hardening'. Case-hardened material is wet under the hard surface and tends to spoil from the inside. The hard, blocked surface does not allow passage inwards of the smoke chemicals, again lowering the preserving effect of the smoking process.

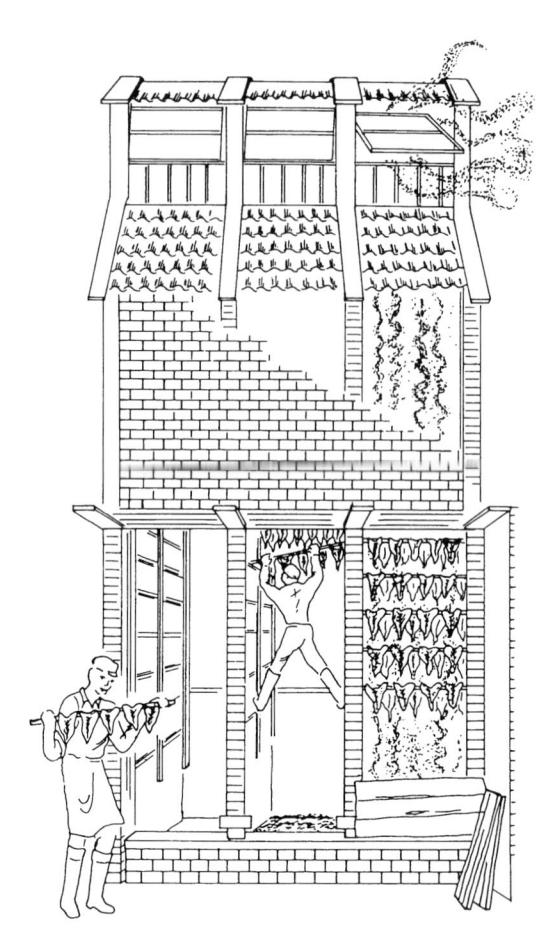

Figure 2.18 Traditional chimney kiln.

Hot air is less dense than cold air so moves upwards carrying the smoke with it. In most smoking kilns, including the traditional chimney kiln shown in Figure 2.18, smoke moves upwards and over the hanging fish. Consequently, temperature is higher and smoke more dense lower down the kiln. Throughout the process, racks of fish have to be moved to different parts of the kiln so that all fish gets the same smoke treatment. This is a difficult job which needs expert judgement.

In many modern kilns the fire box is separated from the smoking chamber. Air movement is across rather than up and smoke is drawn *across* the racks of fish (see Figure 2.19). Still, there is no chance of even smoke or temperature treatment in all parts of the kiln. As the smoke/air mixture passes from one rack to another it becomes:

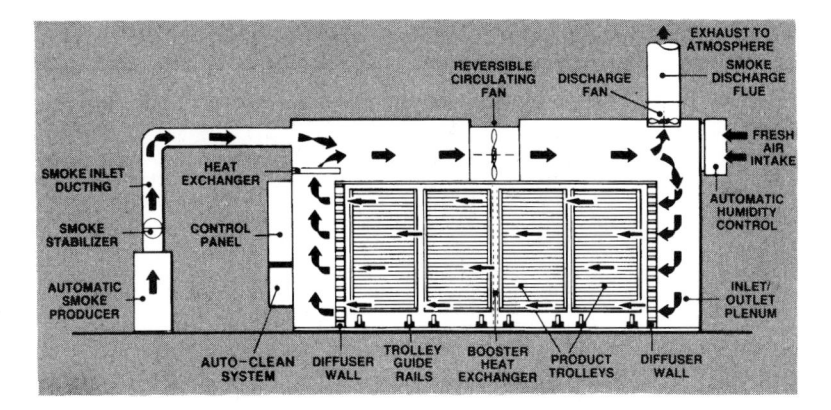

Figure 2.19 Modern mechanical fish-smoking kiln. By courtesy of Afos Limited, Manor Estate, Anlaby, Hull HU10 6RL

- less smoky – because smoke chemicals are taken up by fish from each rack;
- more humid – because water is lost by each rack of fish;
- cooler – as the smoke and air get further from the fire-box.

Again, to achieve an even process for all fish in a batch, it is necessary to keep switching racks of fish from one part of the kiln to another. Although this still requires expert judgement, the job is made easy by racking the fish on wheeled frames. These can be wheeled out of one section and swapped with a trolley in another section in the kiln.

Horizontal flow kilns are often called 'mechanical kilns' because an electric fan, usually in the outlet, draws the smoke along the racks of fish. They can be used for both hot and cold smoking by the use of electrical heaters in the smoke chamber. In large kilns of this type, heaters may be placed at different points in the smoke path to keep the temperatures more even throughout the smoke chamber. Chimney kilns are most often used for cold smoking and squat kilns for hot smoking.

In all types of kiln, there must be an air vent as well as a damper. With the air vent fully open very little air is drawn through the damper. The fire burns less fiercely and the smoke going through the smoke chamber is diluted with air. With the air vent fully closed all the air is drawn through the damper to the fire. This 'draws' the fire to burn more strongly. At first, more smoke is produced but, very soon, the fire becomes so hot that all the smoke chemicals are oxidised to carbon dioxide and water. Therefore, a great deal of smoking process control depends on vent and damper settings. Table 2.6 shows the range of possible smoked products from different fishery produce. There are many others which are not included and even more which have not been tried.

Table 2.6 Examples of smoke processes and products

Species	Smoke process	Product
Mackerel (fillets)	Hot	Smoked mackerel
Herring (gutted and split)	Cold	Kipper
Herring (fillets)	Cold	Kipper fillets
Herring (whole)	Cold	Bloaters
Herring (immature, whole)	Hot	Sild
Sprat (immature, whole)	Hot	Brisling
Haddock (gutted, split)	Cold	Finnan haddock
Haddock or whiting (fillets)	Cold	Golden cutlets
Haddock (nobbed)	Hot	Arbroath smokies
Cod (fillets)	Cold	Smoked cod
Cod roe	Cold/hot	Smoked cod roe
Salmon (gutted, split)	Cold	Smoked salmon
Trout (gutted, split)	Cold/hot	Smoked trout
Eel (gutted)	Cold/hot	Smoked eel
Oysters (brushed with oil)	Cold/hot	Smoked oysters

References

Broughall, J.M., Anslow, P.A. and Kilsby, D.C. (1983), Hazard analysis applied to microbial growth in foods: development of mathematical models describing the effect of water activity. *J. Appl. Bacteriol.* **55**, 101–110.

Brunauer, S., Emmett, P.H. and Teller, E. (1938), Adsorption of gases in multimolecular layers. *J. Am. Chem. Soc.* **60**, 309.

Curran, C.A. and Poulter, R.G. (1983), Isohalic sorption isotherms 3: Application to a dried, salted tropical fish (*Xenomugil thoburni*). *J. Food Technol.* **18**, 739–746.

Cutting, C.L., Reay, G.A. and Shewan, J.M. (1956), *GB Dep. Sci. Ind. Res., Food Invest. Spec. Rep. No. 62.* HMSO, London.

Doe, P.E., Curran, C.A. and Poulter, R.G. (1983), Determination of the water activity and shelf-life of dried fish products. *J. Food Technol.* **17**, 202–208.

Duckworth, R.B. (1963), Diffusion of solutes at low moisture levels. In *Recent Advances in Food Science* (Vol. 3), Hawthorne, J. and Leitch, J.M. (Eds), Butterworths, London, pp. 230–238.

Duckworth, R.B. and Kelly, C.E. (1973), Studies of solution processes in hydrated starch and agar at low moisture levels using wide-line nuclear magnetic resonance. *J. Food Technol.* **8**, 105–113.

Foster, W.W., Simpson, T.H. and Campbell, D. (1961), Studies of the smoking process for foods: the role of smoke particles. *J. Sci. Food Agric.* **12**, 635.

Gal, S. (1972), Die prak tische Bedeutung von Wasserdampf-Sorptions-Messungen in der Lebensmittelindustrie. *Alimenta* **6**, 213.

Gilbert, S.G. (1986), New concepts on water activity and storage stability. In *The Shelf Life of Foods and Beverages*, Charalambous, G. (Ed.), Elsevier Science Publishers BV, Amsterdam, pp. 303–334.

Horner, W.F.A. (1991), Water sorption and shelf-life of dry-cured cod muscle in humid environments. *M.Phil. Thesis*, Loughborough University of Technology.

Jason, A.C. (1958), A study of evaporation and diffusion processes in the drying of fish muscle. In *Fundamental Aspects of the Drying of Foodstuffs*, Jason, A.C. (Ed.), Society of Chemical Industries, London.

Jones, N.R. (1962), Browning reactions in dried fish products. *Recent Advances in Food Science.* (Vol. 2) Hawthorn, J. and Leitch, J.M. (Eds), Butterworths, London, pp. 74–80.

Karel, M. (1973), Recent research and development in the field of low and intermediate moisture foods. *Crit. Rev. Food Technol.* **3**, 329–358.

Kent, M. (1985), Water in fish, its effects on quality. In *Properties of Water in Foods*, Simatos, E. and Multon, J.L. (Eds), Martinus Nijhoff, Dordrecht, Netherlands, pp. 573–590.

Labuza, T.P. (1970), Properties of water as related to the keeping quality of foods. *Proc. 2nd Int. Congress. Food Science and Technol.* Inst. Food Technol, pp. 618–635.

Labuza, T.P., Tannenbaum, S.R. and Karel, M. (1970), Water content and stability of low moisture and intermediate moisture foods. *Food Technol.* **24**, 543–550.

Langmuir, I. (1918), The adsorption of gases on plane surfaces of glass, mica and platinum. *J. Am. Chem. Soc.* **40**, 1361–1365.

Lewis, G.N. (1907), *Proc. Am. Acad. Sci*, **43**, 259.

Masette, M. (1990), Liquid smoke treatment as an alternative method to traditional smoking in fish processing, *CNAA M.Phil. Thesis*, Humberside Polytechnic, Grimsby.

Mossel, D.A.A. (1975), Water and micro-organisms in foods – a synthesis, In *Water Relations of Foods*, R.B. Duckworth (Ed.), Academic Press, London, p. 347.

Reid, D.S. (1973), Water activity concepts in intermediate moisture foods. In *Intermediate Moisture Foods*, Davies, R., Birch, G.G. and Parker, K.J. (Eds), Applied Science, London, pp. 54–65.

Ross, K.D. (1975), Estimation of water activity in intermediate moisture foods. *Food Technol.* **29**, 26–30.

Salwin, H. (1959), Defining minimum % water for dehydrated foods. *Food Technol.* **13**, 594–595.

Scott, W.J. (1957), Water relations of food spoilage micro-organisms. *Advances in Food Research* (Vol. 7), Mark, E.M. and Stewart, G.F. (Eds), Academic Press, London, pp. 84–127.

Tang, S.L. (1978), A chemical method for the investigation of chemical changes in fish during smoking. *MSc Thesis*, Loughborough University of Technology.

Tilgner, D.J., Miler, K., Prominski, J. and Darnowska, G. (1962), Advances in the engineering of the smoke curing process. *International Session, 2nd Gdansk Poland 1960 Technologia Mesa (Special Issue). Yugoslav Inst. Meat Technol.* p. 37.

Williams, J.C. (1976), Chemical and non-enzyme changes in intermediate moisture foods. In *Intermediate Moisture Foods*, Davies, R., Birch, G.G. and Parker, K.J. (Eds), Applied Science, London, pp. 100–119.

Wolf, M. Walker, J.E. and Kapsalis, J.G. (1972), Water vapour sorption hysteresis in dehydrated food. *J. Agr. Food Chem.* **20**, 1073–1077.

Ziemba, Z. (1969), Role of tissue amino compounds and environmental variables in the production of surface colouring during smoke curing of food products. *Roczniki Technologu Chemii Zywnosci*, **15**, 139.

3 Surimi and fish-mince products
G.M. HALL and N.H. AHMAD

3.1 Introduction

The vast growth of surimi-based products in the USA and Europe has engendered a belief that everyone in the food industry knows what 'surimi' is. However, like that other buzz word of the 80s, 'biotechnology', everyone knows what they mean by surimi because they all have their own definitions.

For example, Lee (1984) described surimi as '...mechanically-deboned fish flesh washed with water and mixed with cryoprotectants for good frozen shelf life', while Johnston (1989) said, '...[surimi] was washed, refined mince prepared from fish, which is relatively stable and can be frozen and cold stored and still retain the necessary functional properties for making kamaboko'. Martin (1986) in describing the nomenclature for fabricated seafood products used the following terms, '...an intermediate processed seafood product used in the formulation/fabrication of a variety of finished seafood products... minced fish meat (usually pollack), which has been washed to remove fat and undesirable matters (such as blood, pigments and odorous substances), and mixed with cryoprotectants (such as sugar and/or sorbitol for a good shelf life)'. It is interesting that there appears to be no consumer definition of surimi!

All these definitions contain elements that are true but each has a different function to fulfil and all are valid in their own areas of interest. It is not the intention here to add another definition, but to review the essential parts of the surimi process. The first element to highlight is that surimi is an *intermediate product*, which is later used to make other products, which could be the traditional Japanese 'kamaboko' products or the new fabricated products, collectively called *neriseihin*, that have shown the phenomenal market growth described earlier. It is this growth that has focused the attention of fish technologists in the West on the potential for surimi as a raw material. The consumption of surimi in Japan is declining as more Western-style foods become fashionable so an increase in surimi production depends on new applications elsewhere. This is happening as witnessed by the fact that the USA is now the leading producer of surimi worldwide with Thailand, Korea and South America following Japan as producers. Although most of the production in these countries is for export to Japan some is used in the home-based industry (Ishikawa, 1996). Before looking at the production process in detail, some description of the fish-muscle proteins is appropriate here.

3.2 Fish-muscle proteins

3.2.1 *Nature of muscle proteins*

In this chapter, the structure of native fish muscle is not dealt with in detail as excellent reviews of the subject exist (*e.g.* Love, 1988). Instead the biochemical nature of muscle proteins is concentrated on.

Fish muscle can consist of two main types, dark and white, depending on the life-cycle of the species concerned. Strong-swimming species, such as tuna and mackerel, have a larger proportion of dark muscle than relatively sluggish fish such as cod, haddock and flat fish. The two muscle types are essentially similar in composition but the dark muscle has a higher content of haem pigments, such as myoglobin, for oxygen transport and more non-structural lipids, to provide energy; this reflects its role in active strong-swimming. The significance for surimi production is that the higher levels of lipids and metabolites in dark muscle will affect the flavour and colour of the surimi which, traditionally, should be bland on both these counts; this is discussed later.

The actual amino-acid composition of dark/white fish muscle is roughly the same as in terrestrial species such as cattle, although the proportions of different protein types vary – reflecting the environment in which these creatures live. Three groups of protein can be differentiated and separated by solubility in salt solutions of increasing concentration.

The sarcoplasmic proteins are water soluble and normally found in the cell plasma where they act as enzymes and oxygen carriers. They will comprise anything from 18 to 20% of total muscle protein. The largest proportion of muscle proteins, 65–80% of total protein, consists of the myofibrillar proteins, which give the muscle its fibre-like structure and muscular activity. The major components are myosin, actin, tropomyosin and troponin. These proteins can be extracted by the use of salt solutions up to 0.3 M. The final group of proteins are those making up the connective tissues, surrounding the muscle fibres and in skin; they include collagen and elastin. These proteins, known as the stroma, comprise about 3–5% of the total protein (which is much less than in terrestrial animals) and are easily solubilised by cooking. They are, however, resistant to solubilisation except in strong salt solutions.

The significance of the muscle types for surimi production reflects their nature *in vivo*. The sarcoplasmic proteins, being very soluble, do not contribute to gel formation and may indeed impede the process, for example some proteases may contribute to breakdown of the myofibrillar proteins. Hamann *et al.* (1990) showed the use of hydrolysed beef plasma protein and egg-white proteins to prevent protease activity (during the modori stage at 60°C) during kamaboko gel formation from surimi (Figure 3.1). They attributed the effect to a broad-range enzyme inhibitor in the two additives. The myofibrillar proteins are those that give the gel structure when surimi is further processed by heating to give final products. Recovery, concentration and protection of these proteins is the object of the surimi process. The stroma proteins are not

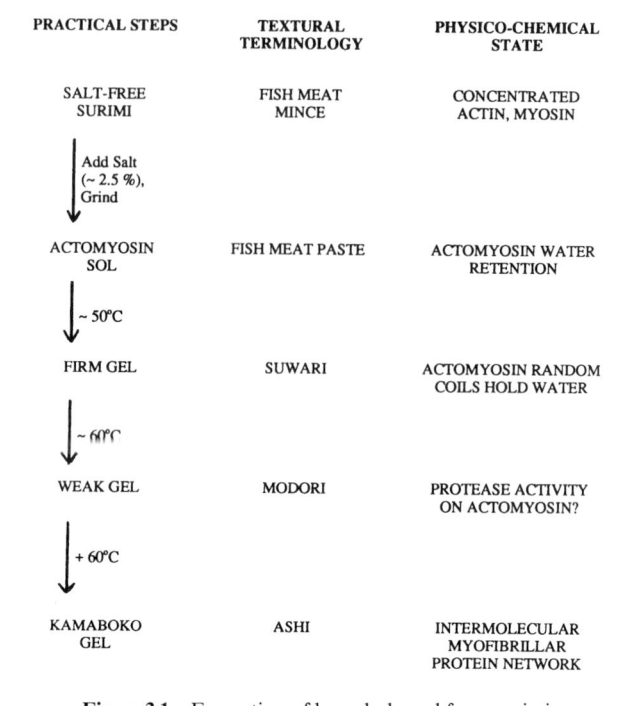

PRACTICAL STEPS	TEXTURAL TERMINOLOGY	PHYSICO-CHEMICAL STATE
SALT-FREE SURIMI	FISH MEAT MINCE	CONCENTRATED ACTIN, MYOSIN
↓ Add Salt (~ 2.5 %), Grind		
ACTOMYOSIN SOL	FISH MEAT PASTE	ACTOMYOSIN WATER RETENTION
↓ ~ 50°C		
FIRM GEL	SUWARI	ACTOMYOSIN RANDOM COILS HOLD WATER
↓ ~ 60°C		
WEAK GEL	MODORI	PROTEASE ACTIVITY ON ACTOMYOSIN?
↓ + 60°C		
KAMABOKO GEL	ASHI	INTERMOLECULAR MYOFIBRILLAR PROTEIN NETWORK

Figure 3.1 Formation of kamaboko gel from surimi.

removed by the surimi process but are solubilised by heat and may be 'neutral' components of the final products.

The protein content of fish muscle varies with season in some species and, where this occurs, the protein content varies inversely with the water content. Higher levels of protein are found in the feeding season and less around the spawning season – there are obvious incentives in using fish with high protein content. In some species of fatty fish, herring and mackerel for example, lipid reserves are also associated with the muscles; again there is a seasonal variation and an inverse relationship between lipid content and water content of the muscle in these species. The convential wisdom is that high lipid content muscle should be avoided in surimi manufacture because of the physical difficulty in removing the lipid and the flavour and colour associated with the residual lipid.

3.2.2 *Properties of actin and myosin*

As these two proteins are the major components of surimi and fish mince, their properties under processing conditions are of importance. Two particular conditions deserve attention: frozen storage and heat processing.

3.2.2.1 *Frozen storage of surimi and fish mince.* The behaviour of actin and myosin in frozen storage is of premier importance as both surimi and minces are stored in this way prior to final processing. The major problem is denaturation of the actin and myosin, which reduces their water-holding properties in subsequent processing. In the final products, the effects of denaturation can be seen as dryness, loss of succulence and poor texture.

The mechanism of freeze denaturation has been studied for many years and several significant papers have been published (Buttkus, 1970; Matsumoto, 1980; Shenouda, 1980; Suzuki, 1981). In general, the denaturation is a consequence of aggregation of the myofibrillar proteins in which hydrogen bonds, ionic bonds, hydrophobic bonds and possibly disulphide bonds are formed. The role of water in these changes is central. As the water is frozen, there is a concentration of solutes; this changes the ionic strength and pH around the proteins, which allows the new bonds to form. The water immediately surrounding the protein molecules is also responsible for maintaining the ordered native state of the protein and, as it is progressively removed by freezing, dehydration occurs, again causing conformational changes. Other reactions taking place in the dehydrated environment, such as lipid oxidation, can give rise to compounds (*e.g.* malonaldehyde) that will react with the proteins giving more crosslinking and denaturation. One way to overcome these problems is through the process itself by rapid freezing to a low temperature, storage at a low temperature and avoidance of cyclic changes in temperature (freeze–thaw) until the moment of thawing.

Another way of preventing denaturation is through the use of cryoprotectants to protect the proteins during frozen storage. The discovery of such compounds was the stimulus that allowed surimi production to move from a cottage industry into a large-volume factory-based process, as discussed later. Initially, a large number of compounds from a variety of chemical groups were tried and, of these, a small group was found to be most effective. The most commonly used have been sucrose, sorbitol and polyphosphates (Suzuki, 1981). Their action seems to be a combination of water-holding to prevent water migration (Matsumoto, 1980), and an increase in surface tension (Arakawa and Timasheff, 1982).

The sweet taste of surimis containing these sugars is unacceptable to some Western consumers and, recently, efforts have been made to assess new cryoprotectants to reduce sweetness and give protection in further processes such as extrusion, which were not envisaged when the original cryoprotectants were introduced. Krivchenia and Fennema (1988a,b) used sodium tripolyphosphates, monosodium glutamate and antioxidant mixtures to protect whitefish (*Coregonus cupleaformis*) and burbot (*Lota lota*) fillets – treated samples were better than untreated samples. Park *et al.* (1988) used polyols and maltodextrin in mixtures with sugars and phosphates to give favourable results in Alaska pollack (*Theragra chalcogramma*). Yoon and Lee (1990) used polyols, crystalline sorbitol (rather than the usual liquid form) and modified starch with red hake (*Urophycis chuss*) and obtained good results, although freeze–thaw cycling reduced the cryprotectant effects. Sych *et al.* (1990) included casein, and fish hydrolysates and hydrocolloids with cod (*Gadus morhua*) surimi, either alone or with sucrose, but these components were not so effective as sugars, either alone or with sorbitol. These recent papers show that a variety of less-sweet cryoprotectants are worthy of investigation, especially with the advent of a new generation of surimi-based products.

Recent work has expanded the range of additives to improve the quality of surimi by different effects. Milk whey was added to surimi from Jack mackerel (*Trachurus murphyi*) as a cryoprotectant (Dondero *et al.*, 1994) whilst bovine plasma was said to act as a gelling agent, when heated (Abe, 1994), or as a protease inhibitor to conserve gel strength (Saeki *et al.*, 1995). Another interesting innovation has been the addition of a microbial transglutaminase, an enzyme, which generates links between the side chain of lysine (ε-NH$_2$) and glutamic acid (γ-COOH) enhancing gel strength (Sakamoto *et al.*, 1995). However, the effect may be fish species dependent (Asagami *et al.*, 1995).

3.2.2.2 *Effect of heat on actin and myosin.* Heat denaturation of proteins is a common phenomenon seen in many classes of proteins. The conversion of surimi into final products is a deliberate transformation brought about by heat. This process is not discussed in detail here but is shown in outline in Figure 3.1. However, during the production of surimi (and crude fish mince) excessive heating is to be avoided as it causes a loss in gel formation. The use of cold water and strict control of mixing/grinding and transfer equipment can eliminate most problems. Douglas-Schwarz and Lee (1988) compared the thermo-stability of red hake (*Urophycis chuss*) with Alaska pollack (*Theragra chalcogramma*) during surimi production and gel formation. Optimum temperatures for washing of the mince and chopping of the surimi were 15°C and 12°C respectively for hake and 10°C and 4°C for pollack. There are cost advantages in being able to use higher temperatures in processing and tropical water species should be particularly temperature-resistant.

3.2.3 *The action of salt*

The presence of salt is essential for the conversion of surimi into final products and in the removal of the sarcoplasmic proteins in surimi manufacture. The latter are water-soluble and usually removed easily by water washing; however, salt can be added to a final wash to remove the last traces and improve water removal. Salt can also be added to the surimi with the cryprotectants (at about 2.5% by weight), this is known as ka-en surimi. This type is less stable in frozen storage as the salt encourages premature gel formation, which separates out; this can be overcome by the addition of higher levels of sucrose.

As indicated in Figure 3.1, salt is added to surimi for conversion into kamaboko products and initiates the process by forming an actomyosin sol from actin and myosin. The level of addition is critical as too much salt leads to salting out of the proteins (Suzuki, 1981). The use of other salts in the wash water, including calcium and magnesium chlorides, has been proposed and said to improve the functional properties of the resulting surimi (Ofstad *et al.*, 1993).

3.2.4 Surimi-based products

As already indicated, surimi-based products are made possible because of gel formation due to the presence of actin and myosin (muscle proteins) which have been concentrated in the surimi process. When heated, often at moderate temperatures, these proteins react to form the actomyosin gel, which gives the textural properties of the products. According to the Japanese, a good, resilient, texture is called 'ashi'. The addition of flavours and colours together with various moulding processes allows the creation of many different products. A bland-flavoured, white surimi to which these components may be added is another traditional quality attribute of good surimi.

The term 'kamaboko' has now become a generic for the Japanese style products (although it actually refers to steamed products), which also include fried products (tempura) and broiled products (chikuwa). In Japan, these cooked and seasoned products are also called 'neriseihin' products. The best description of these products is given by Suzuki (1981).

In the west, products that are shellfish analogues have found greatest acceptance such that, in the USA in 1984, about 94% of all surimi-based products were analogues and, in 1987, the proportion was still 90% (Johnston, 1989). Gwinn (1987) reported 14 plants in the USA with a capacity of 40 million pounds of imitation crab products. The consumption pattern in Europe was similar but at lower volumes. The products could be in the form of flakes, chunks or extruded shapes to resemble crab, lobster and shrimps. The brand names given to these products leave the consumer under no illusion that they are buying the real thing, but have found their own niche in the market. Other products have included sauces, dips, spreads, soups and chowders, sausages, pies, loaves and desserts (Wray, 1987). It has been suggested that the only limit to the use of surimi lies in the imagination of the product developers, although in the authors' view it is really limited by the imagination of the consumer. The use of high pressure as a sterilising process has come out of Japan in recent years and has been effective whilst usually having no deleterious effect on food flavour or texture, unlike heating processes. With surimi, however, the pressure treatment has actually helped to develop the gel character (Chung *et al.*, 1994) and this has been aided by the addition of ammonium salts (Shoji *et al.*, 1994). Miyao *et al.* (1993) showed that pressures of 300–400 MPa did not kill all organisms associated with surimi, especially *Moraxella*, *Acinetobacter*, *Streptococcus* and *Corynebacterium* species.

From the foregoing, it is obvious that the success of these products lies in the gel-forming ability of the muscle proteins and this depends on control of the elements of the surimi process. It is now appropriate to look at the process in detail.

3.3 The surimi process

3.3.1 Basic concepts

Historically, the Japanese produced surimi and converted it into the kamaboko products on a daily basis. This was necessary because in the early days there was no frozen storage capability and surimi itself is not inherently stable. When frozen storage was possible it was found that the gel-forming capacity of frozen surimi was poor on thawing. The discovery of cryoprotectants in 1959 (Matsumoto, 1978) allowed the volume of surimi production to

RAW MATERIAL AND PREPARATION

FRESH FISH
↓
GUT AND CLEAN
↓
FILLET
↓
MINCE
↓

SURIMI PROCESS

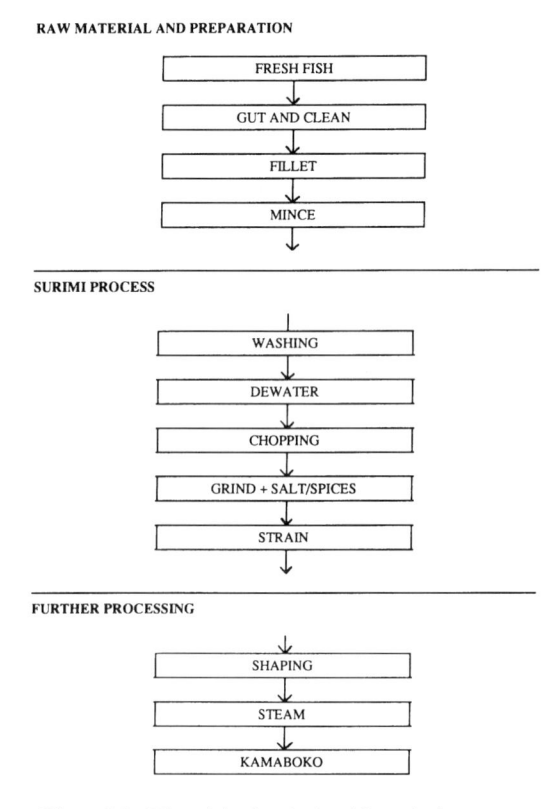

WASHING
↓
DEWATER
↓
CHOPPING
↓
GRIND + SALT/SPICES
↓
STRAIN
↓

FURTHER PROCESSING

↓
SHAPING
↓
STEAM
↓
KAMABOKO

Figure 3.2 The original surimi and kamaboko process.

increase greatly, and called for an increase in the scale of each stage of the process while keeping to the spirit of the original process.

The old process was small-scale and manual in nature and involved the preparation and pre-treatment of fresh fish, the surimi process itself and then the conversion into kamaboko products (Figure 3.2). In the mechanisation and scale-up of this process, it was found that the same factors controlled quality as in the manual process. Figure 3.3 shows a typical modern process, and its genesis from the original process is easily seen. Each aspect can now be discussed in turn.

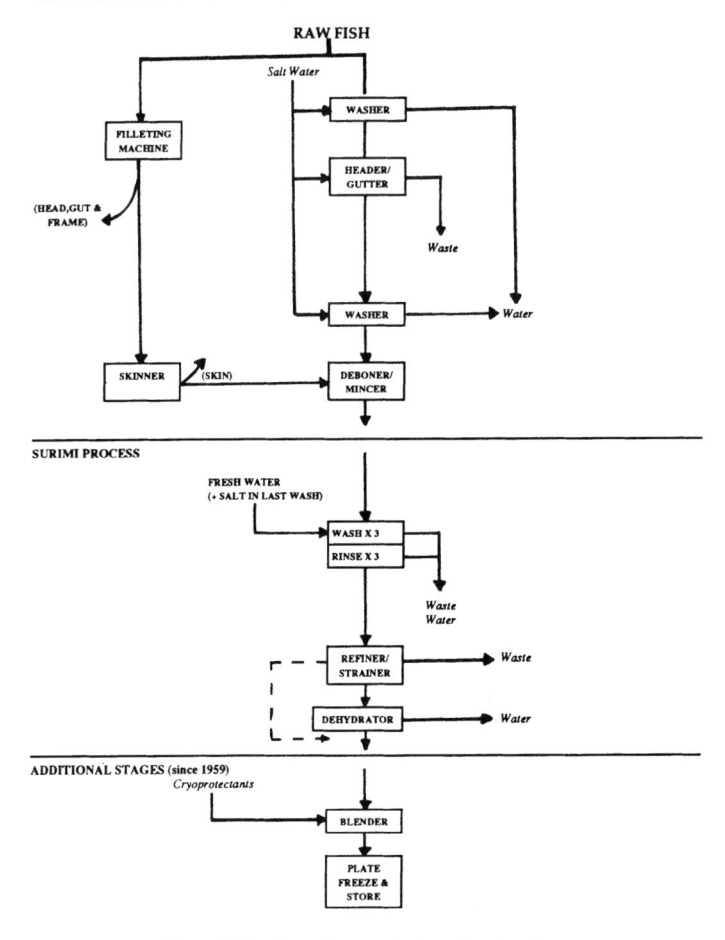

Figure 3.3 A modern surimi production line.

3.3.2 *Process elements*

3.3.2.1 *Raw material.*
The freshness of the raw fish is considered to be of paramount importance such that, in Japan, surimi produced on factory ships is highly valued. Most Western plants are shore-based and here the fish should be *post rigor* but less than 4 days old depending on the storage conditions at sea. However, each species must be treated on its merits, for example MacDonald *et al.* (1990) showed good kamaboko gels from surimi of hoki (*Macruronus novaezelandiae* Hector) stored chilled for up to 10 days before processing. The rate of loss of gel quality with storage with this species was significantly slower than for Alaska pollack, although the authors emphasised the need for strict quality control of the raw material by sensory and chemical assessment. Fresh fish are preferred as they produce less blood and gut residues in the tissues (and so less colour) and experience less autolysis of the muscle proteins giving a better gel. Seasonality is also important as fish in the feeding season have a low water content and high protein content, which means fewer washing cycles and better gel strength, respectively. A low pH at this time also tends to a stronger gel as water is more easily removed in the process. Fish caught during or after spawning show a higher pH, retention of water and hence a softer gel. A uniform size of fish is important for consistent yields from deboning/mincing machines, and fish with a good flesh:frame ratio may give overall better yields of mince.

3.3.2.2 *Fish preparation.*
The transformation of the raw fish into fish mince, which is the immediate raw material for the surimi process, understandably also plays an important role in quality determination. The fresh fish is a whole organism where tissues are delineated and potentially reactive species (structural proteins, enzymes and lipids) are controlled by barriers, even if only in the short term. Once the fish is cut open, the opportunity for biochemical interactions and bacterial contamination leading to loss of muscle protein functionality becomes apparent. Figure 3.3 shows two routes for fish preparation; via filleting/skinning or heading/gutting before the deboner.

The former route is said to give lower yields of better quality mince (a good gel-forming capability and pale colour) requiring few washes and no refining. The latter route gives higher yields of lower-quality mince, particularly contaminated with skin and traces of blood, requiring more wash cycles and refining (Lee, 1986). Proponents of both systems exist and the choice of method may depend on fish species, availability of wash water and final use of the surimi.

The advent of the deboner/mincer machines during the 1970s undoubtedly contributed to the surimi boom, as previously under-utilised fish (too small or bony for traditional use) could be converted into mince. Several versions exist but the most common types involve an adjustable belt and perforated drum, rotating in opposite directions, between which the fish is pressed. The flesh, with some blood and fat, is forced into the drum and the skin and bones rejected. The belt tension and size of the perforations on the drum decide the yield and quality of the mince obtained. Perforation sizes range from 1 to 5 mm (Regenstein, 1986), with smaller sizes giving fine mince with good colour but possibly leading to losses during rinsing in the surimi process. Larger sizes give higher yields that may require refining. Again, experience of the fish species and ultimate use of the surimi is essential when deciding mincing parameters.

3.3.2.3 *The washing process.* The cyclic washing and rinsing of the fish mince with water is the central process in surimi production. Washing removes:

- water-soluble sarcoplasmic muscle proteins which do not form gels;
- enzymes (proteases);
- pigments/blood;
- lipids;
- haem compounds causing lipid oxidation leading to protein denaturation.

In addition to removing these undesirable compounds, washing will concentrate the actin and myosin to give a good gel.

The number of cycles, contact time and wash water:mince ratio will be dependent on the raw material/preparation, as previously noted, and on the level of the removal, and hence surimi quality, of contaminants required. In general, three cycles of 10-minutes contact with water:mince ratios of 3:1 or 4:1 would be adequate for most applications. Lee (1984) gives a detailed account of these factors. The water temperature is usually 5–10°C to prevent muscle protein denaturation, although this will depend on the fish species – tropical fish should be capable of withstanding higher temperatures.

Water quality is important as a high pH will lead to water retention in the mince. Hard water with metals present such as calcium/magnesium and iron/manganese will affect texture and colour, respectively. Salt (up to 0.2%) is usually added to the last wash to remove water but should not be at a level to solubilise the actin and myosin.

Rinsing is performed by rotary perforated drums, where a balance must be struck between potential yield losses due to small mince size, and ease of drainage (where a large mesh is needed).

3.3.2.4 *The refiner/strainer.* This operation can be performed either pre- or post-dehydration. The choice seems to depend on the amount of water retained after dehydration, which may make refining slow and difficult. The purpose of refining is to remove bones, skin and perhaps dark muscle if this is objectionable. Good practice up-stream can obviate the need for this process, which may be another source of yield loss.

3.3.2.5 *Dehydrator.* Traditionally, screw presses (extruders) were used where the extent of water removal depended on the residence time (screw speed) and on the pressure exerted (screw:barrel dimensions). For some species and for non-fresh fish, the addition of brine (up to 0.2%) may be necessary at this stage. The final moisture content of the mince here is 80–85% (wet weight basis). Scroll decanting centrifuges can also be used in this operation. Alfa-Laval have designed a continuous process with a scroll decanter for dehydration, and claims are made for less dry-matter loss and uniform moisture in the refiner feed (Wray, 1987).

Dehydration and/or refining mark the end of the surimi process according to tradition; however, the advent of cryoprotectants for frozen storage stability has led to additional stages.

3.3.2.6 *Addition of cryoprotectants.* Uniform addition is essential and is achieved by the use of silent cutters, which are fast (down to 30 s), or ribbon blenders (3–5 min). Temperature increases must be controlled. The cryoprotectants traditionally used include sucrose (4%), sorbitol (4%) and polyphosphates (0.3%) in various combinations. Their mode of action has been mentioned earlier. This surimi is termed salt-free (or mu-en surimi) but further salt can also be added if desired (ka-en surimi). The final moisture content is now 75–80% (wwb). The surimi can now be packed in trays, frozen in a plate freezer and stored below $-20°C$.

3.3.2.7 *Other considerations.* The large volumes of water used in every stage of surimi production pose problems in terms of the pre-treatment of hard water, if necessary, and as a pollution hazard. A reduction in the volumes used can be made by applying an appropriate water:flesh ratio in the wash cycles and by a counter-current recycle of the water removed in rinsing, except the first rinse, which carries the worst contamination with, for example, blood (Lee, 1986).

The pollution hazard can be reduced by recovery of the water-soluble proteins by flocculation, air flotation or, more recently, by ultrafiltration (Jaouen *et al.*, 1990). The recovered material can be used as animal/fish feeds. Recent work by Lin *et al.* (1995) has shown that proteins, recovered by micro- and ultrafiltration could be included in surimi with no loss in gel strength. The permeate from the filtration was less polluting than conventional effluents and could be recycled. These considerations merit more work as water is no longer a cheap commodity and pollution controls are likely to become more strict in the future. Hastings (1989) compared heat-set gels made from surimi of cod (*Gadus morhua*) with those from unwashed or once-washed fresh or frozen mince. The use of fewer washes would reduce the demand for water. The conventional surimi process gave the best results, of soft and flexible gels, but those from the reduced-wash material would be satisfactory for use as binders and where the white colour of conventional surimi was non-essential. Another approach has been described by Pacheco-Aguilar et al. (1989) where a single wash was used under acid conditions (pH 5.0–5.3) and a reduction in water use and an increase in yield were achieved. The gel properties of the surimi were worse than from a conventional process but may be acceptable for US consumers.

3.3.3 Appropriate species for surimi production

Classically, Alaska pollack (*Theragra charcogramma*) was used by the Japanese, being available in the N.E. Pacific and Bering Sea. Stock reduction and competition has led to a large number of other species being tried. Suzuki (1981) lists croaker (*Argyrosomus argentatus*, *Nibea mitsukurii*), lizard fish (*Saurida undosquamis*), cutlass fish (*Trichiurus lepturus*) and horse mackerel (*Trachurus japonicus*) as other major species, together with various sharks and flounders. Recently, threadfin bream (*Nemipterus tolu*) from Thailand has become an important surimi source for Japan (Ishikawa, 1996).

Any reader wishing to try a local species for surimi production should be encouraged to do so as the gel-forming ability of the muscle proteins can only be assessed by practice. In addition, if the surimi is to be used to produce fish/crustacean analogues, a non-conventional texture/flavour may be acceptable and appropriate to the species available.

Generally, lean white fish are considered to be better than fatty fish (probably because of poor colour and the problem of lipid removal) but any species with actomyosin gel-forming capability should be acceptable. Nine UK species including cod, whiting, dogfish, herrings and sardines have been investigated and the white fish species gave better products than the fatty species (Hastings et al., 1990). Chang-Lee et al. (1990) studied Pacific whiting (*Merluccius productus*) as an example of an under-utilised species in the conventional sense, because of protease activity and the presence of a parasite in the flesh. They showed that the addition of egg white to the surimi was necessary for good gelling in the final products.

Fatty pelagic species such as sardines, herrings and mackerels are a large resource but are difficult to process into conventional surimi due to their dark muscle content, poor stability, high-lipid content and, in some cases, the presence of histidine, which can bring about allergic response in some consumers (Suzuki, 1981). Special techniques were developed to overcome the problems mentioned above and amounted to the use of very fresh fish (Ishikawa et al., 1977) and a variety of techniques to remove dark meat and lipid. Roussel and Cheftel (1988, 1990) studied the surimi and kamaboko of Atlantic sardine (*Sardina pilchardus*) and emphasised the need for adequate suwari gel formation in the production of a good elastic, rather than rigid, final product. Tsukuda (1980) studied changes in lipids during surimi production and storage and reported hydrolysis and oxidation during storage.

As might be expected, much recent research effort has gone into surimi production from fatty species. The European Community under its Fisheries and Aquaculture Research Programme (FAR) has investigated several under-utilised fatty species. Work has also been done on processing fatty species under conditions aimed to reduce oxidation damage (Kelleher *et al.*, 1994) by the addition of antioxidants and exclusion of salt and oxygen.

Freshwater species have been neglected but are worthy of investigation as, again, they are a large resource and readily available when farmed. Two species have been studied: milkfish (*Chanos chanos*), which are found all over S.E. Asia, and tilapia (*Tilapia niloticus*), which is found in many parts of Africa. Milkfish tended to give a rather tough kamaboko gel while the tilapia give an excellent gel but low recovery of mince from the live fish (Suzuki, 1981).

3.3.4 Quality of surimi products

Surimi quality is primarily decided by the quality of the kamaboko products made from it. The Japanese set strict standards for thawed surimi – based on colour/blandness of flavour (affected by impurities) and gel-formation (actin and myosin properties). Other tests performed may be chemical (pH and moisture) and physical (drip loss and viscosity). The gel quality so measured will reflect: (i) the quality of the raw material; (ii) the effect of the washing cycles in concentrating the actin and myosin; and (iii) any changes brought about by frozen storage.

The tests done in Japan were codified by the Tokai Regional Fisheries Research Laboratory (1980) and are done on the raw thawed surimi and on the heat-set gel made from it. The raw surimi is tested for moisture content, pH, presence of impurities such as dark skin and bones, whiteness, and physical attributes such as expressible drip and viscosity.

Gel-forming ability is assessed by use of a standard gelling method after which the gel is tested for strength by a special gel-testing instrument (the Okada gelometer), which has been developed together with sensory testing (Suzuki, 1981). The sensory tests include folding the gel between the thumb and index finger and looking for the extent of cracking, and a bite test. These tests particularly refer to the Japanese products and also to the Alaska pollack as the 'reference' species.

It has been argued (Lee, 1984) that the tests could be modified for other uses and should play the role of specification tests for particular applications. The modifications suggested include a test for gel-forming ability, with and without added starch, and gel properties measured by the Instron universal testing machine, which would presumably give results that are comparable with other familiar materials and aid in choice of application for the surimi. In the authors' opinion, the suggestion of Lee (1984) is valid because, as the applications for surimi increase and more species are investigated to fit them (or vice versa), the quality criteria will change. For example, in certain uses the acceptance of colour/fish flavour or lipid content may be very different from that of the classic surimis.

The advent of new products also leads to consideration of the use of mixed-fish species in a product, and the labelling/identification of these species. To prevent fraudulent claims where shellfish/surimi mixtures are produced, there must be a means of assessing the contribution of each constituent to the product. The identification of raw fish species by electrophoresis has been well established but the identification of species after heat treatment is more difficult. In surimi production, the sarcoplasmic proteins are removed, further reducing the number of proteins for identification. An *et al.* (1989) used sodium dodecyl sulphate-polyacrylamide gel electrophoresis (SDS–PAGE) and urea-gel isoelectrofocusing (IEF) to identify raw and cooked fish fillet, and surimi from Alaska pollack (*Theragra chalcogramma*) and red hake (*Urophycis chuss*). They showed that for the cooked samples it was necessary to use SDS or urea to extract sufficient proteins to make an identification possible.

As indicated in the introduction to this chapter, there may be several definitions of surimi depending on the view of the definer and, in such a rapidly expanding market sector, there is a great need for unambiguous terms that are easily understood by the consumer. Martin (1986) showed that the US consumer was generally well-disposed to surimi-based products, seeing them as wholesome (sea) foods with many recipe opportunities, with even the additives used in the production of final products being acceptable. Such a favourable response to a new product should not be jeopardised by poor labelling or misleading of the consumer. The unique nature of surimi products should be used as a positive element in their marketing.

3.3.5 *Microbial aspects of surimi*

This is a subject which has received less attention compared to the physicochemical aspects of surimi production. On freshly caught fish for surimi production, the sites of bacterial contamination are, as usual, the skin, gills and guts. These are removed in the preparation stage so that with careful filleting skinning, gutting and heading processes the bacterial load on the flesh should be minimal. Deboner/mincers may be possible causes of contamination as traces of skin and blood are forced through the drum with the flesh. The subsequent washes in the surimi process itself may help to remove contaminant bacteria and the addition of salt and cryoprotectants further contribute to their suppression, although some work shows contrasting results (Lee, 1992).

The chill temperature of wash waters may select psychotrophic bacteria including spoilage organisms but frozen storage of the surimi should prevent growth and adequate heat treatment in subsequent product manufacture kill remaining organisms. Bacterial contamination could lead to problems of food poisoning or spoilage. Food poisoning directly attributable to surimi itself rather than unhygienic processing and handling does not seem to be a problem. Food spoilage organisms capable of utilising protein as an energy source might be thought of as a problem. Their activity would lead to distinct aroma/flavour changes and poor texture. These would be serious problems in a product defined by blandness and textural properties. Again there seems to be little evidence of surimi quality being affected by bacterial agents.

The Hazard Analysis and Critical Control Point (HACCP) concept (see Chapter 8 for details) may be particularly applicable to the surimi process which is easily divided into steps. However, the variability in surimi production lines might necessitate different control points for each (Lee, 1992). Finally, a number of reports have appeared recently in which an allergy to fish by individuals could also include surimis made from the fish as well as the fish itself (Helbling *et al.*, 1992; Musmand *et al.*, 1992; Mata *et al.*, 1994).

3.4 Fish mince

3.4.1 *Sources of raw material*

The original source for fish mince was the trimmings from hand or machine filleting operations. These trimmings were used in fish sticks (fingers) and portions which could be enrobed in batters or breadcrumbs. A great impetus to the use of mince came with the development of mechanical flesh–bone separators, which could use whole fish, fillet trimmings or filleting waste as a raw material. Babbit *et al.* (1984) suggested that 32% of the weight of headed/gutted Alaska pollack (*Theragra chalcogramma*) could be recovered as fillets (by mechanical means) and a further 34% by mechanical separation of fillet trimmings and backbones.

The essential difference between surimi and fish mince is that, in the latter, there is no separation of the sarcoplasmic proteins and lipids, and hence the components that lead to instability (enzymes, haem pigments and lipids) are present. This gives textural and flavour changes on storage. The fish-mince products are closer in nature to fillets than are surimi products therefore any change from the norm (*i.e.* fillet-like characteristics) are readily noticed. Since changes in texture have been shown to be detrimental to consumer acceptance, even before changes in flavour were detectable (Wesson *et al.*, 1979), the removal of destabilising compounds in the production of surimi has been considered a considerable bonus.

Obviously, the same fish stocks can be used for mince production as for surimi – after all, the surimi process revolves around washing fish mince. As with surimi, the same steps to preserve raw-material quality are necessary for good mince production. Since most mince is stored frozen, myofibrillar protein denaturation is possible and will be more pronounced than in surimi, or even fillets, because of the intimate mixing of compounds that takes place as a result of the mechanical flesh–bone separation (Babbitt, 1986; Regenstein, 1986). Thomson and Mackie (1982) showed that the proportions of sarcoplasmic, myofibrillar and stroma proteins in mechanically recovered mince were not the same as in the fillet if cod (*Gadus morhua*) skeletons, backbone and belly flaps were the feed. Collagen, a stroma protein, was higher in the mince and it was suggested that some myofibrillar protein may be heat-denatured in passing through the perforated drum. In the authors' opinion, it is possible that the surimi process may obviate some of these effects by removal of released collagen and concentration of the remaining myofibrillar protein so that the denatured part is insignificant.

For the best quality mince, if possible, only single species should be used so that less-stable fish minces do not 'contaminate' the better-quality material. The view that all trash fish may be 'grist to the mill' does not hold if good-quality mince fish products are required. Labelling products with a single species source can also be taken as a sign of a superior product.

3.4.2 Fish-mince products

Frozen blocks of mince can be used to replace fillet blocks in fish sticks and portions as mentioned above but, as they have a poorer texture, they can be considered 'down-market' to the consumer. Attempts to use mince as an extender for extruded crab and shrimp products gave good results but the products were inferior to surimi-based products. The introduction of recipe dishes where the mince appears as fish chunks together with a sauce has led to more productive use; all the consumer needs to provide to make a meal is some cereal. Fish mince can also form the base of a chowder or fish soup. If the fish is pre-cooked, some enzyme-related deterioration can be prevented, while products designed for microwave cooking in the home will receive a short heating time and suffer little textural breakdown.

It is obvious that each product has to fit into a social/cultural context, and the use of binders and flavours can generate a wide variety of products; readers are encouraged to experiment with recipes and fish types. A recent report (Abrahams, 1990) describes the use of surimi from Alaska pollack (*Theragra chalcogramma*) for the production of kosher versions of shrimp, calamari and crab-style legs in Israel, which seems to be a first under the Jewish culinary laws.

3.4.3 Comparison of surimi and fish-mince products

The vast increase in the production of surimi-based products mentioned in the introduction to this chapter has been possible because of the general acceptance of marine foods, initially in the USA, as being healthy, and because the process was seen in the West as being exotic. 'Ethnic' foods have recently formed a greater part of Western diets as people travel more, eat out more and the younger generation is generally a little more adventurous than their parents were.

This combination, notwithstanding the actual quality of the products, has launched the surimi 'bandwagon' and, as mentioned earlier, led to it being one of the food industry buzz words of the 1980s; viewed from Europe the industry might be reaching middle-age, needing some new products and process innovations to revitalise the market. The modern process is highly capital-equipment intensive, requires good quality control and, with the products selling at premium prices, very competitive. So a word of warning is appropriate to those who see surimi as a licence to print money.

The fish-mince products are of an older generation and have taken a back-seat to surimi product development but their relatively simple production process is more widely applicable around the world. Before the surimi process and products find their true level of production and use in Western markets, it may be advantageous to have another look at mince products and use some elements of the former to improve the latter. Some suggestions are (i) the attention to mince quality (colour and texture) rather than yield through raw material selection and processing; (ii) the use of cryoprotectants (which are acceptable to the consumer) to prevent freeze denaturation; and (iii) consideration of the best product lines to include mince. Imitation may be the sincerest form of flattery in some fields, but this is rarely so in the food industry. Mince products need a niche and identity of their own.

References

Abe, Y. (1994), Quality of kamaboko gel prepared from Walleye pollack surimi with bovine plasma powder. *Nippon Suisan Gakkaishi* **60**(6), 779–785.

Abrahams, R. (1990), Kosher innovation. *Food Processing*, October, 30.

An, H., Wei, C.I., Zhao, J., Marshall, M.R. and Lee, C.M. (1989), Electrophoretic identification of fish species used in surimi products. *J. Food Sci.* **54**, 253–257.

Arakawa T. and Timasheff, S.N. (1982), Stabilisation of protein structure by sugars. *Biochemistry* **21**, 6536–6544.

Asagami, T., Oguwara, M., Wakameda, A. and Noguchi, S.F. (1995), Effect of microbial transglutaminase on the quality of frozen surimi made from various kinds of fish species. *Fish Science* **61** (2), 267–272.

Babbitt, J.K. (1986), Suitability of seafood species as raw materials. *Food Technol.* **40**, 97–100.

Babbitt, J.K., Koury, B., Groninger, H. and Spinelli, J. (1984), Observations on reprocessing frozen Alaska pollack (*Theragra chalogramma*). *J. Food Sci.* **49**, 323–326.

Buttkus, H. (1970), Accelerated denaturation of myosin in frozen solution. *J. Food Sci.* **35**, 558–562.

Chang-Lee, M.V., Lampila, L.E. and Crawford, D.L. (1990), Yield and composition of surimi from Pacific whiting (*Merluccius productus*) and the effect of various protein additives on gel strength. *J. Food Sci.* **55**, 83–86.

Chung, Y.C., Gebrehiwot, A., Farkas, D.F. and Morrissey, M.T. (1994), Gelation of surimi by high-hydrostatic pressure. *J. Food Sci.* **59**(3), 523–524.

Dondero, M., Carvajal, P. and Cifuentas, A. (1994), Cryoprotective effect of milk whey on surimi from Jack Mackeral (*Trachurus murphyi*). *Revista Espanola de Ciencia, Techologia de Alimento* **34**(3), 285–300.

Douglas-Schwarz, M. and Lee, C.M. (1988), Comparison of the thermostability of red hake and Alaska pollack surimi during processing. *J. Food Sci.* **53**, 1347–1351.

Gwinn, S. (1987), International perspective: market and economic situation. *Proc. Atlantic Canada Surimi Workshop*, Newfoundland, Canada.

Hamman, D.D., Amato, P.M., Wu, M.C. and Foegeding, F.A. (1990), Inhibition of modori (gel weakening) in surimi by plasma hydrolysate and egg white. *J. Food Sci.* **55**, 665–669.

Hastings, R.J. (1989), Comparison of the properties of gels derived from cod surimi from unwashed and once-washed cod mince. *Int. J. Food Sci. Technol.* **24**, 93–102.

Hastings, R.J., Keay, J.N. and Young, K.W. (1990), The properties of surimi and kamaboko gels from nine British species of fish. *Int. J. Food Sci. Technol.* **25**, 281–294.

Helbling, A., Lopez, M. and Lehrer, S.B. (1992), Fish allergy – is it a real problem with surimi-based products? *Int. Arch. of Allergy and Immunology* **99**(2–4), 452–455.

Ishikawa, Y. (1996), World surimi market outlook. *Infofish International* **1/96**, 16–21.

Ishikawa, S., Nakamura, K. and Fujii, Y. (1977), Fish-jelly product (kamaboko) and frozen minced meat (frozen surimi) made of sardine. I. Freshness and handling of the fish material affecting its kamaboko forming ability. *Bull. Tokai Regional Fish Res. Lab.* **90**, 59–66.

Jaouen, P., Bothorel, M. and Quemeneur, F. (1990), Membrane processes utilisations in fishing industries and aquacultural farming. *Proc. 5th World Filtration Congress*, Nice, 5–8th June, 1990.

Johnston, W.A. (1989), Surimi – an introduction. *Eur. Food Drink Rev.* **4**, 21–24.

Kelleher, S.D., Hultin, H.O. and Wilhelm, K.A. (1994), Stability of mackerel surimi prepared under lipid-stabilising processing conditions. *J. Food Sci.* **59**(2), 269–271.

Krivchenia, M. and Fennema, O. (1988a), Effect of cryoprotectants on frozen whitefish fillets. *J. Food Sci.* **53**, 999–1003.

Krivchenia, M. and Fennema, O. (1988b), Effect of cryoprotectants on frozen burbot fillets and a comparison with whitefish fillets. *J. Food Sci.* **53**, 1004–1008.

Lee, C.M. (1984), Surimi process technology. *Food Technol.* **38**, 69–80.

Lee, C.M. (1986), Surimi manufacturing and fabrication of surimi-based products. *Food Technol.* **40**, 115–124.

Lee, J.S. (1992), Microbiological considerations in surimi manufacturing. In *Surimi Technology*, Lanier, T.C. and Lee, C.M. (Eds), Marcel Dekker, New York, pp. 113–121.

Lin, T.M., Park, J.W. and Morrissey, M.T. (1995), Recovered protein and reconditioned water from surimi processing waste. *J. Food Sci.* **60**(1), 4–9.

Love, R.M.(1988), *The Food Fishes, Their Intrinsic Variation and Practical Implications*. Farrand Press, London.

MacDonald, G.A. Lelievre, J. and Wilson, N.D.C. (1990), Strength of gels prepared from washed and unwashed minces of hoki (*Macruronus novaezelandiae*) stored in ice. *J. Food Sci.* **55**, 976–978.

Martin, R.E. (1986), Developing appropriate nomenclature for structured seafood products. *Food Technol.* **40**, 127–134.

Mata, E., Faview, C., Moneretvautrin, D.A., Nicholas, J.P., Ching, L.H. and Gueant, J.L. (1994), Surimi and native codfish contain a common allergen identified as a 63-kDa protein. *Allergy* **49**(6), 442–447.

Matsumoto, J.J. (1978), Minced fish technology and its potential for developing countries. In *Proceedings on Fish Utilisation Technology and Marketing*, (Vol. 18, Sect. III) Indo-Pacific Fishery Commission, Bangkok, p. 267.

Matsumoto, J.J. (1980), Chemical deterioration of muscle proteins during frozen storage. In *Chemical Deterioration of Proteins*, Whitaker, J. and Fujimaki, M. (Eds), *Amer. Chem. Soc. Series* (Vol. 123), American Chemical Society, Washington, DC, pp. 95–124.

Miyao, Y., Shindoh, T., Miyamori, K. and Arita, T. (1993), Effects of high pressurisation on the growth of bacteria derived from surimi (fish paste). *J. Jap. Soc. Food Sci. Tech.* **40**(7), 478–484.

Musmand, J.J., Helbling, A., Eldahr, J., Haydel, R. and Lehrer, S.B. (1992), Surimi – a hidden, potentially serious cause of fish allergy. *Chemical Research* **40**(4), A805.

Ofstad, A., Grahlmadsen, E., Gundersen, B., Lauritzen, K., Solberg, T. and Solberg, C. (1993), Stability of cod (*Gadus morhua* L.) surimi produced with calcium chloride and magnesium chloride added to wash water. *Int. J. Food Sci. Technol.* **28**(5), 419–427.

Pacheco-Aguilar, R., Crawford, D.L. and Lampila, L.E. (1989), Procedures for the efficient washing of minced whiting (*Merluccius products*) flesh for surimi production. *J. Food Sci.* **54**, 248–252.

Park, J.W., Lanier, T.C. and Green, D.P. (1988), Cryoprotective effects of sugar, polyols, and/or phosphates on Alaska pollack surimi. *J. Food Sci.* **53**, 1–3.

Regenstein, J.M. (1986), The potential for minced fish. *Food Technol.* **40**, 101–106.

Roussel, H. and Cheftel, J.C. (1988), Characteristics of surimi and kamaboko from sardines. *Int. J. Food Sci. Technol.* **23**, 607–623.

Roussel, H. and Cheftel, J.C. (1990), Mechanisms of gelation of sardine proteins: influence of thermal processing and of various additives on the texture and protein solubility of kamaboko gels. *Int. J. Food Sci. Technol.* **25**, 260–280.

Saeki, H., Iseya, Z., Suguira, S. and Seki, N. (1995), Gel-forming characteristics of frozen surimi from chum salmon in the presence of protease inhibitors. *J. Food Sci.* **60**(5), 917–921, 928.

Sakamoto, H., Kumazawa, Y., Toiguchi, S., Seguro, K., Soeda, T. and Motoki, M. (1995), Gel strength enhancement by addition of microbial transglutaminase during onshore surimi manufacture. *J. Food Sci.* **60**(2), 300–304.

Shenouda, S.Y.K. (1980), Theories of protein denaturation during frozen storage of fish flesh. *Adv. Food Res.* **26**, 275–311.

Shoji, T., Saeki, H., Wahemada, A. and Nonaka, M. (1994), Influence of ammonium salt on the formation of pressure-induced gel from walleye pollack surimi. *Nippon Suisan Gakkaishi* **60**(1), 101–109.

Suzuki, T. (1981), *Fish and Krill Protein: Processing Technology*. Applied Science Publishers, Barking, Essex.

Sych, J., Lacroix, C., Adambounu, L.T. and Castaigne, F. (1990), Cryoprotective effects of some materials on cod-surimi proteins during frozen storage. *J. Food Sci.* **55**, 1222–1227.

Thomson, B.W. and Mackie, I.M. (1982), Technical note. The content of sarcoplasmic, myofibrillar and connective tissue proteins in mechanically separated tissue from filleting offal of cod (*Gadus morhua*). *Int. J. Food Sci. Technol.* **17**, 767–770.

Tokai Regional Fisheries Research Laboratory (1980), *Standard Procedure for Quality Evaluation of Frozen Surimi*. Tokai Regional Fisheries Research Laboratory, Tokyo, Jan 31.

Tsukuda, N. (1980), Lipids in kamaboko of sardine and mackerel flesh. *Bull. Tokai Region. Fish Res. Lab.* **103**, 99–104.

Wesson, J.B., Lindsay, R.C. and Stuiber, D.A. (1979), Discrimination of fish and seafood quality by consumer populations. *J. Food Sci.* **44**, 878–882.

Wray, T. (1987), Surimi: a protein for the future. *Food Manufacture* **62**, 48–49.

Yoon, K.S. and Lee, C.M. (1990), Cryoprotectant effects in surimi and surimi/mince-based extruded products. *J. Food Sci.* **55**, 1210–1216.

4 Chilling and freezing of fish
G.A. GARTHWAITE

4.1 Introduction

4.1.1 *Relationship between chilling and storage life*

Fish and shellfish are considered to be among the most perishable of foodstuffs; even when held under chilled conditions the quality quickly deteriorates. Generally, it is desirable to consume fish and shellfish as soon as possible after catching in order to avoid undesirable flavours and loss of quality due to microbial action. There are, of course, exceptions to this rule, where fish are normally kept for a short period in order to eliminate undesirable qualities, for example elimination of unpleasant flavours from plaice.

Loss of quality in fish is brought about initially by autolytic deterioration due to the action of enzymes that are present in the gut and in the flesh of the fish. This is followed by the growth of micro-organisms on the surface of the fish, which manifests itself as a slime developing on the surface. The bacteria then invade the flesh of the fish, causing breakdown of the tissues and a general deterioration of the product.

Generally, the rates at which both autolytic and microbial spoilage take place are dependent upon the temperature at which the fish is stored. Deteriorative processes are retarded at reduced temperatures and, when the temperature is low enough, spoilage can almost be stopped.

Normally, to keep fish cool, packing in ice is used; this avoids the possibility of the temperature dropping too low with the concomitant freezing of the flesh of the fish and resultant loss in quality on defrosting if slow freezing takes place. Keeping the fish cool thus extends the high-quality life (HQL) of the fish. High-quality life may be defined as 'the shelf-life of a product where the quality is of a high standard'. The standard is set much higher than the level at which the food becomes inedible. This provides a safety margin of time which assures that the fish is still edible if minor problems in distribution are encountered, such as delays or *slight* rises in temperature. Good chilling practices on board the fishing vessel and on shore result in better-quality fish which, on landing, command higher prices at auction and greater acceptance by the consumer.

The storage life of chilled fish will vary from species to species, and fish caught in tropical waters and subsequently iced will last longer than similar species caught in cooler, temperate waters. Similarly, the storage life will be affected by breeding condition, time of year and area where the fish is caught. For example, herring caught in the months of summer when the fish is feeding and the fat content is increasing may last up to 4 days on ice whereas the same fish caught in winter may last as long as 12 days on ice (Shewan, 1977). Relative rates of spoilage at different temperatures are often useful when estimating the quality of the fish of known temperature history, although, as indicated later, this applies only to fish above 0°C.

The purpose of freezing fish is to lower the temperature even further thus slowing down spoilage such that when the product is thawed after cold storage it is virtually indistinguishable from fresh fish.

4.1.2 Relative spoilage rates

Since ice melts at 0°C under normal conditions, a reference temperature of 0°C is normally used when comparing storage times of fresh fish and shellfish. A development from the Arrhenius reaction rate equation allows us to calculate 'relative spoilage rates' for fish and shellfish for temperatures above 0°C (Gorga et al., 1988). The relative spoilage rate (R) is given by:

$$R = (1 + 0.1T)^2$$

where T is the temperature of the fish measured in °C.

Where the temperature history of a fish after catching is known, the quality of the fish may be estimated by integrating the relative spoilage rates with time, resulting in an equivalent length of time of storage at 0°C. This will give an indication of the quality of the fish but assumes that the fish is in prime condition upon catching, and that handling is of a sufficiently high standard to avoid problems such as bruising. Poor handling would result in much more rapid spoilage than the calculated equivalent 'length of time on ice' would indicate.

The relative spoilage rate has been incorporated into microprocessors attached to temperature-measuring devices. The temperature probes can be placed in the centre of boxes of fish and by integration, the 'equivalent age on ice' is shown at any point in time using a suitable display mechanism. Such an instrument is of obvious advantage to the fish technologist or fisheries officer when 'trouble-shooting'.

As the distances over which fresh fish is transported increase, the maintenance of quality and extension of shelf-life become more important. Also, the consumer is demanding a higher quality product at the point of retail sale and the 'relative spoilage rate' provides a quantitative method of monitoring quality. It has already been noted that the natural spoilage mechanisms of enzyme and microbial activities may be reduced by bringing the temperature down to 0°C. In order to reduce the spoilage rates still further, packing of the product in controlled gas mixtures has been developed.

4.2 Modified-atmosphere packaging (MAP)

4.2.1 Introduction

Early work in the 1930s demonstrated that 10–20% carbon dioxide (CO_2) in an atmosphere surrounding a foodstuff will suppress the growth of *Pseudomonas* spp. and certain other spoilage organisms, provided that the temperature is maintained at or below 4°C. This physiological effect of CO_2 provides the fish technologist with a method for controlling the growth of *Pseudomonas* spp. in chilled fish, and consequently of increasing the shelf-life of chilled fish.

4.2.2 Modified-atmosphere packaging systems

Packing fish in special mixtures of gases may extend the shelf-life by up to 30%, provided that the temperature is kept below +2°C (SFIA, 1985). This is of value to supermarket chains and other distribution systems in extending the high-quality life (HQL) of the fish. It is normal when packing fish in this manner to use a mixture of three gases: carbon dioxide (CO_2), oxygen (O_2) and nitrogen (N_2).

The proportions of these gases are accurately controlled at the time of packing giving rise to the term 'controlled atmosphere packaging' (CAP). This may be considered a misnomer, as, once the pack has been sealed there is no control of the mixture of gases from that point onwards. Thus the term modified atmosphere packaging (MAP) is more accurate. In its simplest form, MAP is achieved by placing the fish in a plastic bag or sleeve, which is flushed with the gas mixture immediately prior to sealing. The plastic bag or sleeve must have a low permeability to the gases used. It is more usual to see MAP machinery that produces thermo-formed base trays from a continuous roll of plastic film into which the fish is placed (Figure 4.1). After fish is placed in the tray, it then moves along a conveyer to a section where a vacuum is drawn in the tray and the void is filled with the appropriate mixture of gases. A film lid is then heat-sealed to the top edge of the tray, completing the process.

The composition of the gas mixture will vary, depending upon whether the fish in the pack is lean or oily fish. Oxygen sustains basic metabolism and minimises the possibility of anaerobic spoilage; carbon dioxide inhibits bacterial and mould activity; and nitrogen is chemically inert and prevents rancidity, mould growth and insect attack by displacing oxygen. For lean fish, a ratio of 30% O_2:40% CO_2:30% N_2 is recommended. Higher levels of CO_2 are used for oily and smoked fish with a comparable reduction in level of O_2 in the mixture, with the gas-mixture ratio becoming 0% O_2:60% CO_2:40% N_2. By excluding oxygen, the development of oxidative rancidity in the fish is slowed.

The gas supply to the machine is normally taken from cylinders of pure gas and mixed in the correct packing ratio on the machine itself (Figure 4.2). The

Figure 4.1 Packing fish into thermo-formed base trays for modified atmosphere packing. Photograph by courtesy of UB Chilled Foods.

Figure 4.2 View of gas flushing and sealing head of MAP line. Note the reel of top web material to provide the lid for the trays and, above the reel, the control panel for the gas mixture showing flow meters. Photograph by courtesy of UB Chilled Foods.

mixture of gases can be checked by sampling at regular intervals (hourly intervals normally suffice) and analysing using gas chromotographic techniques. Portable machines for this are readily available and can be connected to the machinery on the packaging line for direct sampling if necessary. Concentric columns of Porapak Q for CO_2 (outer) and Molecular Sieve 5A (inner) for N_2 and O_2 with a thermal conductivity detector, enable all three gases to be estimated in one injection. It is important to control both the ratio of the gas mixture, and the volume ratio of fish to gas. A minimum ratio of 1:3 (fish:gas) is recommended.

CO_2 permeates packaging films up to 30 times faster than N_2 and is more fat-soluble and water-soluble, its solubility increasing as temperature is lowered. These factors lead to a reduction of pressure within the pack, resulting in a tendency for the pack to collapse. (This manifests itself as a concave surface to the lid of rigid-base packs.)

Dissolution of CO_2 into the surface of the fish muscle can reduce the pH sufficiently to lower the water-holding capacity of the proteins. This results in unsightly drip within the pack which is often absorbed by placing a cellulose pad beneath the product. The resultant pack has the advantages of retaining any drip and fishy aromas within the package while allowing the customer to view the fish prior to purchase.

The use of MAP for shell-on crustacea appears to inhibit the development of blackening of the shell (black spot), at higher chill temperatures of 5–10°C (Cann et al., 1985).

MAP-chilled fish is a most attractive proposition both to the retailer (with extended shelf-life in store) and the consumer (with cleanliness and convenience); however, the quality is dependent upon very carefully-controlled temperatures throughout the production, transportation and storage of the raw material and product. The generous use of ice on the fresh fish, accurately-controlled chill stores (less than 2°C) and air-conditioned packing rooms are essential to the achievement of maximum shelf-life extension.

The following list of recommendations highlights some of the important factors when using MAP:

- Use only fresh fish.
- Ensure fish temperature is below 2°C prior to packing.
- Pack under cool conditions and move finished pack to chill store (less than 2°C) as soon as possible after packing.
- Check that the gas mixture being used is suitable for the fish in the pack.
- Check the gas mixture on a regular basis.
- Use refrigerated transport capable of holding the product between 0 and 2°C during distribution.
- Check that the product temperature is between 0 and 2°C on arrival at the depot or retail store.
- Store at 0–2°C in display cabinets (or chill store), which should be monitored regularly to ensure this temperature range is achieved.
- Ensure that the shelf-life particulars on the label, for example 'sell by' and 'use by' are within the achievable limits for that particular product.

Although this list is not exhaustive, it attempts to highlight the main areas concerned in this increasingly popular method of preservation and distribution of fish by chilling.

4.3 Freezing

4.3.1 General aspects of freezing

Fish contains about 75% by weight of water. The process of freezing converts most of the water into ice. Water in fish contains dissolved and colloidal substances, which depress the freezing point below 0°C; the extent of freezing-point depression is proportional to the concentration of the solutes. A typical freezing point of fish is −1°C to −2°C. During freezing, water is gradually converted to ice and the concentration of dissolved organic and inorganic salts increases, depressing the freezing point continuously. Even at a temperature of −25°C, only 90–95% of the water is actually frozen. This does not include bound water (i.e. water chemically bound to specific sites such as carbonyl and amino groups of proteins, and hydrogen bonding), as it is never available for freezing. Nevertheless, the largest part of the water (about 75–80%) is frozen between −1°C and −5°C. This temperature range is known as the critical or freezing zone (Heen and Karsti, 1965; Dyer, 1971; Leniger and Beverloo, 1975).

During the freezing process, the temperature of fish flesh first falls rapidly from the initial temperature to just below 0°C. The temperature then falls very slowly until most of the water has changed state to become ice. Once the critical zone is passed, the temperature drops again quite quickly, the thermal diffusivity of ice being much higher than that of water. Since spoilage continues fairly rapidly at temperatures just below 0°C, it is important to pass the critical range quickly.

When the temperature of a biological system is reduced below 0°C, the solution first 'super cools' then solutes start crystallising or precipitating out of the solution and the formation of ice crystals occurs in two stages. Firstly, there is nucleation, that is small insoluble particles suspended in the liquid, or random aggregation of water molecules to a critical size. Secondly, there is crystal growth. Structurally, the ice crystals consist of layers of hexagonally orientated hydrogen-bonded water molecules with the oxygen of each water molecule at the centre of a regular tetrahedron, having four of the water molecules at its vertices. It is known that slow removal of heat results in slow freezing, producing ice crystals comparatively large in size and few in numbers, which may cause rupture of the cell walls and result in fluid loss and textural changes on defrosting. In contrast rapid removal of heat results in fast freezing, producing large numbers of small crystals, thus reducing the possibility of shrinkage and rupture.

In fish, however, the cell walls may be considered sufficiently elastic to withstand excessive damage from the growth of large crystals, therefore this does not account for the drip loss evident on thawing the frozen fish. In fact, most of the water is bound in the protein structure and would not be lost as drip on thawing. This binding of water may be demonstrated by squeezing fresh fish muscle in the hand when virtually no liquid will be expressed.

Despite this, on thawing of any fish product there is a loss of fluid from the flesh, which is explained through denaturation of the protein during the freezing process causing the protein to lose its water-binding capacity. Protein denaturation is dependent upon concentration of enzymes (and other compounds) and temperature. Increased concentration of the enzymes increases the rate of denaturation, whereas the rate is reduced as the temperature falls. Of course, as the temperature falls, more of the water is converted to ice and the concentration of enzymes in solution increases, therefore below the freezing point of water, concentration and temperature are very closely related.

The optimum temperature range for denaturation by this means is $-1°C$ to $-2°C$. Thus, in order to reduce thaw-drip to a minimum, the time spent in this temperature zone during freezing must be as short as possible. Protein denaturation due to dehydration also occurs during cold storage.

'Quick freezing' is a general term applied to most freezing processes and gives rise to the term 'IQF' or 'individual quick-frozen'. Quick freezing is difficult to define and although in the UK it is recommended that all the fish should be reduced from $0°C$ to $-5°C$ in 2h or less, it is generally acknowledged that a period of 2h would be too long for products such as tails of scampi (*Nephrops norvegicus*). The alternative is to compare the rate of movement of the ice front through the foodstuff (Table 4.1).

The rate of movement of the ice front will, of course, depend upon factors such as the shape and thermal properties of the fish, and the type of freezer used.

During the freezing process, the temperature of the fish should be lowered to $-30°C$ before transfer to the cold store. Most commercial freezers operate at temperatures of $-40°C$ to $-35°C$ and a rough guide to attaining the temperature of $-30°C$ would be to lower the thermal centre of the fish to a temperature of $-20°C$ prior to removal from the freezer. The time taken to lower the temperature of the thermal centre to $-20°C$ is normally termed the freezing time.

As already indicated, lowering of the temperature reduces reaction rates. Furthermore, as the water in the fish freezes it becomes bound, thus reducing

Table 4.1 Rate of movement of ice front through food during freezing

Rate of movement of ice front (cm h^{-1})	Type of freezer
0.2 to 0.4	Cold store
0.5 to 3.0	Plate freezers and air-blast freezers
5.0 to 10.0	Air-blast freezers and fluidised bed freezers
10 to 100	Liquid nitrogen and carbon dioxide freezers

the water activity (a_w) and hence bacterial growth. Thus it may be said that freezing preserves fish by a combination of reduction of temperature and lowering of water activity.

4.3.2 Prediction of freezing times by numerical methods

The freezing of any foodstuff is a complex, unsteady-state heat-transfer situation. Not only is the temperature at any point changing with time, but there is also a change of state at the freezing temperature. This, in turn, results in change of thermal properties between the thawed and frozen material making accurate estimation of freezing time difficult.

As a consequence of this, it is normal to estimate freezing times for particular circumstances, for example cod fillets frozen using a Torry Freezer, by experimental methods rather than calculations based on mathematical models. However, there are some circumstances where mathematical modelling is required. In such models, the accuracy of the calculation depends upon the accuracy of the data incorporated in the equations used; thus thermal characteristics such as heat-transfer coefficients at the surface and thermal conductivity of the foodstuff need to be known with a fair degree of accuracy. In the case of fish, while thermal conductivity is known for some species where work at research stations has centred around those species, the condition of the fish (e.g. oil content of fatty fish) will affect the thermal properties. Also, there are many species where these properties are unknown and estimates of data from similar species must be used.

Notwithstanding the above limitations, mathematical models can be used in situations where indications of the effect on freezing time are required when process parameters may change.

4.3.2.1 Plank's equation (1941); cited in Ede (1949).
This mathematical model assumes steady-state conduction in the frozen section with the product initially at its freezing point. The time θ_f for the slowest cooling point (thermal centre) to be completely frozen is given by:

$$\theta_f = \frac{L\rho}{T_f - T_m}\left[\frac{Pd}{h} + \frac{Rd^2}{k}\right] \qquad (4.1)$$

and, where the food is contained in packaging materials, by:

$$\theta_f = \frac{L\rho}{T_f - T_m}\left[Pd\left(\frac{1}{h} + \frac{x}{k_1}\right) + \frac{Rd^2}{k}\right] \qquad (4.2)$$

where θ_f = freezing time (s)
  d = diameter of sphere or cylinder, thickness of slab, smallest dimensions of brick (m)
  T_f = freezing point of food (°C)
  T_m = temperature of freezing medium (°C)

h = surface heat transfer coefficient (W m^{-2} K^{-1})
k = thermal conductivity of frozen material (W m^{-1} K^{-1})
k_1 = thermal conductivity of packaging material (W m^{-1} K^{-1})
x = thickness of packaging material (m)
P = 1/6 for sphere, 1/4 for cylinder, 1/2 for slab
R = 1/24 for sphere, 1/16 for cylinder, 1/8 for slab
L = latent heat of freezing
ρ = density of material.

P and R are factors representing the shortest distance from the surface to the centre of the food.

The application of Plank's work is limited by the assumption that the material must be at the freezing temperature uniformly throughout, the cooling medium is at a constant temperature, the material is homogeneous and both freezing point and latent heat of fusion can be defined accurately.

Work by Nagaoka *et al.* (1955) on freezing fish resulted in a suggested modification of Plank's basic formula. The revised equation takes into account the initial and final temperatures, thus:

$$\theta_f = [1 + 0.0044(T_1 - T_f)]\left[\frac{Z}{V(T_f - T_m)}\right]\left[\frac{Pd}{h} + \frac{Rd^2}{k}\right] \qquad (4.3)$$

where: T_1 = initial temperature of fish (°C)
V = specific volume of ice (m^3 kg^{-1})
Z = heat to be removed from the fish lowering it from its initial temperature to the final temperature

$$Z = C_{p1}(T_1 - T_f) + L + C_{p2}(T_f - T_2) \qquad (4.4)$$

where: C_{p1} = specific heat above freezing point
C_{p2} = specific heat below freezing point
T_2 = final temperature of fish.

4.3.2.2 *Neumann equation (1959).* Neumann proposed an equation for calculating the temperature distribution throughout a mass in which a change of state is occurring. The temperature is expressed as a function of time and position in an infinite slab while it is being frozen:

$$v_1 = \frac{T_f}{\text{erf}\,\lambda}\,\text{erf}\,\frac{X}{2\left[\frac{(k\theta)}{\rho_1 c_1}\right]^{1/2}} \qquad (4.5)$$

$$v_2 = T_0 - \frac{(T_0 - T_f)}{\text{erfc}\,\lambda\left(\frac{k/\rho_1 c_1}{k_2/\rho_2 c_2}\right)^{1/2}}\,\text{erfc}\,\frac{X}{2\left(\frac{k_2\theta}{\rho_2 c_2}\right)^{1/2}} \qquad (4.6)$$

where: v_1 = difference in temperature between frozen section and surface temperature (°C)
v_2 = difference between temperature at which freezing occurs and surface temperature (°C)
k = thermal conductivity of frozen material (W m^{-1} K^{-1})
k_2 = thermal conductivity of thawed material (W m^{-1} K^{-1})
ρ_1, ρ_2 = densities of frozen and thawed material
c_1, c_2 = specific heats of frozen and thawed material
X = distance from surface of slab
θ = time
erf = error function
erfc = co-error function
λ = factor determined from equation 4.7:

$$\frac{e^{-\lambda^2}}{\operatorname{erf}\lambda} - \frac{k_2\left(\dfrac{k}{\rho_1 c_1}\right)^{1/2}(V-T_f)\exp\left[-\dfrac{k\lambda^2}{\rho_1 c_1}\bigg/\dfrac{k_2}{\rho_2 c_2}\right]}{k\left(\dfrac{k_2}{\rho_1 c_2}\right)^{1/2} T_f \operatorname{erfc}\left[\lambda\left(\dfrac{k}{\rho_1 c_1}\bigg/\dfrac{k_2}{\rho_2 c_2}\right)^{1/2}\right]} = \frac{\lambda L \pi^{1/2}}{c_1 T_f} \qquad (4.7)$$

where: L = latent heat of freezing
V = initial temperature of thawed material.

Equation 4.7 must be solved by trial and error to determine λ. The error function and co-error functions are obtained from tables.

4.3.2.3 *Prediction for irregular shapes.* Both Plank's and Neumann's solutions are limited by use of regular shapes for the objects to be frozen. Cleland and Earle (1982) proposed a simple model, with an accuracy of ±10%, that takes the shape of the product into account. The model is based upon Plank's equation but incorporates a concept of equivalent heat transfer dimensions (EHTD). Three dimensionless groups are used:

(i) the Plank number (Pk):

$$\mathrm{Pk} = \frac{C_L(T_i - T_f)}{\Delta H} \qquad (4.8)$$

(ii) the Stephan number (Ste):

$$\mathrm{Ste} = \frac{C_s(T_f - T_a)}{\Delta H} \qquad (4.9)$$

(iii) the Biot number (Bi):

$$\mathrm{Bi} = \frac{hd}{k} \qquad (4.10)$$

Modification to the terms P and R in Plank's equation are made as follows:

$$P = 0.5[1.026 + 0.5808\,Pk + \text{Ste}(0.2206\,Pk + 0.1050)] \qquad (4.11)$$

$$R = 0.125[1.202 + \text{Ste}(3.410\,Pk + 0.7336)] \qquad (4.12)$$

Then:

$$t_f = \frac{\Delta H}{(T_f - T_a)(\text{EHTD})}\left(P\frac{d}{h} + R\frac{d^2}{k}\right) \qquad (4.13)$$

The value of EHTD may be taken as follows: (i) for infinite slabs EHTB = 1; (ii) for infinite cylinders EHTD = 2; and (iii) for spheres EHTD = 3. For other shapes, EHTD may be found experimentally and substituted in Equation 4.13 where the conditions have similar Biot numbers.

The mathematical models given in this section indicate the complexity of the heat transfer during the freezing process. It is not within the scope of this section to expand on the above models, but calculation of freezing times is extensively covered elsewhere (Ede, 1949; Nagaoka *et al.*, 1955; Neumann, 1959; Cleland and Earle, 1977, 1979, 1982; Cleland 1990).

Any mathematical solution gives at least a close indication of the actual freezing time; practical confirmation of such predictions is normally essential.

4.3.3 Freezing systems

There are three basic methods available for freezing fish, the choice of which will depend upon cost, function, and feasibility governed by factors of location and type of product. The three methods are:

- air-blast freezing – where a continuous stream of cold air is passed over the product;
- plate or contact freezing – where the product is placed in direct contact with hollow, metal, freezer plates, through which a cold fluid is passed;
- spray or immersion freezing – where the product is placed in direct contact with a fluid refrigerant.

All three types are used both on shore-based and ship-board freezing systems in the fish industry.

4.3.3.1 Air-blast freezers.
The main advantage of air-blast freezers is their versatility. However, because of this, they are frequently used incorrectly and inefficiently. They are suitable for irregular-shaped, different-sized and non-deformable foods. Hence they are advantageous for producing individual quick frozen (IQF) products of different crustaceans, fish fillets and added-value products such as breaded fish portions or fish sticks. Moreover, they can be operated on either a batch or continuous basis. Uniform freezing is achieved only if the temperature and air velocity over the product are constant. A velocity of about $5\,\text{m s}^{-1}$ is usually economical, although a higher

velocity of 10–15 m s^{-1} may be justified for continuous-belt freezers. Such higher air velocities are also used on board factory vessels since noise is rarely a problem and speeds of 15–25 m s^{-1} are used.

However, air-blast freezers have a slightly slower freezing rate compared with immersion freezing and need frequent defrosting. A further disadvantage is that a slight increase in air velocity demands a large increase in power. Prolonged higher air velocities may dehydrate the product surface severely, causing freezer burn. This is especially the case if the product is not chilled prior to entering the freezer (although it has been shown that chilling prior to freezing may extend the time taken to pass through the critical zone). High air speeds may also result in the product being blown across the belt. This is particularly the case where the surface of the belt is smooth and the coefficient of friction between the product and the belt is very low. In the extreme case, this results in the product being blown off the belt inside the freezer resulting in product loss.

During the continuous operation of blast freezers, there is a build-up of frost on the conveyor belt. This can be a particular problem where a mesh conveyor belt is used. The build-up of frost blocks the holes in the mesh, reducing air flow and thus increasing freezing time. The problem is overcome by incorporating a belt washer into the system.

One of the most recent developments in the design of continuous air-blast freezers is the Torry Freezer (Figure 4.3), which was developed at the Ministry of Agriculture, Fisheries and Food Research Station in Aberdeen. This uses a continuous, flat belt made from stainless steel and, as such, has the advantage of the high heat-transfer rates that are achieved in contact freezers coupled with versatility. It is particularly suited to the freezing of individual fish fillets where the bottom surface is flat and the top surface is irregular (Figure 4.4).

Figure 4.3 Loading cod fillets into a Torry Freezer. Photograph by courtey of Findus (UK) Ltd.

Figure 4.4 Unloading from the Torry Freezer shown in Figure 4.3. Note the rigid shape of the frozen fillet causing it to detach from the curving belt. Photograph by courtesy of Findus (UK) Ltd.

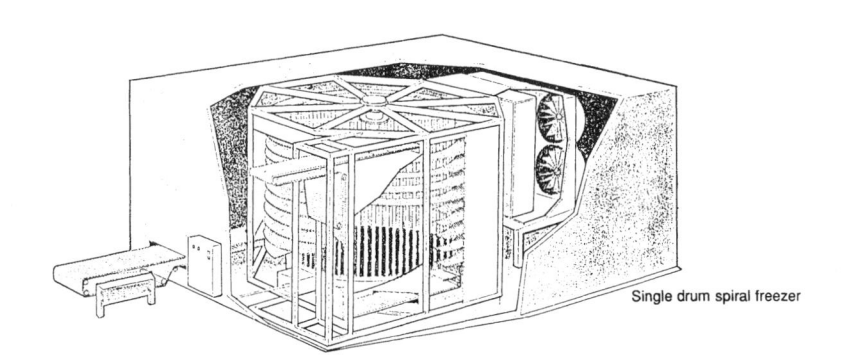

Figure 4.5 A spiral air-blast freezer. Diagram by courtesy of APV Baker Ltd, Freezer Division.

Where high capacity or extended freezing times are required using air-blast freezers, the conveyor length becomes excessive. This problem has been overcome by the use of multiple-pass systems where the product is transferred inside the freezer from one belt to another, and travels backwards and forwards along the length of the freezer before being discharged. In this type of freezer, three passes are normal. A more usual method of overcoming the same problem is to use spiral freezers. The single, continuous belt can be operated on a single- or twin-drum application in ascending or descending combinations. The whole system is enclosed in an insulated chamber, the floor of which is usually constructed to form a tank, which is usually sloped to drain and thus aids cleaning (Figure 4.5).

4.3.3.2 Plate or contact freezers. Plate freezers are used for freezing fish into blocks but are not so versatile as blast freezers. They may be of a vertical or horizontal type, according to the arrangements of the plates. The plates are now made from extruded aluminium, which, in cross-section, shows channels through which the liquid refrigerant is passed. Heat transfer takes place through the upper and lower surfaces of the plate, and freezing is accomplished by direct contact between the cold plates and the product. Pressure applied to the plates on each side of the product will improve the contact and, in this way, increase the heat-transfer coefficient between the plate and the product. The pressures applied are between 1 and 10 bar using hydraulic pressure to hold the plates closed. It is worth noting that the design of plate freezers for shore-based and ship-based systems differs slightly because of the limitations of ceiling height between decks.

The vertical type of plate freezer is particularly suitable for freezing whole fish at sea and for bulk freezing. The thickness of the blocks may be 25–130 mm depending upon the product. To operate these freezers, the openings at the bottom and at the vertical edges between the plates are enclosed. The product is dropped into the box thus formed and, when full, the plates are closed and the refrigeration process commences. The product is loaded with the plates in a 'defrost' condition because if they were pre-cooled it is likely that the product would stick to the plates instead of falling to the base of the box. A packaging material between the fish and plates is sometimes used to prevent sticking to the plates; however, with the use of warm refrigerant to defrost the plates and facilitate removal of the frozen blocks, packaging is not always necessary. Unloading of such freezers may be by discharging to the top of the plate or from the base of the plate or to the front of the plate, the direction of discharge suiting the user's requirements.

Horizontal-plate freezers are used for the production of laminated fish blocks, which may be 25–100 mm thick. They are also used to freeze pre-packed retail flat cartons of fish products such as packs of prawns (both with the shell on and shell off) and other fish products. The product is usually wrapped in plastic film and then placed in cartons or directly on to aluminium freezing trays, which are, in turn, placed on the freezer shelves.

4.3.3.3 Spray and immersion freezers. These are mainly used for IQF products. They find limited application in the fish-processing industry and tend to be used only for specialised or high-value products, such as scampi, and added-value or seasonal products, such as meats of bivalve molluscs.

4.3.3.4 Immersion freezers. Use of immersion freezers ensures intimate contact between the surface of the fish and the freezing medium, so would appear to ensure good heat transfer. The freezing medium used is normally sodium chloride brine, which has a eutectic point of $-21.2°C$. Because of this, brine temperatures of about $-15°C$ are used in the freezing process. Further reduction in temperature must be achieved by transferring the product to cold storage.

Freezing of large tuna in brine may take up to 3 days to achieve complete freezing. Using modern blast freezers operating at temperatures as low as -50 to $-60°C$, the fish may be frozen in less than 24 h. Brine freezing, once popular in the tuna fishing industry, is being replaced by air-blast freezers for preservation of the catch.

4.3.3.5 *Spray freezers.* Spray freezers are also classified as cryogenic freezers. In cryogenic freezing a very rapid rate of freezing is achieved by exposing the products, either unpacked or thinly packed, to an extremely cold refrigerant with a low boiling point. In this method, the refrigerant is sprayed onto the product and heat removal is accomplished during a change of state by the refrigerant.

4.3.3.6 *Carbon dioxide.* With this type of freezer the liquid carbon dioxide is sprayed on the product as it passes through the tunnel. The fish moves through the tunnel on a conveyor belt and, as it passes under the nozzles, the surface is sprayed with carbon dioxide in liquid form. The carbon dioxide changes state on leaving the nozzles and in so doing absorbs large quantities of heat resulting in rapid cooling of the product. In some systems a layer of solid CO_2 (dry ice) is laid down on the conveyor belt and the product is placed on top. Liquid carbon dioxide is then sprayed overhead and, as sublimation of dry ice occurs at $-78°C$, it is possible to freeze to at least $-75°C$. Freezing under these circumstances is very rapid and drip losses are reduced to less than 1%.

4.3.3.7 *Liquid nitrogen.* In the case of liquid nitrogen freezers, the liquefied gas is also sprayed onto the product as it travels through a tunnel on a moving conveyor. The nitrogen gas passes countercurrent to the conveyor, thus the fish is pre-cooled before reaching the liquid nitrogen sprays. At atmospheric pressure liquid nitrogen boils at $-196°C$, therefore it is essential to incorporate a pre-cooling stage in the tunnel to avoid stress cracking of the product as a result of too rapid cooling. After the spray nozzles, a region for tempering the product is allowed before the conveyor exits from the freezing tunnel. This enables the temperature gradient from the outside to the centre of the product to equilibrate to some extent. It is normal for complete equilibrium to take place once the product has been moved into the cold storage area.

Both carbon dioxide and nitrogen may be used in specially designed blast freezers normally of the spiral type.

In any given commercial situation, the type of freezer employed to produce frozen-fish products must be capable of producing a product that is of the quality demanded by the customer. Certain types of freezers are simply not compatible with the production of the end-product. For example, it is essential that a plate freezer is used to produce frozen fish blocks that are to be used in the production of fish portions and fish sticks. The use of any other type of freezer would result in blocks that have non-parallel sides, and this would lead to greatly reduced yields on further processing. For most applications, however, the use of blast freezers offers the best means of achieving the desired results. Liquid refrigerants, such as carbon dioxide, are used mainly for high-value products such as prawns, lobsters, and so on, or for seasonal products where the reduced capital cost of the liquid freezing tunnel outweighs the higher running costs of such freezers. Freezers on board fishing vessels or factory ships are normally of the plate or air-blast types only, with immersion freezing being used in the tuna fishery.

4.4 The application of freezing systems in fish processing

4.4.1 Freezing on board

Much of the frozen fish traded throughout the world is in block form, frozen in plate freezers as already mentioned; such freezers are extensively used on fishing vessels. Vertical plate freezers are used for freezing whole fish that will be re-processed onshore.

Fish should be graded to ensure that blocks are of a single size grade, as the purchaser normally requires a particular grade for subsequent shore-based operations. Nor should the spaces between the plates be overloaded, as this will increase freezing time and may cause other problems such as difficulty unloading. Some standard block sizes used on the world market are shown in Table 4.2.

Where larger fish such as mackerel are being frozen in bags, water is placed in the bag first to cushion the fall of the fish, which otherwise may rupture the base of the bag. If the fish are frozen without the addition of water, the frozen block may be packed in a polythene bag and possibly fibreboard outer after freezing to aid handling (McDonald, 1982).

Fish-fillet blocks are also produced on factory vessels or freezer trawlers, and onshore. In each case, the procedure is similar. A variety of blocks may be produced, ranging from laminated fillet blocks to blocks of minced fish recovered by using flesh–bone separators.

Table 4.2 Some standard block sizes used on the world market

Weight (lbs)	Dimension (inch)	Dimension (mm)
13.5	21 × 11.5 × 1.5	533 × 292 × 381
16.5	19 × 10 × 2.5	482 × 254 × 63.5
18.5	19 × 11.375 × 2.5	482 × 292 × 63.5

Figure 4.6 Jumble-packed fish fillets being loaded into mould. Note that the mould rests on an aluminium tray and is lined with polythene-lined card. Photograph by courtesy of Findus (UK) Ltd.

Laminated skinless fillet blocks are the most labour-intensive block to produce. The block is produced by placing the fish fillets in a waxed board or polyethylene laminated board carton, which is held in a rigid frame. The fillets are laid out either lengthways or across the carton but all fillets are arranged parallel to each other. The corners are filled first to avoid any voids, after which a fixed weight of fillets is filled into each carton. Bad corners or voids produce non-uniform blocks resulting in wastage in subsequent user-processing operations. Figure 4.6 shows jumble-packed fish fillets being loaded into a mould.

Once filled with fillets, the lid of the carton liner should be folded down to cover the upper surface, ensuring the tabs on the lid fit outside the walls of the liner but inside the walls of the frame. This prevents the fish becoming embedded in the frozen block. The frame and contents, which are normally on a thin aluminium sheet, are placed in horizontal plate freezers (see Figure 4.7).

It is important to ensure that any condensation is removed from the plates before loading, as droplets of water may freeze *in situ* on the plates, producing an irregular surface to the block and reducing heat-transfer rates.

Other types of block produced in a similar manner have randomly arranged fillets or mixtures of fillets and fish mince. IQF fillet blocks may be produced by laminating layers of fillet with layers of polyethylene. Such blocks produce separate (IQF) fillets if pressure is applied to the block to break it, as fracture occurs on the lamination rather than across the fillets themselves. The polyethylene is coloured (blue, red or green) in order to facilitate inspection for any that is frozen into folds in the fillet.

Pelagic fish, such as herring and mackerel, will be frozen whole (gut in) and the addition of water to fill the voids between fish is a fairly standard procedure. The use of polyethylene or polyethylene-lined paper bags as liners

Figure 4.7 Loading block in frames into plate freezer for freezing. Photograph by courtesy of Findus (UK) Ltd.

for the stations of the freezers, while adding to the cost of production, facilitates release of the frozen blocks from the moulds and, like the water, acts as a protectant against dehydration and rancidity development during subsequent storage.

The freezing of 'whole' white fish such as cod is carried out after evisceration. The head may or may not be removed before freezing and, as this process is usually mechanical due to restrictions on space and labour, the final yield of fillet flesh from the fish may be lower than one would expect from a hand-filleting operation. With large catches, the problems of fish going into rigor before freezing are particularly problematic – especially where the ambient air temperature is high, which accelerates the onset of rigor. Where possible, fish should be frozen before the onset of rigor as this reduces the possibility of severe muscle contraction resulting in 'gaping' in the fillet. Fish should never be frozen while in rigor as this will result in broken, gaping fillets on defrosting, also the density of the fish in the block is reduced.

Where ambient temperatures are high, it is usual for freezer trawlers to have refrigerated seawater-holding tanks fitted for storing the catch prior to preparation for freezing. Ideally, the catching rate should match the freezing capacity of the vessel but this is rarely achievable.

Storage prior to freezing using ice or refrigerated sea water will not affect different species the same way (Table 4.3).

The loading of the fish into the plate freezer should be such that the fish are packed head to tail, ensuring that the head is towards the outside of the block to reduce damage subsequent to freezing.

Air-blast freezers are fitted in some larger freezer trawlers and factory ships. These freezers are tunnel freezers, which may incorporate conveyor systems or use moveable racking systems, both with trays into which the fish is loaded for

Table 4.3 Chilled storage for various fish species prior to freezing

Species	Maximum holding time under chill conditions prior to freezing
Cod	3 days
Haddock	2 days
Hake	1 day
Plaice	5 days
Herring	1 day
Mackerel	1 day

freezing. Electrical heaters may be used to release the frozen fish from the tray and the block is subsequently glazed using water sprays and packed prior to storage in the hold. The use of air-blast freezers is advantageous for larger fish that do not lend themselves to plate-freezing techniques, such as in the tuna-fishing industry where the fish may be conveyed through the freezing tunnel hanging by the tails.

4.4.2 Onshore processing

Where fish landed fresh is to be frozen 'whole', the use of plate freezers is very similar to on-board methods. However, crustacea such as prawns and scampi (*Nephrops norvegicus*) are often frozen using liquid nitrogen freezers.

Where catches of shrimps and prawns are large, the use of fluidised bed freezers may be considered, but such freezers are expensive and require throughputs of at least $2 t h^{-1}$ for economy.

Most shore-based freezing of fish is in the production of added-value fish products. The simplest form of adding value is to remove the fillets from the fish. Fillets may be frozen in blocks as described in section 4.1 or using the continuous stainless-steel belt of a Torry Freezer. If the Torry Freezer is used, fillets should be placed on the belt with the thicker (head) end leading (see Figure 4.3); this reduces problems of breakage of the fillets due to curvature of the belt. Slated belts on cryogenic freezers may also be used for the freezing of fillets.

The coating of fish in batter with or without breadcrumbs is practised throughout the world. The 'fish finger' (fish stick) and 'fish portion' are both portion-control products where shape and weight must be controlled accurately. Consequently, they are produced from frozen fish-fillet blocks, which allow regular-shaped portions to be cut, using a bandsaw, with a minimum of waste.

The batters used are basically a suspension of flour in water and are of two types:

- adhesive
- tempura or puffing.

Depending upon the application, the adhesive batters may be used in various degrees of viscosity. Thin batters are used for the first coating of crumb on a multi-pass line, whereas thicker batters are used for a single-pass line. Complete cover of the fish by the coating is essential to prevent the coating blowing-off on subsequent frying operations. Improved adhesion to the fish, especially for irregular shaped products, is achieved by incorporation of gums or modified starch into the batter or by pre-dusting with dry batter mix or fine crumb. Batter viscosity is a major factor in the control of pick-up of crumb in the subsequent operation, and most production lines use an automatic batter viscosity control system.

Incorporation of sodium bicarbonate and leavening acid in the batter produces a puffed batter on frying due to the CO_2 produced. Such batters are termed 'tempura' batters. The product is always flash-fried prior to freezing, and too much leavening agent may cause blow-off of the coating during frying, resulting in a reject product. Flash-frying of enrobed products prior to freezing, although not always essential, is advisable as it 'sets' the starches and proteins in the batter and imports mechanical strength to the coating, which reduces losses in subsequent handling. In the case of tempura-type products, the flash-frying is essential. The frying process lasts only 20–30 s during which time the fish, if originally frozen, remains in that condition. Subsequent freezing takes place immediately after the frying stage as any delay would result in defrosting of the fish with subsequent loss in quality both of the fish and the coating. Continuous air-blast freezers with mesh conveyor belts are normally used for such operations.

Where a factory is producing a variety of frozen products all of which must pass through the same freezer, other types of freezer belt may be employed. An example is prawn-processing plants where the raw meat would freeze to a mesh belt and be difficult to remove in one piece. Such plants may be reforming flesh from prawn tails (by extrusion) and would require a flat surface on which to freeze the product. Whether freezing whole prawns, prawn tails, breaded/batter prawns or re-formed products, using an air-blast freezer, air speeds must be controlled to avoid blowing the product off the belt.

The regular shapes of fish sticks and rectangular portions do little to help the natural image demanded by today's consumer. A recent development is a moulding press, which takes tempered frozen portions placed in a mould and applies very high pressures, thus moulding the frozen block to the shape of the mould. Such a technique opens up a whole range of possibilities for new frozen products.

Production of products incorporating sauce with fish is well-developed but gaining in sophistication. With the increase in use of microwave ovens, packaging of such products is geared towards this market (microwaveable material, avoidance of sharp corners). Production is often form-fill packing on lines already discussed under MAP (section 1.2). Because of the regular nature of the packaging, plate freezers are often used and high-volume lines incorporate continuous-plate freezers with automatic loading and unloading.

Cook–freeze and cook–chill fish products for catering purposes may incorporate cooked fish, whereas in the frozen, added-value retail packs this is rarely the case. Where the fish is to be pre-cooked, the fish should be slightly under-cooked allowing the re-heating procedure to complete the cooking process, for example in airline catering. This will avoid loss in quality due to over-cooking. The packaging material used must be compatible with the end usage. In the case of airline catering, the re-heating is usually in forced convection hot-air ovens where aluminium trays are used, the excellent heat conduction properties of aluminium also assisting in the freezing process.

In all freezing processes, microbiological quality must be monitored to ensure safety to the consumer. This is particularly important in cook–chill products where temperature control during distribution is of paramount importance.

4.5 Changes in quality on chilled and frozen storage

4.5.1 Chilled storage

The shelf-life for different fish products will vary considerably and is primarily dependent on the temperature of storage and initial condition of the fish. The fish species, treatment and method of packaging and storage will also affect the high-quality life (HQL) of the product.

The initial quality of the fish dictates the maximum possible shelf-life and may be affected by its feeding habits, breeding condition, handling on board, and so on. Once caught, the quality deterioration is highly temperature dependent. The quality during storage may be assessed by analysing the products of protein degradation. Measurement of hypoxanthine, di-methylamine and tri-methylamine to give an indication of freshness has been used to give an objective assessment of fish quality.

The K-value index is gaining in popularity in certain parts of the world and relates freshness to the autolytic degradation of nucleotides in fish muscle after death (Elira et al., 1989). Adenosine triphosphate (ATP) degrades catabolically as:

Adenosine triphosphate (ATP) → adenosine diphosphate (ADP) → adenosine monophosphate (AMP) → inosine monophosphate (IMP) → inosine (HXR) → hypoxanthine (HX)

The nucleotides are separated and collected by means of anion-exchange column chromatography and measured using a spectrometer.

$$\text{The } K \text{ value } (\%) = \frac{HX + HRX}{ATP + ADP + AMP + IMP + HX + HXR} \times 100$$

Standards may be set for different species. The Japanese set a standard of K (%) equal to 20 maximum for fish destined for consumption in the raw state.

The ATP is degraded to IMP very soon after death and consequently a K_i value may be used which does not use determination of ATP, ADP or AMP.

$$K_i = \frac{\text{hypoxanthine} + \text{inosine}}{\text{IMP} + \text{inosine} + \text{hypoxanthine}}$$

$$= \frac{\text{HX} + \text{HRX}}{\text{IMP} + \text{HXR} + \text{HX}}$$

Where the K_i value is used, with some species it increases very rapidly and then remains constant. This is due to the accumulation of a large quantity of inosine relative to the quality of hypoxanthine produced. Whilst this gives a reasonable indication of freshness in terms of quality parameters (which may refer to raw seafood consumption), it does not accurately reflect the freshness of some species of fish in other parts of the world (where the fish is normally cooked before consumption).

An example of the latter is North Atlantic cod, which can accommodate inosine rapidly. For this species, the K_i value is considered inadequate as a quality indicator. Instead, the G value is used (Burns *et al.*, 1985) where:

$$G\% = \frac{\text{hypoxanthine} + \text{inosine}}{\text{AMP} + \text{IMP} + \text{inosine}} \times 100$$

$$= \frac{\text{HX} + \text{HRX}}{\text{AMP} + \text{IMP} + \text{HRX}} \times 100$$

Pacific cod quality has been assessed using the H value (Luong, *et al.*, 1992) where:

$$H\% = \frac{\text{hypoxanthine}}{\text{IMP} + \text{inosine} + \text{hypoxanthine}} \times 100$$

$$= \frac{\text{HX}}{\text{IMP} + \text{HRX} + \text{HX}} \times 100$$

The work on these two species does imply that the use of the K_i value as an indicator of freshness must be verified and adapted where appropriate, depending upon the species in question. There are also indications that the location of the sampling point on the fillet from which the sample is taken for analysis can affect the percentage of nucleotide degradation observed.

4.5.2 *Frozen storage*

Once the fish is frozen, bacterial spoilage caused by the exogenous enzymes of the bacteria ceases, and autolytic changes caused by the breakdown of the chemical constituents of the fish by the intrinsic enzymes tend to be relatively slow and contribute little to loss of quality. The major causes of quality loss during frozen storage and on thawing are dehydration, drip loss, protein denaturation and discoloration and cold-store odour.

The surface of frozen fish may dry slowly in cold storage due to migration of water by sublimation caused by, even, small fluctuations in cold-store temperature. Such fluctuations alter the water-holding capacity of the air in the store or pack. Consequently, water is absorbed from the ice in the product and condensed to form ice crystals at the evaporator coils, or the inner-surface of the packaging.

Frozen fish that have suffered severe drying in cold store develop a chalky white, dry and wrinkled appearance on the surface which, in extreme cases, discolours to produce yellow or brown discoloration in white fish fillets. This is associated with the tough texture that is characteristic of the condition known as 'freezer burn'. When thawed, the product is dry and spongy due to denaturation of the protein.

Weight loss by dehydration is directly proportional to the exposed surface area and can be reduced by two methods; covering the surface with packaging material, or surrounding the product with a sacrificial layer of ice. The use of close-contact packaging materials for uniform shapes, (*e.g.* fish-fillet blocks) or vacuum packing for irregular shapes (*e.g.* smoked fillets or crustacea), may be employed. Crustacea with their often spiny exoskeleton must be packed in especially strong film to give puncture resistance. In Canada, the use of surlyn (ionomer) and nylon (polyamide) laminate provides a barrier to oxygen and gives the required toughness and puncture resistance for packing lobsters. Lobsters may be packed in tubes filled with brine solution prior to freezing, resulting in a better-quality product on defrosting.

The use of ice glaze for small and irregularly shaped fish products may be considered essential where the fish is stored without packaging or is packed in 'pillow packs', for example small prawns and shrimp (IQF). The glaze is applied by dipping the frozen product into chilled water or passing between spray nozzles situated above and below a mesh conveying belt, along which the frozen fish passes. The belt speed may be adjusted to give the correct dwell time to achieve the desired glaze pick-up. By making the conveyor dip into a water trough, thus omitting the lower spray nozzles, savings on water consumption may be achieved. The water must be of drinking quality. The amount of glaze pick-up is dependent upon:

- fish temperature
- water temperature
- product size and shape
- glazing time.

The fluctuation in temperature during storage mentioned above causes re-distribution of the ice crystals, resulting in the growth of a very few extra-cellular ice crystals at the expense of many small intracellular ones. Once thawed, the water from these large crystals cannot diffuse back into the cells (Giddings and Hill, 1978). Denaturation of protein adversely affects its water-holding capacity and adds to thaw-drip. Concentration of solutes in the bound water during freezing (resulting in them crystallising out) produces localised areas of mechanical damage of the tissue cells.

The initial quality of the raw material may also affect drip loss (Hebbar and Hiremath, 1980).

Colour plays a major role in the acceptance of food by the consumer (Simpson, 1982). Loss of the characteristic pink colour in crustacean shellfish results from changes in the cartenoid pigments:

$$\begin{array}{ccc} \beta\text{-carotene} & \text{astaxanthin} & \text{astacene} \\ \text{(red)} & \xrightarrow{} \text{(pink)} & \xrightarrow{} \text{(orange-yellow)} \end{array}$$

Enzymic oxidation results in the yellow discoloration of frozen lobster meat, particularly in the lipid-rich region of the tips of the claws.

4.5.3 Thawing

Thawing of fish for re-processing is of considerable importance in the maintenance of fish quality. The thawing system used should avoid: (i) localised overheating of the fish (the protein from cod caught in arctic waters denatures at about 30°C); (ii) excessive drip loss; (iii) dehydration; and (iv) bacterial growth.

When thawed in air, the temperature should not be allowed to rise above 20°C and it is normal to use air saturated with water vapour, which is circulated by fans at speeds of $8-10 \text{ m s}^{-1}$. Blocks of frozen fish are placed on deeply grooved trays to ensure good circulation of air around the block (Figure 4.8).

Thawing in water, while simple and inexpensive may cause the fish to lose quality in terms of flavour and appearance.

Vacuum thawers consist of airtight chambers into which the fish is loaded using trolleys. A vacuum is drawn and water in a tray on the base of the chamber is heated, filling the chamber with water vapour. The vapour condenses on the cold surface of the fish where the latent heat of vaporisation is absorbed by the fish. The water usage is low and the defrosting rate is similar to forced-circulation air defrosters. However, care must be taken that gases released from the fish as thawing proceeds do not cause rupturing of the flesh (*e.g.* belly-burst in herring, mackerel).

Faster thawing of fish is achieved by the use of microwave, dielectric or electrical resistance heating. Microwave heating is expensive and as the energy is absorbed at the surface, localised overheating is a problem and the risk of cooking the surface is very real. Dielectric thawing, while more expensive, takes only 20% of the time for air or vacuum thawing.

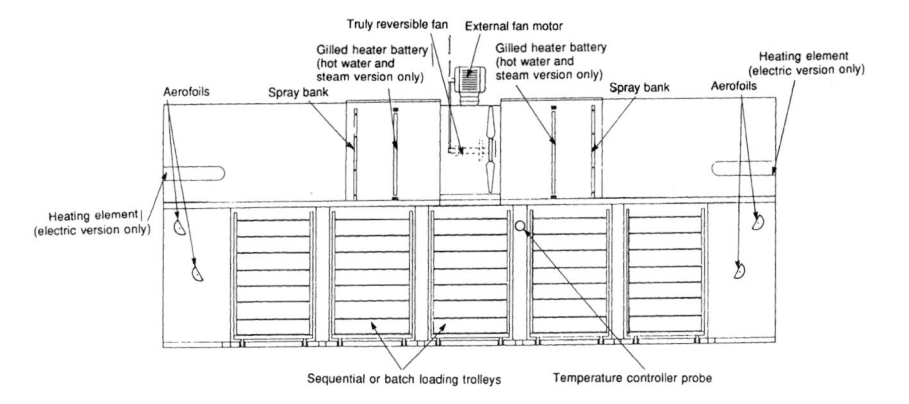

Figure 4.8 Thawer. By courtesy of AFOS.

Electrical resistance thawing requires the fish to be warmed to about $-10°C$ by conventional means such as immersion in water. Above this temperature the fish is made an electrical conductor by placing it between two metal plates, forming the electrical contacts, and an alternating current at low voltage is applied. The electrical frequency causes dipoles in the water to oscillate as the electrical-field direction changes and the frictional energy produced causes the fish to warm. Ideally, the blocks of fish should be uniform with parallel, flat surfaces providing good contact for the electrical contact plates.

Electrical methods for defrosting are expensive and require good control procedures; however, if correctly applied, they can result in good-quality thawed fish.

References

Burns, G.B., Ke, P.J. and Irvin, B.B. (1985), Objective procedure for fish freshness evaluation based on nucleotide changes using a HPLC system. *Canadian Technical Report of Fisheries and Aquatic Science* **1373**, 35 pp.
Cann, D.C., Houston, N.C., Taylor L.Y., Strand, G., Early, J. and Smith, G.L. (1985), *Studies of Shellfish Packed and Stored Under a Modified Atmosphere.* Torry Research Station, Aberdeen.
Cleland, A.C. (1990), *Food Refrigeration Processes, Analysis, Design and Simulation.* Elsevier Applied Science.
Cleland, A.C. and Earle, R.L. (1977), A comparison of analytical and numerical methods for predicting the freezing times of foods. *J. Food Sci.* **42**, 1390–1395.
Cleland, A.C. and Earle, R.L. (1979), Prediction of freezing times for foods in rectangular packages. *J. Food Sci.* **44**, 964–970.
Cleland, A.C. and Earle, R.L. (1982), Freezing time prediction for foods – a simplified procedure. *Revue Int. Froid* **5**, 3.
Dyer, W.J. (1971), Speed of freezing and quality of frozen fish. In *Fish Inspection and Quality Control,* Kreuzer, R. (Ed.), Fishing News Books, Farnham, pp. 75–78.

Ede, A.J. (1949), The calculation of the freezing and thawing of foodstuffs. *Modern Refrig.* **52**, 52–55.

Elira, S. and Uchiyama, H. (1989), The methods for estimating the freshness of fishes. *Int. J. Agric. Fish Technol.* **1**, 287–291.

Giddings, G.G. and Hill, L.H. (1978), Relationship of freezing preservation parameters to texture-related structural damage to thermally processed crustacean muscle. *J. Food Proc. Pres.* **2**, 249–264.

Gorga, C. and Rousivalli, L.J. (1988), *Quality Assurance of Seafood*. AVI, Westport, Connecticut.

Hebbar, K.S. and Hiremath, G.G. (1980), Effect of freshness on the drip loss and other quality parameters of frozen prawns. *Mysore J. Agric. Sci.* **14**, 584–588.

Heen, E. and Karsti, O. (1965), Fish and shellfish freezing. In *Fish as Food (Vol. 4 Part 2)*, Borgstrom, G. (Ed.), Academic Press, London, pp. 355–418.

Leniger, H.A. and Beverloo, W.A. (1975), *Food Process Engineering*. D. Reidel, Dordrecht.

Luong, J.H.T., Male, K.B., Masson, C. and Nguyen, A.L. (1992), Hypoxanthine ration determination in fish extract using capillary electrophoresis and immobilized enzymes. *J. Food Sci.* **57**, 77–81.

McDonald, I. (1982), In *Fish Handling and Processing (2nd edn)*, HMSO, London, pp. 79–87.

Nagaoka, J., Tagagi, S. and Hotani, S. (1955), Experiments on the freezing of fish in an air blast freezer. *Proc. Ninth Int. Congr. Refrig. Paris* **2**, 4.

Neumann, B.S. (1959), In *Conduction of Heat in Solids*, Carslaw, H.S. and Jaeger, J.C. (Eds), Clarendon Press, Oxford.

SFIA (1985), *Guidelines for the Handling of Fish Packed in a Controlled Atmosphere*. Sea Fish Industry Authority.

Shewan, J.M. (1977), The bacteriology of fresh and spoiling fish and the biochemical changes induced by bacterial action. In *Handling, Processing and Marketing of Tropical Fish*, Tropical Products Institute, (New N.R.I. Chatham), London, pp. 51–66.

Simpson, K.L. (1982), In *Chemistry and Biochemistry of Marine Food Products*. AVI Publishing Company, Westport, Connecticut, pp. 115–136.

5 Canning fish and fish products
W.F.A. HORNER

5.1 Principles of canning

The contents of cans are generally ideal growth media for a vast array of micro-organisms. In particular, and in contrast to other prepared food products for retail sale, they will readily support the growth of anaerobic over aerobic organisms. Since the most familiar signs of food spoilage reflect the growth of aerobes, this could lead to contaminated can contents becoming toxic before becoming noticeably spoilt. Canning, therefore, is a technology where *mistakes cost lives*... a fortunately rare situation in the world of food processing but one that, through neglect of due diligence, continues to destroy human lives, livelihoods and businesses.

In the search to discover means through which the quality of canned fish products can be improved by the judicial manipulation of processing conditions, the technologist must never compromise product safety. There are three essential maxims of cannery safety (applying equally to bottled and retortable pouched products):

- *Container seal integrity* – the vacuum in the can will tend to draw fluids (and the microbes they contain) through a faulty seal recontaminating the sterile contents.
- *Adequate thermal process lethality* – the times of exposure, to given high temperatures, required to effectively eliminate the most dangerous and heat-resistant pathogens, particularly *Clostridium botulinum*, are accurately known. Thermal processes are calibrated in terms of the product centre's equivalent time at 121.1°C even though the process itself is unlikely to be conducted at a temperature as high as this. This 'thermal process lethality' time is called the F_0 value.
- *Scrupulous post-process hygiene* – while the can is still hot and wet, after the sterilisation process, it is most vulnerable to leakage inward through the seal. Cooling water is, therefore, measuredly chlorinated, as are all surfaces with which the can comes into contact, and wet cans are *never* handled.

5.1.1 Thermal destruction of fish-borne bacteria

5.1.1.1 *Selection of heat process.* For the purpose of canning, food is divided into three pH groupings:

High acid (below pH 4.5). Fish marinades and pickles containing acetic, citric or lactic acids will not support the growth of spore-forming, human pathogenic micro-organisms. Those organisms that can grow in such acid conditions are destroyed by relatively mild heat treatments, such as elevating the contents temperature to 90°C at the coldest point followed by immediate cooling, or even by the temperatures to which the fish and liquor are heated for hot filling and sealing.

Medium acid (pH 4.5 to pH 5.3). Many canned fish products packed in tomato sauce will fall into this category and consequently require the full heat sterilisation process (often based upon the destruction of *Clostridium botulinum* spores) that is designated to this pH category for safe storage.

Low acid (above pH 5.3). Most canned fish products, other than those mentioned above, will have a close-to-neutral pH and require a full heat sterilisation process, as do the medium acid group. Additionally, it may be necessary to take into account the possibility that some extremely heat-resistant, spore-forming, thermophilic organisms could survive such a process. For example, the thermophile *Bacillus stearothermophilus* has been found responsible for the flat-sour spoilage of canned foods. However, since the heat processing required to effectively eliminate the spores of this organism is so severe that the fish would be grossly overcooked if subjected to such a process, it is better to avoid the use of raw materials, like untreated herbs and spices, which might contain such organisms and/or post-process conditions that might encourage their spores to germinate. When, for example, large diameter cans are cooled naturally, that is, without the water and compressed air of pressure cooling, it can take longer than a day for the can centres to pass through the organisms growth temperature range, thus allowing germination of spores and thermophilic spoilage.

5.1.1.2 *Thermal resistance of micro-organisms.* It is thought that the destruction of micro-organisms is due to the coagulation of their proteins, specifically those that form part of their metabolic enzyme systems, although there is great variability in the degree to which different micro-organisms will resist heat. Indeed, the heat resistance of a single micro-organism type may vary widely according to the environment in which it finds itself. Already noted is the effect of pH on micro-organism viability but, additionally, the effect of salt and nitrite, particularly upon the heat resistance of *Clostridium botulinum*, was drawn to our attention by Perigo and Roberts (1968). Water activity (a_w) and the presence of organic acids and antibiotics, such as nisin, which is particularly active against *Clostridia* spp. (see Boone, 1966), also have their effects on microbial heat resistance. However, until these effects have been conclusively validated in terms of lethality at different temperatures, no reduction in the severity of the heat-sterilisation process should be contemplated.

In studies of the thermoresistance of bacteria, it has been necessary to define the death of a micro-organism as 'its incapacity to reproduce in its optimum environmental conditions'. Vegetative cells of bacteria, yeasts and moulds are destroyed almost instantaneously on exposure to 100°C. Bacterial spores, however, are more heat-resistant than their vegetative cells and some resist prolonged periods of boiling.

The technique where bacteria suspensions are repeatedly heated to temperatures that are adequate only to destroy the vegetative cells, interspersed by periods of resuscitation to encourage the outgrowth of surviving bacterial spores, is referred to as 'Tyndallisation' after its proponent, John Tyndall (1877). Unfortunately, this less thermally-damaging means of commercially sterilising food in sealed containers, founders on the unpredictability of spore germination and, therefore, process safety.

If a suspension of identical bacterial spores were subjected to a period of exposure to a constant elevated temperature above that species' growth range and samples were to be withdrawn and inoculated into a resuscitation medium, a progressively smaller percentage of the spores would regenerate as the time of exposure to the high temperature increased. A graph plotted to show this trend would resemble that shown in Figure 5.1.

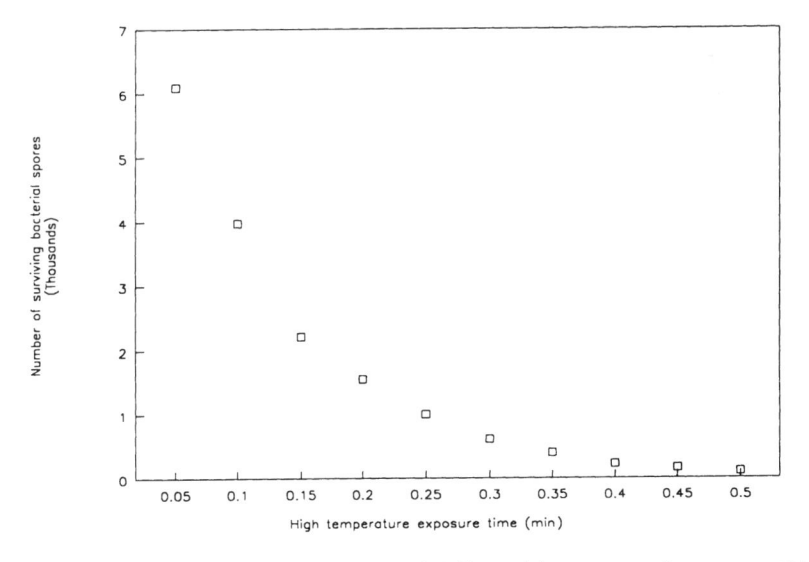

Figure 5.1 Thermal death curve (TDC) for a species of bacterial spore exposed to a constant high temperature.

122   FISH PROCESSING TECHNOLOGY

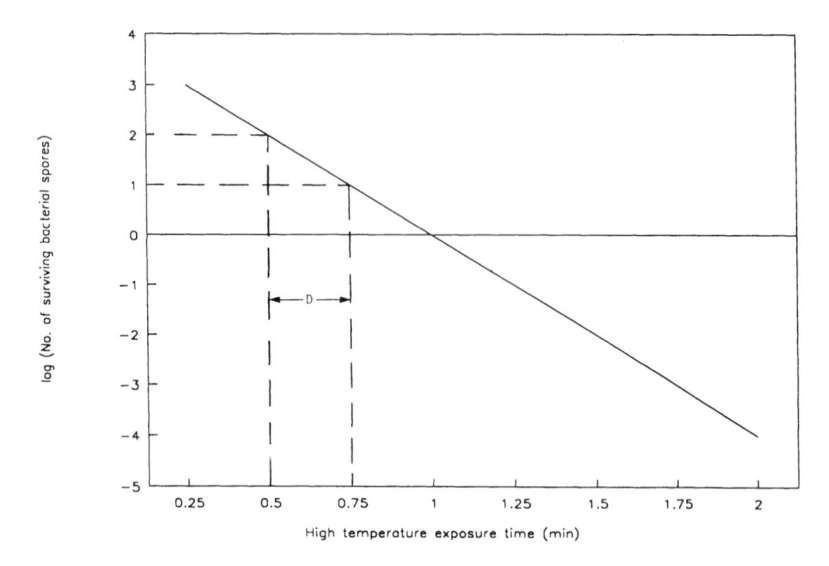

Figure 5.2 Semi-logarithmic interpretation of the TDC shown in Figure 5.1.

The plot approaches the x-axis asymptotically, which suggests that a straight line could be obtained by taking the logarithm of the number of organisms and plotting this against time as in Figure 5.2. In fact, most species of bacterial spore behave in this fashion, only differing in the slope of the straight line describing their destruction.

5.1.1.3 *D value.* Zero on the $\log_{10}$ scale corresponds to one surviving spore and it might be asked 'How is it possible to have less than one spore?' To answer this, it is necessary to look upon the scale below zero in terms of the odds against one spore surviving a given heat process. Thus -1 corresponds to 1 chance in 10, -2 as 1 chance in 100 of a surviving spore and so on.

The time taken for this graph to traverse one logarithmic cycle, that is, to reduce the number of spores from 100 to 10 for example, is called the 'decimal reduction time' or 'D value'. If the spores are subjected to a single specific temperature (121.1°C), the time required to decimally reduce the population is denoted D_0. Different micro-organisms and their spores have different D_0 values as shown in Table 5.1.

5.1.1.4 F_0 *value.* From Table 5.1 and the initial number of spores inside the sealed container, an idea of the severity of the heat process required to reduce the spore population to a pre-determined level, N_t, can be calculated. For example, taking D_0 for *B. stearothermophilus* to be 5 min and the initial number (N_0) in the container to be 10 000, if it was required to reduce this number to one in the heat process, four decimal reductions would be needed.

Table 5.1 Some micro-organisms and their D_0 values

Organism	D_0(mins)
Spores of *Bacillus stearothermophilus*	4–5
Spores of *Clostridium thermosaccharolyticum*	3–4
Spores of *Clostridium nigrificans*	2–3
Spores of *Clostridium botulinum* types A & B	0.1–0.25
Spores of *Bacillus coagulans*	0.01–0.07
Non-spore forming mesophilic bacterial yeasts and modulus	0.5–1.0

The time at 121.1°C required to accomplished this would be $4 \times D_0$ that is $4 \times 5 = 20$ min. Similarly, if the spores were of *C. botulinum* the required time would be $4 \times 0.25 = 1$ min at 121.1°C. Putting this in mathematical terms: $N_0 = 10\,000$; and $N_t = 1$. The number of decimal reductions required is given by:

$$\log N_0/N_t = \log N_0 - \log N_t$$
$$= 4 - 0 = 4$$

This $\log N_0/N_t$ is sometimes referred to as the 'order of process' factor 'm', and the value of the product of m and D_0 is called the 'process value' or 'F value', that is:

$$F_0 = mD_0$$

Obviously a process designed to reduce the spore population to one is inadequate. What, then, would be regarded as a commercially safe heat process? Ideally, 'sterility' of the can contents should be the microbiological objective of the canner but, on the basis of the logarithmic death curve shown by bacterial spores, this is not an achievable target. 'Commercial sterility' is, therefore, taken as a compromise whereby the initial bacterial load is taken through sufficient decimal reductions to reduce the possibility of a single organism surviving to an acceptably low level. What constitutes an 'acceptably low level' depends upon the organism the process is designed to destroy.

Clostridium botulinum is usually taken as the organism whose destruction is the process objective because, during its growth, which is particularly favoured by the low oxygen conditions inside an hermetically sealed container, it releases a substance that is extremely toxic to humans and responsible for the often lethal disease, botulism. Since its spore is very heat-resistant and the organism is widely distributed in nature, including marine and fresh water environments, it may be present in many foods, including fish, which are canned. An 'acceptably low level' in the context of this dangerously pathogenic organism means less than one in a billion (*i.e.* 10^{-12}) chance of survival.

Thus, if there were one organism in the container initially,
$$m = \log N_0/N_t = \log 1 - \log 10^{-12} = 12$$
$$F_0 = mD_0 = 12 \times 0.25 = 3$$
Since, in this case, the reference organism is *Clostridium botulinum*, and the reference temperature is 121.1°C, this is more correctly written:
$$F_{121.1°C}^{C.\,botulinum} = 3,$$
which is usually shortened to $F_0 = 3$. F_0 values commonly used by canners for medium- and low-acid products range from 6 to 14 to give an additional safety margin to compensate for temperature-measurement inaccuracies.

An 'acceptably low level', when applied to an organism which is a non-pathogenic spoilage organism may be subject to a more compromising interpretation than that for a pathogen like *C. botulinum*. However, if this organism had a D_0 value of 5, like *B. stearothermophilus*, the reduction to one can per million, given an average initial contamination per unit of 10, with this particular organism, gives $m = 7$ and still requires $F = 35$. Fortunately, most circumstances do not require that *B. stearothermophilus* be taken into consideration, but there are mesophilic spoilage organisms with D_0 values greater than that for *C. botulinum*, and the compromise between least spoilage rate and best eating quality (optimum cooking) must be struck.

5.1.1.5 *Heat transfer in canned fish.* In the fish itself, heat transfer is predominantly conductive and, therefore, it takes a great deal of time to elevate the thermal centre or 'cold spot' temperature of a solid pack of fish from 20 to 120°C in a can 145.5 mm diameter by 168 mm height. By comparison, if it can be arranged that most of the in-pack heating is convective, the same temperature rise could be achieved in as little as 20 min.

To avoid fish closest to the can walls overcooking, and to accelerate heat transfer to the cold spot, oil, sauce or brine is added to the contents. Also, convective heat transfer is promoted by agitating the cans in the retort. End-over-end rotation is more efficient in this respect than the axial rotation.

Radiation heating plays no part in conventional retort processes but both ohmic (resistance) and microwave heating are used to cook foods prior to aseptic filling.

Most fish packs, being solids suspended or packed in liquid, exhibit both conductive and convective heat transfer through their contents, and the location of the cold spot is not simply the geometric centre of the container but the geometric centre of the thickest piece of fish in the pack, wherever this is located, since conductive heat transfer is considerably slower than convective heat transfer.

5.1.1.6 *Measuring process lethality.* By taking a series of readings over the length of the process, the temperature history of the cold spot can be revealed.

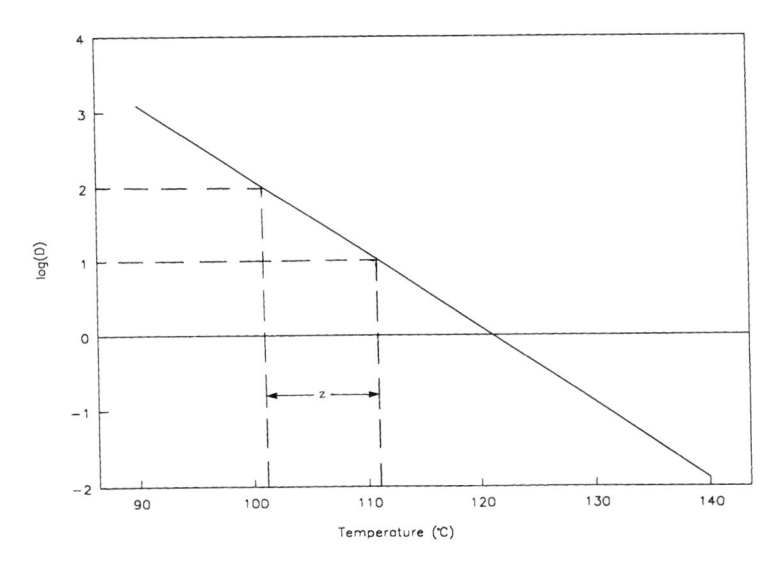

[Figure 5.3: plot of log(D) vs Temperature (°C) showing a linear decreasing line, with z-value indicated between 100 and 110 °C]

Figure 5.3 Temperature dependency of D-value: z-value.

It now becomes necessary to relate this temperature history to the destruction of a selected micro-organism. Hence the D value for this organism at different temperatures must be known. The graph of the D value plotted on a logarithmic scale against temperature on a normal arithmetic scale, Figure 5.3, is approximately linear over the limited range of temperatures used in conventional retort processing. (See Cowell's (1968) criticism of Jones' (1968) paper on thermal process evaluation.) The temperature interval over which the graph passes through 1 log cycle is called the 'z value'. Thus, we can obtain a D value corresponding to each temperature measurement.

Different research workers have put the z value for the different strains of *C. botulinum* between 8 and 11°C; in Table 5.2 it is taken to be 10 as with the hypothetical organism. Several other spore-forming bacteria of significance in canning have z values close to 10°C but it is by no means universally applicable. *Bacillus subtilis*, for example, has a z value of 6.5°C at common processing temperatures.

For the purpose of heat-process determination with respect to their lethality towards specific micro-organisms, the reciprocal of the D value called the lethal rate, L, is used. So with $L = 1/D$, Table 5.2 becomes as shown in Table 5.3. If now during the process, instead of recording temperatures, the corresponding lethal rates are plotted against time, the area enclosed by the graph and the ordinate represents the F value for the process. That is:

$$\int_0^t L \, \mathrm{d}t = F$$

Table 5.2 D values of a hypothetical organism and *C. botulinum* at various temperatures

Temperature (°C)	D value (min) Hypothetical organism	D value (min) C. botulinum
131	0.1	0.025
121	1	0.25
111	10	2.5
101	100	25
91	1000	250

Table 5.3 Lethal rates of a hypothetical organism and *C. botulinum* at various temperatures

Temperature (°C)	L value (1/min) Hypothetical organism	L value (1/min) C. botulinum
131	10	40
121	1	4
111	0.1	0.4
101	0.01	0.04
91	0.001	0.004

The heat process required to meet the condition of commercial sterility can now be specified mathematically in terms of lethal rates which are readily available for a number of micro-organisms (although *C. botulinum* is by far the most used):

$$\int_0^t L \cdot dt > mD_0$$

5.1.1.7 Predicting the process. Heat-process determinations should be carried out before deciding upon a process for any given product, when a change in the process (including pre-process treatment) or product formulation is contemplated, and as a routine quality control check on the evenness of heat treatment throughout one retort or between different retorts. It may, however, be useful to be able to predict the effect of a change in process time on the F value, or to predict the process time required to deliver a given F value. This may be done with foreknowledge of two other values 'j' and 'f_h', which relate to the heat transfer through the contents.

The graph of temperature against time for any given heat process is an asymptote towards the set retort temperature. If cold-spot temperatures are converted to temperature deficits (*i.e.* retort temperature – cold spot tempera-

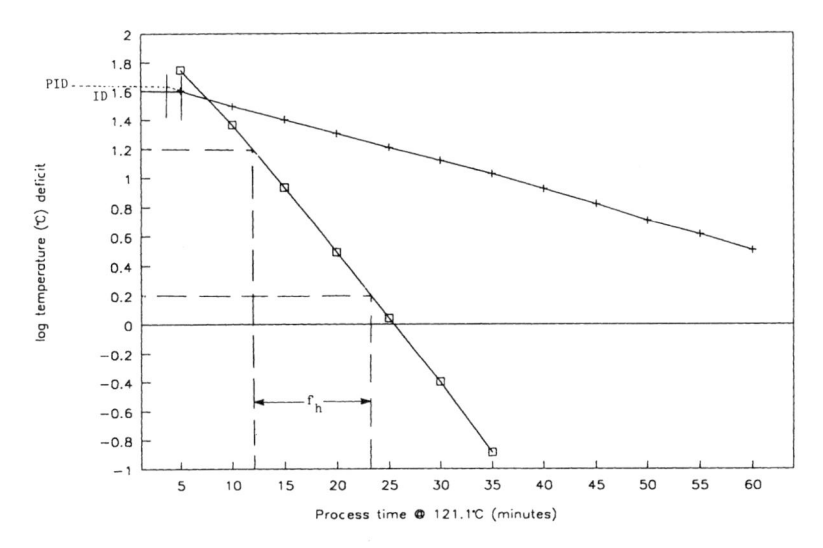

Figure 5.4 Log deficit temperature histories recorded at the thermal centres of convection and conduction heating packs: (□) convection; (+) conduction.

ture) and plotted on a logarithmic scale against time on an arithmetic scale, a straight line should be obtained for a convective pack, and a straight line with an initial, almost horizontal, lag period for a conductive pack (Figure 5.4).

The time taken for such a graph to traverse one log cycle is called the 'f_h' value. For purely convective packs, this may be as little as 5 min, while for purely conductive packs it could be 15 or more times greater than this.

The j value is given by the pseudo-initial temperature deficit (PID) divided by the actual initial temperature deficit (ID) where the pseudo-initial deficit is obtained by extrapolating the rectilinear portion of the graph back to the point where it intersects a vertical line drawn on the heat penetration graph 0.40 of the distance from a vertical line (representing the time at which the retort temperature was reached) to another vertical line representing the time when the steam began to enter the retort. (The reason for this complication is that much of the steam used initially heats the retort rather than the can contents. Ball (1923) empirically determined that about 40% of the time between steam on and process temperature being reached, that is, the come-up time, should be considered as process time at the process temperature.)

For purely convective packs, it follows that, since there is no lag period, PID and ID coincide, so that $j = 1$. In purely conductive packs, conversely, a value of $j = 2$ should be obtained. Most fish packs, therefore, being mixed convective/conductive, will reveal j values between 1 and 2.

Knowing the f_h and j values for any given pack of fish, together with the retort temperature to be used during processing and the initial temperature of the cold spot as heating commences, it is possible to draw the graph for the process as in Figure 5.4 and predict the F value for a given processing time or, more likely, the processing time required to arrive at specified F values.

Having decided upon a target F value for a commercial sterilisation process and knowing the f_h and j values for the specific container and contents to be processed, it is possible to compute the process time, B, from Ball's (1923) equation:

$$B = f_h(\log jI - \log g)$$

where I is the initial temperature deficit $(T_r - T_0)$ and g is the temperature deficit after the process time B. The required F value depends upon the value of g and can be found using $f_h/U:g$ tables (Table 5.4). U (which is defined as the time required at the actual retort temperature to accomplish the same amount of bacterial destruction equivalent to the F value of the process) must be calculated from the required F value via the relationship:

$$U = F \cdot 10^{(121.1 - T_r)/z}$$

The process time B, found by substituting the obtained values of f_h, j, I and g into the penultimate equation, includes 40% of the 'come-up' time and should therefore be reduced by this amount if we are defining process time strictly as the time over which the retort remains at the selected process temperature. Note that this mathematical method of predicting the process time to achieve a specific F value should only be used for products showing an unbroken heating curve (i.e. those whose f_h value remains constant over the full retort temperature range).

Process determinations that include the lethality accumulated in the cooling as well as heating process first require establishment of the time constant f_c and lag factor j_c from the empirically obtained cooling curve. These may differ substantially from f_h and j_h obtained from the heating curve (Pham, 1990).

Table 5.4 $f_h/U:g$ table (assuming $z = 10°C$). Values of $g \times 10^2$ are expressed for various j values

				$g \times 10^2$			
j	0.80	1.00	1.20	1.40	1.60	1.80	2.00
F_h/U							
0.30	0.126	0.133	0.141	0.148	0.155	0.163	0.170
0.50	2.63	2.81	2.99	3.17	3.34	3.52	3.69
0.70	9.78	10.5	11.2	12.7	13.4	14.2	—
0.90	20.6	22.2	23.8	25.4	27.1	28.7	30.3
1.0	26.9	29.1	31.2	33.3	35.4	37.6	39.7
2.0	85.0	92.2	115.	123	130	138	145
3.0	169	181	193	204	216	228	239
4.0	230	245	260	274	289	304	319
5.0	282	300	317	335	353	371	388

5.1.1.8 *Integral F values.* Quoting the *F* value of a process based on the foregoing theory refers to the thermal centre of the pack, as if *all* the contaminating organisms were located precisely at that point. Of course, these organisms are distributed throughout the rest of the container and receive a proportionately more severe process the further is their position from the thermal centre. To this end, Stumbo (1953) proposed that the integration of the lethal effects over the whole can contents be used in determination of process time temperature requirements. He imagined cans of food as an infinite number of containers resting, like Russian dolls, inside the can.

The surface of each imaginary container represented all points within the container subject to the same process lethality. By integrating the heat process lethality over the full volume of the can, a more accurate evaluation of total lethality was attained – the *integrated F value*, F_s. Its calculation simplified to:

$$F_s = F_0 + D_0[1.084 + \log(F_x - F_c)D_0]$$

where F_c is the *F* value at the thermal centre and F_x is the *F* value received at the surface of one of the imaginary containers where $j_x = 0.5j$ at the centre and $g_x = 0.5g$ at the centre. The $f_h/U:g$ table can be used to calculate a value for U_x and hence F_x from the equation:

$$U_x = F_x 10^{(121.1 - RT)/z}$$

It should be reiterated that the use of integrated *F* value allows a small time-saving in sterilisation processes for mainly conduction-heating packs. Such a saving, however, is very worthwhile if one considers that a 1-min saving in process time may increase a large canner's production by up to 2%.

5.1.2 *Thermal processing: quality criteria*

5.1.2.1 *Effect of heat processing on fish.* The softening of tissues and loss of volatile components, associated with heating fish, would be advantageous if the sterilisation treatment, associated with canning, coincided with the agreeable changes associated with cooking. In practice, most improvements in canning technology have centred upon minimising over-cooking, during sterilisation, by increasing the rate of heat transfer through to the cold spot. White fish, particularly, with its delicate flavour and structure, is rendered virtually unmarketable by conventional canning processes.

Processing damage to nutritional value with respect to proteins was reviewed by Bender (1972). The dietary value of protein seems not to be significantly affected by exposure to canning time/temperature processes; in fact, some proteinaceous components that would otherwise be plate-waste, like salmon and sardine bones, are softened enough to become edible. The heat denaturation of protein causes water losses varying from 9 to 28%, dependent on the severity of the process/pre-process, species, pH and other physiological factors. It is necessary to limit such losses in the can, known as 'cook-out', otherwise there is an unsightly, curdled appearance to the contents on opening the container. Pre-processes such as curing, pickling, smoking and cooking minimise 'cook-out', but encompass a loss of soluble proteins, which could be collected and concentrated in similar fashion to the 'stickwater' from fishmeal production.

Slight losses of B-group vitamins, thiamin, riboflavin, nicotinic acid, folic acid and cyanocobalamine have been revealed in comparisons between fresh and canned fish.

Flavour changes, which occur in fish during the long exposure to high temperatures, limit the acceptability of the product but are otherwise difficult to define. Where possible, they are masked by the use of sauces and seasoning.

Textural changes during the process are also inevitable but, as mentioned already, may be advantageous to a limited degree. Excessive protein denaturation and the accompanying decrease in the water-holding capacity of the structural components yield a product with a dry, chewy mouthfeel like damp cotton-wool. Oily-fleshed fish suffer less, in this respect, due to the restrictive effect of the lipids on water migration. Choice of raw material is important in this context, with less fresh fish losing more water and, therefore, showing greater textural deterioration after processing.

Spoiling fish flesh, which becomes subject to excessive autolysis, may yield a heat-processed product with a pitted or honeycombed texture, although a restricted degree of proteolysis, prior to processing, may result in a desirable softening of the texture of the finished product.

Colour changes may also be more pronounced when poorer-quality raw material is used. The discoloration in canned tuna known as 'greening', relates to the trimethylamine oxide, myoglobin, cysteine concentration and the cooking operation itself and, as Ali Khayat (1978) found, can be reduced by additions of combinations of these substances or antioxidants. Nagaoka et al. (1971) have developed a method whereby determination of combined TMAO and TMA content of the raw fish can be used to indicate the probability of 'greening' occurring during the heat process.

The colour of canned salmon is extremely important and greatly affects its market value: red salmon (*Oncorhynchus nerka*) commands a higher price than medium-red salmon (*Oncorhynchus kisutch*), which in turn, commands a higher price than pink salmon (*Oncorhynchus gorbuscha*). Although species is the main colour-controlling factor, the use of poor-quality fish or unnecessarily severe heat processing can cause colour changes sufficient to make the consumer believe that the product has been misrepresented.

Browning in canned fish is commonly associated with the 5-carbon reducing sugar, ribose. This is released increasingly in spoiling fish by the action of riboside hydrolase on ribonucleic acid. However, with ribose being soluble, pre-cooking the fish and decanting the cook-out can help avoid the problem. It has also been suggested that *Lactobacillus pentoaceticus* would, in 2 days at

0°C, remove all the ribose. Another browning phenomenon, occurring in marinated fish when packed together with onion, may be due to amino acids reacting with 2,5-diketogluconic acid, released by bacterial action on onion. This is another non-enzymic, carbonyl-amino browning reaction.

Undesirable colour changes in shellfish during canning often involve metal ions, for example, the blue discoloration of crab meat involves iron, whereas a black discoloration in prawns relates to copper content. Eels, abalone and albacore tuna all occasionally suffer discoloration, on processing, due to the high iron content of the raw material. Discoloration of this type is increased when the material is held in frozen storage prior to canning, because of the build-up of free sulphur in the tissue. Iron and free sulphur react together, during heat processing, precipitating black iron sulphide on the sides of the container, in the fish itself and, especially, in any free liquid.

'Struvite' crystals, occasionally found in heat-processed packs of crustacea, salmon, tuna and scombroids are often mistaken for glass by the consumer. They are, however, crystals of calcium or magnesium ammonium phosphate. Yet, because the discovery of these crystals is more distressing to the consumer than, perhaps, any other form of foreign body contamination, utmost precautions should be adopted to ensure that they do not occur. Inclusion of sodium hexametaphosphate or citric acid, which sequester free calcium and magnesium ions, or lowering the pH to avoid precipitation of struvite crystals, are common preventive measures.

Curd formation and the tendency of fish pieces to adhere to the side of the can are made more probable if raw, previously frozen fish are canned and less probable if pre-sterilisation treatments include brining, pre-cooking and inclusion of tartaric acid.

5.1.2.2 *Cook values.* The changes that occur in fish during heat sterilisation processes are desirable in so far as they correspond to what is individually perceived as 'cooked'. Many developments in canning technology have centred on the improvement of product quality by restricting overcooking while maintaining the objective of commercial sterility. However, while process value (or F value) may be defined with respect to the effect of exposure time at a specific temperature on a specific organism, 'cooking' encompasses a multiplicity of physical and biochemical changes. They range from finite chemical changes, such as the de-activation of an enzyme or destruction of thiamin, to subjectively perceived physical changes of colour, flavour and texture. All such changes, however, can be assigned z values under standard conditions, and from these an overall z value for 'cooking' may be computed allowing for the relative importance of each change in the product's quality as expected by the consumer. Thus, an analogy can be drawn with the concept of 'decimal reduction time' D, at a steady temperature θ given by

$$D = 10^{-(121.1 - \theta)/z}$$

where D for 'cooking' would symbolise a 10-fold change in one or a combination of quality parameters. Similarly, the 'process value' F, given by:

$$F = \int_0^t \frac{1}{D} \cdot dt$$

would become the 'cook value' (C_g) and the latter equation would become

$$C_g = \int_0^t 1/[10^{-(121.1-\theta/z)}] \cdot dt$$

Thus, Thijssen et al. (1978) have proposed a procedure for calculating optimum quality retention while maintaining acceptable sterility from the process. Those factors comprising 'quality change' during heat processing include single-step reactions like protein denaturation (including enzyme deactivation) and micro-organism destruction, and multiple-step reactions like non-enzymic browning and fat oxidation. Activation energies (E) from the Arrhenius equation are given by:

$$E = 2.3(RT^2)/z$$

where R is the universal gas constant and T the absolute temperature. These energies show that increasing process temperature is more effective in increasing micro-organism destruction than enzyme inactivation and more effective on the latter than on the acceleration of non-enzymic browning and thiamin destruction.

The temperature sensitivities of these various quality change criteria are denoted by their z values as shown in Table 5.5.

It would seem, from Table 5.5, that higher-temperature/shorter-time processes favour the objective of maximum process safety with minimum quality deterioration. This approach can, however, be pursued too far as sterilisation processes much above 125°C may be insufficient to inactivate some enzymes also associated with spoilage (see the review by Adams 1991, on the subject). Thus enzyme inactivation might become the process completion criterion in aseptically canned fish products. Nevertheless, the use of higher temperatures

Table 5.5 Temperature sensitivities of various quality change criteria as denoted by their z values

Quality change criterion	z(°C)
Thiamin destruction	31
Riboflavin destruction	28
Enzyme inactivation	26
Non-enzymic 'Maillard' browning	22[a]
'Cooking'	18[a]
C. botulinum spore destruction	10

[a] Aggregated z-value for multiple-stage reactions.

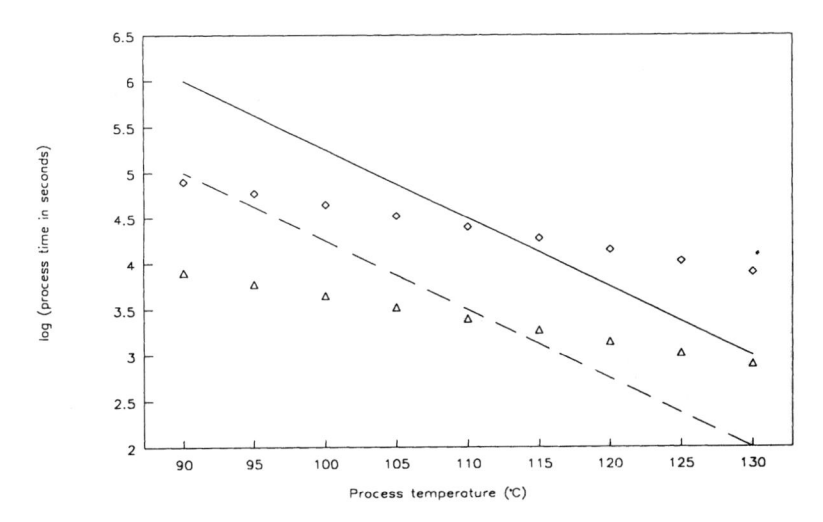

Figure 5.5 Process time and temperature for equivalent thiamin loss and *C. botulinum* spore destruction. (———) 10D* (*C. botulinum*); (– – –) D* (*C. botulinum*); (◇) 90% thiamin loss; (△) 10% thiamin loss.

and shorter process times yield a commercially sterile product that is less brown and contains more vitamins than products yielded by lower temperature/longer time processes of equivalent lethality as exemplified in Figure 5.5.

5.1.2.3 *Aseptic canning.* Heat sterilisation processes, whereby the ingredients are sterilised outside the container and then aseptically filled and seamed, have been used for liquid foods and solids in suspension. Their advantage lies in the shorter exposure time resulting in less overcooking, improving particularly the finished product quality where heat-delicate food components are involved. Fish products have rarely been processed thus. One of the few, however, has been 'tuna salad', composed of flaked tuna and vegetable strips in a mayonnaise sauce. This product would not be feasible using any conventional canning process as the long process times would over-soften the vegetables and de-stabilise the oil-in-water emulsion of the mayonnaise, causing separation of the phases.

The high cost of the specialised equipment and control procedures needed for aseptic canning and the limited range of products to which it can advantageously be adapted has restricted its exploitation.

Using process temperatures above 130°C, therefore, allows commercial sterilisation while destroying less than 10% of the thiamin content, where commercial sterilisation at 100°C would destroy most of the thiamin content. Gaze and Brown (1988) confirmed the adequacy of using a D_0 value of 0.2 min and z value of 10°C over the 120–140°C range for *C. botulinum* spores.

Similarly, with browning, sterilisation processes above 130°C reduce browning more than 100 times compared with that caused by 100°C sterilisation processes.

Ohlsson (1979) found z values for various sensory changes associated with the cooking of fish pudding to range between 23 and 29°C and, on this basis, proposed the optimisation of time/temperature processes, based on both F_0 and C_g values.

5.1.3 Storage of canned fish

With certain canned-fish products, a long storage period is thought to favour the development of desirable characteristic flavours. Time alone, however, has no selective control over the chemical and physical reactions continuing to affect the container and its contents. Storage periods of several years, however, must increase the likelihood of consumers having cause to complain over contents discoloration, container corrosion, crystal formation, curdling of sauces and off-flavour development.

High storage temperatures (above 35°C) must, at all costs, be avoided to prevent the outgrowth of thermophilic spores, which might survive the usual 'botulinum process'. In addition to this risk, Durand and Thibaud (1980) found that corrosion de-tinning of the internal surfaces of tin-plate cans containing sardines and mackerel products that were packed in oil or sauces was much more severe over a 2-year period at a storage temperature of 37°C compared with that when the same products were stored at 20°C. Lead content was also shown to be significantly higher in acid-marinated canned fish (where soldered lidded-cans had been used) that were stored at the higher temperature.

It would seem, then, that if canned-fish products are to be stored for a long period of time, or at tropical temperatures, aluminium cans would be more suitable containers even though, it has been shown, there is some weakening of fish and sauce flavour compared with the same products stored in tin-plated cans.

5.1.4 Choice of heat process

The factors to be taken into consideration when attempting to select the most appropriate heat preservation process for any given canned-fish product may be summarised thus:

- $\log_{10}(N_0/N)$ sets the safety target of the heat process.
- The character of the product (its pH, its salt content, its fat content (since fat has a protective effect on proteins exposed to high temperatures it also tends to raise micro-organism D values) and whether it contains bacterial D-value lowering components such as spices, nitrates or nitrites) will modify the heat process requirements necessary to achieve the target.

- $\log_{10}(X_0/X)$ where X_0 is the initial thiamin content, colour index or sensory score and X is the acceptable level of the same after the heat process, sets the quality (cook value) target of the heat process.
- The post-processing storage conditions determine whether certain thermoduric spore-forming bacteria should be used, in addition to *C. botulinum*, in setting the process lethality target.

5.2 Design of packaging for fish products

An ever-increasing range of containers is becoming available to replace the standard open-top tin cans. To qualify as a suitable container, two essential features are that: (i) it should be possible to hermetically seal the container such that it will withstand pressure differences between inside and outside, during and after the process, without danger of leakage in or out; and (ii) it should withstand the high temperatures involved in the heat process, without danger of melting or reacting with the contents.

5.2.1 *Glass jars*

Nicholas Appert's pioneering work in the early nineteenth century, which established canning as an alternative method of food preservation, was, in fact, conducted using glass jars sealed with corks. In spite of their drawbacks, namely that they take longer to process, are subject to thermal shock if the cooling stage of processing is not carefully controlled, and that they break easily when subject to mechanical shock, glass containers are still used by tradition for fish pastes and spreads and where it is considered an advantage to display the product or prevent it reacting with the container as for example, with pickled shellfish and herring.

Closures feature a malleable, polymer sealant-ring, which is tightened against the jar rim in the closing operation. These are easily removed by the consumer, but require even more care during the processing stage. 'Pry-off' closures, especially, will blow off during the cooling phase of the process, if the pressure inside the container ever exceeds the pressure outside, leading to the loss of a complete processing batch. Most 'twist-off' caps have a 'tamper' indicator – the centre portion of the lid flips outward once the headspace vacuum has been broken and customers are advised not to buy such 'flip-lidded' jars.

5.2.2 *Flexible containers*

Conductive heat transfer has been described by the Fourier equation, which relates the heat flow in any given direction, Q, to the heat-transfer coefficient, U, the surface area perpendicular to the direction of heat flow, A, and the temperature difference, ΔT, between heat source and the point most distant from it inside the container (the 'cold spot') thus:

$$Q = UA\Delta T$$

The overall heat transfer coefficient, U, depends upon the surface heat-transfer coefficient, h, the thermal conductivity of the material across which heat is being transferred, k, and its thickness, x, as follows:

$$\frac{1}{u} = \frac{1}{h} + \frac{x}{k}$$

The use of a pouch in place of a can, can reduce the amount of heat processing required to impart sterility by maximising A and minimising x. Since the late 1960s, flexible pouches have been available as substitutes for the food can. Essentially, they are rugged polymer laminates consisting of a heavy-gauge outer polymer layer, which may be coated to take a decorative logo and printing, which is then bonded to an inner polymer layer that can be heat-sealed. Most frequently, an intermediate layer of aluminium foil is incorporated for its barrier properties, additional strength and light-proofness. An example of such a construction would be, from the inside, polyvinylidene chloride, aluminium foil and polyvinylidene chloride.

Although these containers have become quite popular in Japan and the USA and have been used in Europe for the more expensive recipe-type fish products, they have not gained a substantial share of the canned-fish products market for the following reasons:

- They are more expensive than conventional cans, even without an outer cardboard carton, which is regarded as necessary for protection and display. The latter may bring the unit cost to two or three times that of the conventional can.
- On-line filling and sealing rates may only be 60 units/min compared with perhaps 20 times that rate for conventional cans.
- Heat processing of pouches must be carried out in specially designed retorts using steam and compressed-air mixtures to create an over-pressure sufficient to balance the pressure developed in the pouch and therefore prevent strain on the seals and bursting; hence, change to a pouch line involves heavy capital expenditure on equipping new processing lines.
- 'Leaker' detection is more complicated than with conventional cans. For these reasons, the use of pouches for heat-processed fish products is likely to remain confined to the up-market, luxury products, which can render insignificant the extra costs involved in their higher retail prices and which can be shown to benefit significantly, quality-wise, from shorter processing times.

5.2.3 Rigid metal containers

These include tin-plate and aluminium cans, which may be produced in a wide range of shapes and sizes to conform with traditions and markets for canned-fish products.

Until the early 1970s, aluminium cans, being drawn from a plug of metal, had the advantage over tin-plate containers in having a single seam (at the processors end) compared with the three seams on conventional tin-plate cans. However, this technology has been extended to tin-plate cans to produce the 'drawn, walled and ironed' (DWI) and the 'drawn and re-drawn' (DRD) cans, also featuring the single ('processors-end') seam. Both may also feature an easy-open, ring-pull end, although aluminium, being more malleable, requires less effort in its opening.

Tin-plate is produced mainly by the electrolytic deposition of tin on mild steel plate. The thickness of the steel influences the strength of the container, although very light-gauge corrugated tin-plate can match the strength of the plain-sided heavier-gauge plate. The thickness of the tin coating influences the container's resistance to corrosion. Fish proteins, especially those of the skin, adhere to untreated metal, to aluminium more than to tin-plate, producing an unattractive pack. For this reason, and because the very high price of tin has led to a reduction in the thickness of the tin coating, increasing the danger of reaction between contents and the mild steel, lacquers are applied to the internal surfaces of finished cans. In some cases, where the risk of external corrosion is also great, fish canned at sea or in exposed coastal regions, for example, lacquer is also applied to the external surfaces of fish cans. The lacquer may be applied and baked onto the tin-plate sheets before the end and body blanks used for can forming are cut, or they may be sprayed on internally after forming.

Cans manufactured from pre-lacquered blanks, however, require the application of a further strip of lacquer over the side seam, post-forming, to cover any breakages in the lacquer caused during its formation.

Fishery produce normally has a close-to-neutral pH, unless packed in tomato sauce or marinated, in which case, an acid-resistant lacquer is required. However, many fish products (especially shellfish) release hydrogen sulphide due to the degradation of some of their sulphur proteins during the heat process. This hydrogen sulphide will react with exposed areas of iron, to produce unsightly black precipitates of iron sulphide on the container, and with ferrous ions in solution to discolour the contents.

$$Fe + H_2S \rightarrow FeS + H_2$$

The extent of the hydrogen production can reduce to zero the vacuum inside the container and even cause 'hydrogen swells'. To prevent this phenomenon, a special lacquer incorporating zinc oxide or carbonate is applied to the inner surface of the container. The zinc acts sacrificially, being higher in the electrochemical series than iron, and is converted to white zinc sulphide, which remains embodied in the lacquer, thus producing an acceptable uniform, white appearance to the internal walls of the container.

$$ZnO + H_2S \rightarrow ZnS + H_2O$$

The internal vacuum is not affected in this case, since the other product of the reaction is water. The zinc oxide, or carbonate, is included in an 'inorganic' lacquer whose other constituents are Na_3PO_3, $Na_2Cr_2O_4$ and a wetting agent. Crustacea, particularly, evolve hydrogen sulphide during heat processing but other canned fish products may use cans with organic lacquers as follows:

- Vinyls: these are good barriers, but soften on heating and may cause fish flesh to stick to them.
- Phenolics/phenolformaldehyde lacquers: these are sulphur-resistant but can cause taint and occasionally crack on thin corrugated plate due to inflexibility.
- Oleoresins: these are very popular and cheap but have limited barrier properties against acid and sulphur.
- Epoxide lacquers: these are acid- and sulphur-resistant but expensive.

Apart from lacquers, cans have been lined with vegetable parchment and the contents processed 'dry' to avoid sticking, particularly for small expensive packs like crab-claw meat and lobster.

5.2.4 Rigid plastic containers

Thermoformed from multilayered coextruded plastics like polyvinylidene chloride and EVOH, the 'plastic can' may be lidded with a ring-pull metal end, double-seamed onto the plastic body, or have a foil laminate that is heat-sealed to the rim. The advantage of the latter over the conventional metal can is that, once the metal laminate top is removed, it is microwaveable.

Plastic cans are filled in the same way as metal cans and sealed under vacuum to minimise pressure imbalance between inside and outside during the retorting process. The plastics used have been developed to withstand normal canning processes up to 121°C, but require an over-pressured retorting regime similar to that for retortable pouches. They are available in a range of shapes and sizes, attractive enough to eat from, and, being non-corrodible, can be re-lidded and kept in a refrigerator if the contents are not consumed on first opening.

The main disadvantages of this type of container compared with conventional metal cans are: (i) the higher incidence of seal failure (although it is considered that this can be improved upon at least to the level of that tolerated in metal cans); and (ii) the requirement for a slightly longer heat process to achieve the same process lethality as would be achieved in a metal can of the same dimensions (Berry and Bush, 1988).

5.2.5 Labelling

A product labelled with the common name of a specific fish species, 'red salmon' for example, must contain nothing more than that species, although an allowance is made for salt and seasonings such that 95% raw-fish equivalent is the minimum permitted. Appropriate designations, according to the Labelling of Food Regulations, 1970, for commonly marketed varieties of canned fish in the UK, are given in Table 5.6.

5.3 Process operations and equipment

5.3.1 Pre-processing operations

The sterilisation or pasteurisation heat treatment used in canned-food production is referred to as 'the process' and so all preparatory operations are 'pre-processing'.

The international canned-fish market, like most canned-food markets, expects that the contents of the can should leave no 'plate-waste' from the consumer. To comply with this expectation, it is necessary to subject the raw material to a variety of pre-processes to bring the fish into the form required in the finished product.

Separation of parts, which, even after the prolonged heat exposure of retorting, remain inedible, may take place before any pre-retorting heat treatment. Fish are de-headed ('knobbed') eviscerated and trimmed of fins, scales and other inedible parts before any heat treatment because such handling after cooking would cause the fish to break up and would therefore lower its value. Nevertheless, much canned fish is still hand-filled, and the extent to which the flesh tends to break during this operation is used to adjust the severity of this pre-cooking operation.

5.3.1.1 Skinning.
Although perfectly edible, the skin is removed from many species, especially tuna and mackerel, for presentation purposes. A chemical skinning process is used where fish are briefly immersed in 70–80°C sodium hydroxide solution with a pH of 14. On emerging, water-jet sprays remove the loosened skin and then immersion in pH 1 hydrochloric acid neutralises residual alkali on the fish.

5.3.1.2 Filleting.
Many canned fish, from sardines to salmon, are not filleted for canning, since their bones after the retorting process are soft enough to eat. Also, filleting, apart from deviating from the traditional presentation,

Table 5.6 Appropriate designations of fish that are commonly canned[a]

Marine fish		Salmon and freshwater fish		Shellfish	
Appropriate designation	Species	Appropriate designation	Species	Appropriate designation	Species
Anchovy	All species of *Engraulis*	Cherry salmon	*Oncorhynchus masou* (Walbaum)	Abalone or Ormer	All species of *Haliotis*
Brisling	*Sprattus sprattus* (L)			Cockle	All species of *Cardium* (includes *Cerastoderma* or *Parvicardium*)
Herring	*Clupea harengus* (L) and sub-species	Chum salmon or keta salmon	*Oncorhynchus keta* (Walbaum)		
Mackerel	All species of *Scomber*			Crab	All species of *Cancer*
Pilchard	*Sardina pilchardus* (Walbaum)	Medium red, coho or silver salmon	*Oncorhynchus kisutch* (Walbaum)		All species of *Lithodes*
California pilchard	*Sardinops sagax caerula* (Girard)	Pink salmon	*Oncorhynchus gorbuscha* (Walbaum)		All species of *Paralithodes*
Chilean pilchard	*Sardinops sagax sagax* (Jenyns)	Red or sockeye salmon	*Oncorhynchus nerka* (Walbaum)		All species of *Callinectes*
					All species of *Geryon*
Japanese pilchard	*Sardinops sagax melanostica* (Schlegel)	Spring, king or chinook salmon	*Oncorhynchus tschawytscha* (Walbaum)		All species of *Chionoecetes*
					Erimacrus isenbeckii (Brandt)
					Maia squinado (Herbst)
South African pilchard	*Sardinops sagax ocellate* (Poppe)			Crawfish or Spiny Lobster	All species of *Palinurus*
					All species of *Palinurus*
					All species of *Jasus*
Sardine	Small *Sardina pilchardus* (Walbaum)			Crawfish	All species of *Cambarus*
Scad	All species of *Trachurus*				All species of *Astacus*
Sild	Small *Clupea harengus* (L)			Lobster	All species of *Homarus*
Smelt or sparling	All species of *Osmerus*			Norway Lobster or Dublin Bay prawn or scampi	*Nephrops norvegicus* (L)
Sprat	*Sprattus sprattus* (L)				
Tuna or tunny	All species of *Thunnus* except *Thunnus alalunga*			Mussel	All species of *Myttilus*
	All species of *Neothunnus* (Bonaterre)			Octopus	All species of *Octopus, Polypus* and *Eledone*
Albacore tuna	*Thunnus alalunga* (Bonaterre)			Oyster	All species of *Ostrea* except *Ostrea edulis* (L)
Bonito tuna	All species of *Sarga*				All species of *Crassostrea* except *Crassostrea angulata* (Link)

Skipjack tuna	All species of *Euthynnus* and *Katsuwonus pelamis* (L)
Native oyster	*Ostrea edulis* (L)
Portuguese oyster	*Crassostrea angulata* (Link)
Prawn	Large *Pandalus borealis* (Kroyer)
	Large individuals of:
	all species of Palaemonidae
	All species of Penacidae
	All species of Pandalidae
Scallop or Escallop	All species of Pectenidae
Queen scallop or Queen escallop	*Chlamys* (or *Acquipecton*) *opercularis* (L)
Shrimp	*Pandalus montagui* (Leach)
	All species of Palaemonidae
	All species of Penaeidae
	All species of Pandalidae
Squid	All species of *Loligo*, *Sepia* all *Ilex*
Whelk	All species of *Buccinum*
	All species of *Neptunea*
Winkle	All species of *Littorina*

[a] Useful references when dealing in the international fish market are Aitken and Dunbar's (1990) revision of J.J. Waterman's Torry Advisory Note 55, which appears as Torry Advisory Note No. 96 'Fish Names in the European Community' and the OECD (1978) *Multilingual Dictionary of Fish and Fish Products*.

might weaken the structure sufficiently for the fish to break up during retorting. Other species, however, should be filleted because their bones, even after retorting, remain inedibly hard. Fillets of very oily fish tend to be subject to considerable damage during the filling operation, although the dehydration, associated with the pre-retort cold-smoking operation applied to kipper fillets, makes the flesh firmer and less fragile for easier hand filling. Mackerel are pre-cooked at 90°C until the flesh separates cleanly in two halves from the spine without breaking up.

Pre-process operations that are not essentially separation exercises have been applied to modify the sensory aspects of the product and, thus, involve brining, dry salting, marinating, smoking and cooking. Apart from the modification of the sensory characteristics, all of these pre-processes cause the denaturation of proteins and loss of water, which otherwise would be released during the retorting process. Hence, the proteinaceous exudate, which forms an unsightly curd in the surrounding liquor during processing, is minimised.

5.3.1.3 *Brining.* The main object of brining is the enhancement of flavour in the final product. Normally, the process is short and part of a continuous line from gutting, heading and other separation processes, to cooking, drying or smoking, which may also be continuous processes feeding onto can-filling and seaming lines. As such, there is little water removal from the fish during brining. Indeed, in weaker than 80° brines (21% weight to volume NaCl) there may be a net gain in weight. In the short time of brine immersion, the denaturation of sarcoplasmic, let alone myofibrillar, proteins is insignificant. However, such proteins, solubilised in this period, are conveyed to the surface, as water evaporates in the post-brining stage. There they form a shiny, attractive gloss, which, none the less, permits further moisture loss and inward diffusion of volatile components in a subsequent smoking process. Brines may also carry other sensory enhancing substances like colours, smoke flavour or acetic acid. (The latter toughens the skins of sild and brisling and prevents them sticking to the can sides upon retorting.)

5.3.1.4 *Dry salting.* Removal of water is much faster than with brining but salt penetrating beyond the bounds of taste acceptability limits the time of exposure. Large fish canned in the form of 'cuts' or 'steaks' may be dry salted. Whatever form of pre-process is used, it is unlikely to completely inhibit loss of water and sarcoplasmic proteins from the deepest parts of the muscle, thus such products are often presented in their natural juices.

5.3.1.5 *Marinating.* Most often applied as a main process for seafood products subsequent to being sold as chilled foods, marinating is also used in the preservation of various shellfish meats in glass jars prior to pasteurisation or sterilisation by heat.

5.3.1.6 *Smoking.* Whether 'hot' or 'cold', smoking is essentially a process to impart more flavour to some fatty fish prior to canning. The drying that occurs during the process also denatures and 'sets' the proteins so there is little danger of exudation into the surrounding liquor during the heat process.

The initial composition of fatty fish for smoking and canning greatly affects the quality of the end-product. Low-fat content raw material tends to lose more water during smoking and, although firm and easily handled at the filling stage, nevertheless remains tough-textured after the heat-sterilisation process. On the other hand, high-fat content raw material tends to break up easily during the hand-filling operation and yields a finished product that is too soft.

Sild and brisling, which are immature herring (*Clupea harengus*) and sprat (*Sprattus sprattus*) respectively, that are subjected to brining and hot-smoking before canning, indicate that a market exists for the many sardine-like pelagic species in canned form. More frequently now, rather than smoking, smoke flavour is incorporated into the brine, the fish pre-cooked in their cans and the cook-out decanted prior to adding oil or sauce, or sealing and retorting. This method yields a more satisfactory product than when used for canned kipper fillets.

5.3.1.7 *Cooking.* The heat sterilisation process applied when cans of fish are retorted is more than adequate to cook the contents; indeed, non-fatty, white fish are grossly overcooked in rendering them commercially sterile and are, therefore, generally unsuitable for canning.

5.3.2 *Exhausting*

The removal of gases from cans prior to sealing is necessary:

- to prevent large in-container pressures developing during high-temperature sterilisation due to the expansion of headspace gases; and
- to reduce oxidation of the contents and internal corrosion of the container.

Inequality between internal and external pressures during processing of tin cans causes strain upon the seams, which may cause them to leak. Leaker spoilage is, by far, the commonest source of canned-food spoilage by microorganisms. That there have been pressure imbalances during retorting may or may not be obvious on inspection of the finished product. With large pressure differences and/or large can diameters, greater internal than external pressure causes a permanent tin-plate distortion (usually of the end close to the seam) that is called 'peaking', whereas greater external than internal pressure causes inward collapse of the can body called 'panelling'.

In other containers excessive internal pressure during retorting could lead to lids blowing off during retorting, or pouch seals softened by the heat process being pulled apart. In consequence, with both of the latter examples, efforts are made to ensure an over-pressure in the process medium.

In the traditional small flat cans, the structural strength of the container is sufficient to resist quite large interior/exterior pressure differences without deformation; moreover, with such cans there is no requirement for a headspace to allow for expansion of such small quantities of contents during processing, although every effort is made to eliminate air by tightly packing them with fish and injecting oil or sauce to fill any air gaps. Larger cans need a headspace to allow for expansion of contents during the heat process, thus requiring some means of eliminating air from the headspace immediately prior to sealing. Three methods of achieving a headspace vacuum are: (i) hot filling/hot sealing; (ii) sealing under steam; and (iii) vacuum sealing.

5.3.2.1 *Hot filling/hot sealing.* When packing fish into the container hot and injecting hot oil, brine or sauce, one can leave less ullage (can volume – fill volume) for subsequent expansion and some of the air in the top of the can is expelled by steam from the hot contents. Sealing must follow immediately before cooling and contraction can occur. In a 440 g can of finished product this method of exhausting may achieve a moderate vacuum around 25 mm of mercury on testing with a can vacuum gauge but the handling of hot, delicate fish may be impossible by hand or mechanically. However, prawns and other shellfish meats packed in brines or vinegar are successfully treated in this way. Similarly, those fish that are cooked in the can followed by decanting, hot-liquid injection and sealing (like mackerel in sauce) achieve satisfactory headspace vacuums.

Solid, chunk, flake and fish-paste packs cannot be satisfactorily exhausted, although attempts have been made to fill solids cold and inject hot liquid. The resultant headspace vacuum in such cases may be lower than 5 mm of mercury so that there is a danger that the can ends flip outwards. Such cans are referred to as 'flippers' or, when one end flips out and can be pushed inwards causing the opposite end to flip out, 'springers'. This condition can also be caused by the production of gases inside the can by internal corrosion of the tin plate, ('hydrogen swells') or, more dangerously, by microbiological activity from organisms that have survived the process or gained access subsequent to processing via a leak. For this reason, cans with convex rather than concave ends are correctly rejected by the consumer and condemned by public health authorities.

A long-established exhausting process is where cans are filled and placed in a hot-water bath with their lids 'clinched' on (that is where the end has been fixed to the body by the first operation of double seaming only) to prevent contamination of contents, while still permitting air to escape through the incomplete double seam, followed immediately by completion of the seam. This is a reliable method of achieving a good headspace vacuum but, in being very slow, does not fit well as an operation into modern, high-speed, continuous lines.

5.3.2.2 *Sealing under steam.* Containers that are conveyed through a steam-exhausting chamber and sealed as they emerge have their headspace air replaced by steam, which condenses in the closed container. This leaves a partial vacuum, the extent of which depends upon the extent to which air is excluded from the chamber, which, in turn, depends upon the pressure of steam supplied to the chamber.

5.3.2.3 *Vacuum sealing.* The most reliable method of achieving a constant headspace vacuum is to seal the can in an evacuated chamber. However, the speed at which lines can run may be retarded by time needed to exhaust the cans as they enter the chamber. Laminated foil pouches must be vacuum-sealed to avoid any pressure build-up within, during heat processing. Line speeds are accelerated by having many sealing stations, but pouch lines may run at only 60 units min^{-1} whereas conventional canning lines may run at over 1000 units min^{-1}.

5.3.2.4 *Seal integrity.* It is desirable that neither flesh nor liquid be trapped in the seal during the sealing stage as trapped material (solids in particular) may provide a route for post-process contamination and, in the case of glass pry-offs, which rely upon internal vacuum to hold on the lid during processing, total failure of the seal.

The canned fish processor is responsible for the integrity of the double seam he makes in sealing his product. Both filling and sealing are operations that require close control to ensure seam integrity. Great care should be exercised at the filling stage to avoid air spaces, which, if sealed into a can of fish, would expand during processing causing undue stress on the seams and thereby increasing the likelihood that they leak.

The amount of headspace in a product that affects the heat-transfer characteristics and the final vacuum depends upon the ullage left by the filling operation which, in turn, depends upon the bulk density of the contents, assuming that weight control is within narrow limits. In many fish-canning operations, however, the smallness of the container, the required tightness of the fill and the speed of the production line make the provision of a headspace vacuum impracticable. Such cans of fish are generally processed under water under an excess pressure head (achieved by using compressed air) to balance can internal pressures.

By far the majority of incidences of canned-food spoilage by micro-organisms are traceable to re-infection of the contents after processing usually by ingress through the double seam. This is called 'leaker spoilage'.

Non-destructive visual and tactile examination of can seams may detect a fault in the seaming operation. Such faults and their detection are described in can-makers' manuals. 'Droop', 'spurs', 'cutover', 'jumped seams' and 'false seams' are common descriptions of faults arising from either the use of damaged cans or the erroneous setting of the seaming equipment. Occasional faults, due, for example, to entrapment of product in the seam are unlikely to be detected unless all the finished cans were to be examined. (It would require the examination of 4000 cans to be reasonably confident of detecting a fault that occurred once in every 1000 seaming operations).

Destructive examinations can be made in two ways. The more modern method is to make two parallel cuts about 2 cm apart across the can seam, and press the portion of the seam between the cuts inwards to reveal a cross-section of the seam for inspection on a calibrated can-seam projector.

However, this method suffers from the disadvantage that, with only one point on the seam circumference being examined, it may miss faults like 'jumped seams' and 'end-hook wrinkles'. The older method whereby the seam is 'torn down' to reveal the full circumference of end body hooks is more reliable in this respect. End-hook wrinkles are more likely to occur on larger diameter cans but specifications on 'wrinkle rating' exist for all can sizes.

In the destructive examination of can seams, measurements of dimensions are taken at three points on the seam circumference using a seam micrometer. Pieces are then cut from the body plate and end plate using tinsnips so that 'body plate thickness' (tb) and 'end plate thickness' (te) can be measured using the seam micrometer or, more accurately, using a plate-thickness gauge.

The efficiency of the seaming operation is indicated by the seam thickness (T) minus the five thicknesses of plate it comprises, thus demonstrating how loosely or tightly the seam has been made, and the extent to which body (BH) and end hooks (EH) overlap each other in the seam length (L). Thus:

$$\text{'Freespace'} = T - (2\text{tb} + 3\text{te})$$

and

$$\text{'Actual overlap'} = \text{EH} + \text{BH} + 1.1\text{te} - L$$

From a cross-section of the seam a direct measurement of the 'actual overlap' and the 'body hook butting' can be made. The latter is the extent to which the body hook 'butts' into the sealing compound in the end hook and is the most direct, logical indication of the success of the double-seaming operation. It may also be computed as percentage body hook butting from measurements taken during the tearing down procedure thus:

$$\% \text{ Body hook butting} = \frac{\text{BH} - 1.1\text{tb}}{L - 1.1(2\text{te} + \text{tb})} \times 100$$

This should be greater than 70%. (For more detailed seam inspection information a can-maker's manual should be consulted.)

5.3.3 *Heat processing and heat-processing equipment*

5.3.3.1 *Steam-pressure retorts.*
The most common medium in use for heat-processing canned foods to commercial sterility, is saturated steam under pressure. The greater the pressure inside the retort, the greater will be the

Table 5.7 Commonly used conditions for conventional retorts

Pressure (bar)	Condensation temperature (°C)
1.5	111.4
1.7	115.2
2.0	120.2

temperature at which the steam condenses on the external walls of the can. Commonly used conditions in conventional retorts are shown in Table 5.7.

Pressure vessels are subject to stringent safety laws and must be tested frequently. Tested pressure (TP) and safe working pressure (WP) are stamped onto each vessel, which must be equipped with a safety valve set to open and vent the retort if the WP is exceeded. Safety devices are now mandatory for ensuring that the retort can neither be opened while the retort is under pressure nor while the steam entry valve is open.

Two types of batch retorts still in common use are the static vertical and horizontal types shown in Figure 5.6.

The horizontal type has the advantage that it can be loaded with containers on trolleys, whereas vertical retorts must be loaded by block and tackle. Also, many vertical retorts are mounted in wells, which can pose a hygiene hazard. However, one advantage that the vertical possesses over the horizontal retort, is that it tends to favour a more uniform internal steam distribution. Uniformity of steam distribution in horizontal retorts, in an attempt to overcome cold pockets and, therefore, localised under-processing within a batch, is favoured by introducing the steam through multi-orifice sparge pipes along the length of the retort floor. Orifices, however, tend to become blocked if the sparge pipes are not regularly cleaned.

5.3.3.2 Steam-pressure retort operation. The operation of these simple batch retorts although, increasingly, being converted to automatic control, follows the following sequence:

1. Closing and securing the lid or door(s).
2. Admitting steam with all exit vents and drain open.
3. Closing the drain when the volume of steam condensate emerging has fallen to a quantity that can be efficiently removed via the condensate valve.
4. Allowing the retort to reach 100°C internally and flushing through steam for a fixed interval (depending upon retort size) to ensure the expulsion of all air pockets inside the retort, which could lead to under-processing. This is called *venting* the retort.

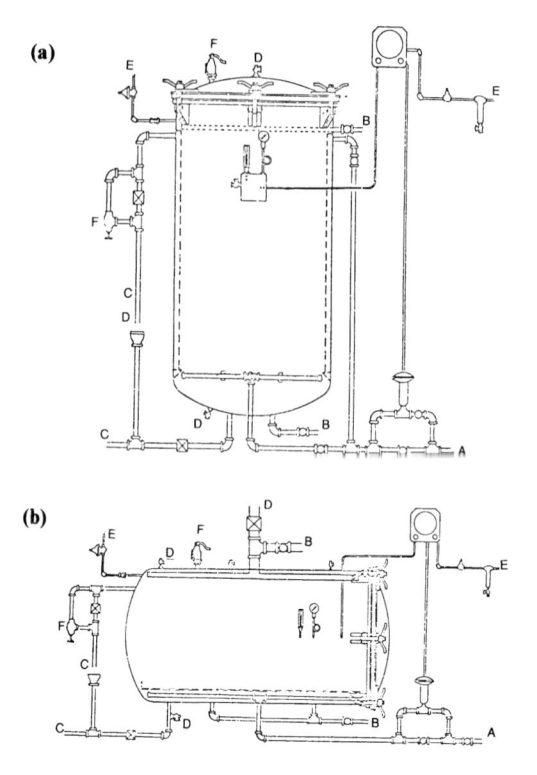

Figure 5.6 Static (a) vertical and (b) horizontal retorts: A = steam; B = water; C = drain, overflow; D = vents, bleeders; E = air; F = safety valves, pressure relief valves, manual valves; (○) globe; (☒) gate.

5. Closing the main steam exit valve so that pressure builds up in the retort to a value, pre-set on the steam regulator valve, corresponding to the required processing temperature.
6. On the attainment of this temperature, the 'process' begins and conditions are maintained for the time or process required. Pet-cocks, kept open throughout the process, ensure movement of steam over the instruments and containers, and expulsion of any air that may enter with the steam.
7. On expiry of the process time the cooling sequence comes into operation.
8. Simultaneous closing of the main steam valve and opening of the air valve, admitting compressed air to maintain the pressure inside the retort.
9. Activation of the cooling water pump followed by opening of the cooling water valve admitting chlorinated cooling water to the retort.

10. Opening the drain, allowing circulation of cooling water through the retort and back to a holding reservoir for re-chlorination and re-use, while gradually reducing the retort pressure to balance the pressure inside the containers, as the latter reduces with temperature.
11. Halting the cooling cycle, draining the retort, opening the retort and withdrawing of the containers at a suitably low temperature.

5.3.3.3 *Heat processing of pouches and heat-sealed plastic containers.* Retortable pouches and heat-sealed plastic cans are processed in specially adapted batch retorts. If, at any time during their process, the pressure inside the containers were to exceed that outside, total loss of the batch may be the result as strained, hot seals burst open. For this reason, it is normal to ensure a fair margin of over-pressure in the retort, compared with that expected in the containers, at any one temperature.

If no vacuum is achieved within the container during the sealing process, a pressure around 0.26 Mn m^{-2} could develop at a contents temperature of 115°C. Thus, a retort pressure greater than 0.3 Mn m^{-2} would be needed to give an adequate safety margin against pouch bursting. The temperature corresponding to 0.3 Mn m^{-2}, if it were to be achieved purely by steam, would be 133.5°C. Such high temperatures, apart from causing heat damage to the product would, inevitably, cause higher temperatures and therefore higher pressures in the container. Consequently, higher retort pressures must be applied without raising the selected processing temperature. Practically, this is accomplished by processing pouches under water, heated to the processing temperature by pressurised steam, with the requisite excess pressure achieved by compressed air above the water level. Alternatively, processing under a homogenised mixture of steam and air, requiring specialised processing equipment but obviating the need for processing under water, is widely practised in countries where pouches have achieved considerable market popularity.

Although it is more difficult to locate and record the temperature during the retorting process for the pouch 'cold-spot', Govaris and Scholefield (1988) found good agreement between computer models for evaluating thermal process safety in pouches and experimental results.

5.3.3.4 *Continuous retorts.* Continuous processing retorts require that the container be introduced to the heating chamber without releasing the internal pressure.

Figure 5.7 shows cans being transferred into a continuous rotary cooker through which the cans roll, at velocities determined by the required process time, on a spiral pathway, around the periphery of the retort shell. Another transfer valve conveys the cans from pressurised cooking vessels to pressurised cooling vessel, where they travel along a similar scroll to the exit-transfer valve. Cans emerge from the latter still warm enough to have an internal pressure

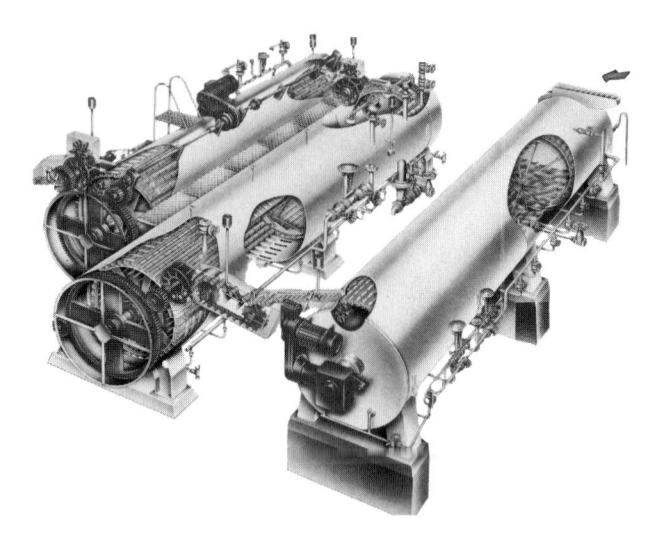

Figure 5.7 Typical rotary cooker with transfer valves. Reproduced by kind permission of FMC Food Machinery, North America.

sufficient to resist panelling due to excessive difference between pressure inside and outside the can. Larger diameter cans, requiring a greater degree of cooling, may be transferred to a second cooling vessel at a lower pressure than the first, before passing to can drying, labelling and bulk packaging/unitisation lines.

The continuous agitating retort uses the same principle as the continuous rotating retort, in that cans are rotated axially on their passage through the retort, but uses a conveyor system back and forth inside the retort, rather than a spiral scroll around the inside retort perimeter. The agitation is achieved by the partitioned conveyor rolling the cans along a fixed track.

Hydrostatic retorts (Figure 5.8) save factory floorspace by carrying the cans along vertically ascending and descending pathways enclosed in a tower several storeys high. Essentially, they are continuous static retorts, the cans not being agitated but conveyed though heating, holding and cooling limbs of the tower. The steam in the central dome is held under constant pressure, and, therefore, temperature, by a head of water in the outer heating and cooling limbs. The process time is once again controlled by the speed of the conveyor and can be adjusted for different can sizes and products.

5.3.3.5 *Other heat-processing equipment.* Steam is the usual heat-transfer medium in canned food processes but other media have been tried and successfully commercialised. The French 'Steriflamme' continuous cooker/cooler employs naked gas flames over which cans roll axially along

CANNING FISH AND FISH PRODUCTS 151

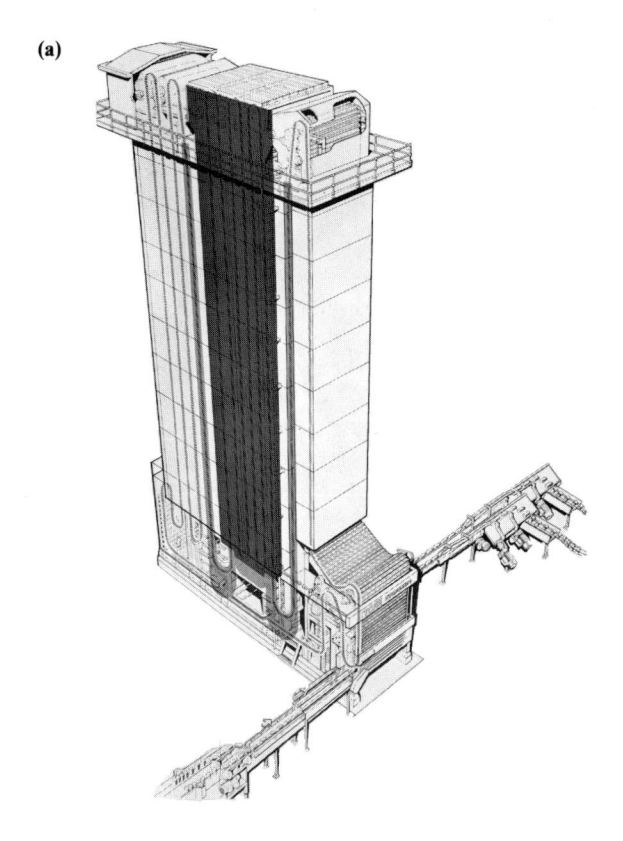

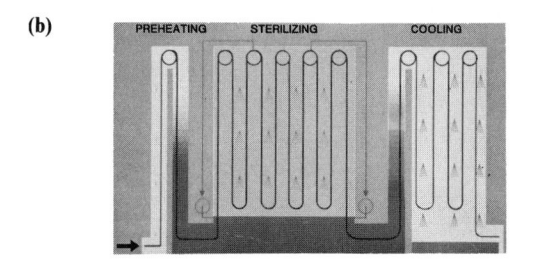

Figure 5.8 (a) An hydrostatic retort with (b) diagrammatic cross-section. Reproduced by kind permission of Stork Amsterdam BV.

tracks running back and forth through the length of the cooker. Similarly, cooling is achieved by the cans rolling under water sprays. The system lends itself to high-temperature short-time (HTST) processes but is unsuitable for can diameters greater than 105 mm because the internal pressures, building up and exceeding the constant atmospheric pressure in the open-sided cooker, might cause larger cans to peak during the process. However, since the cooking and cooling sections are entirely open and accessible, problems like container jamming can be quickly spotted and corrected. The review by Richardson (1987) indicates the suitability of the method for canned-fish products and the large energy savings compared with steam retorting.

Patents also exist for can-processing equipment using hot fluidised sand as the heat-processing medium and such a system has been successfully exploited for processing retortable pouches.

Microwave canning, as a means of sterilising without over-cooking, is a technical possibility with all-plastic rigid or pouch containers, provided that pressures developed inside the container can be balanced with equivalent external pressure. Continuous systems have been developed and have been used for in-container pasteurisation processes. Apart from energy costs, the major obstacles to the use of microwaves in canned food sterilisation are the unpredictability of the temperature distribution in a complex product that is subjected to microwaves and the reliance on slow, conventional methods of cooling.

5.3.4 *Post-process operations*

After the cold spot has received a heat treatment adequate for commercial sterilisation, it is usually desirable to cool it as rapidly as possible to avoid too great a degree of over-cooking. Additionally, a long, slow cooling process (by, for example, allowing the can contents to cool naturally after removing the heating medium) can lead to thermophilic spores (surviving a process designed to eliminate *C. botulinum*) germinating and multiplying to cause spoilage. There is a danger, particularly with large-diameter containers, where it might take more than 24 h for the cold spot to pass through the temperature range suitable for thermophilic growth. The substitution of the heating medium with a cooling medium must be finely controlled. Above all, the pressure in the medium surrounding the containers should, initially, be maintained at that which was applied during processing. The balance of pressure between the inside and outside of the can, and consequences of imbalance, at all times during the retorting process, have been considered in section 5.3.2. (With respect to cooling cans in conventional steam-heated retorts, this is accomplished by replacing the steam with compressed air.) Hence, the cold spot temperatures should be monitored throughout cooling, to determine the optimum pattern of falling pressure in the retort during the post-process period. The water used to cool the processed containers must be chlorinated because, while the containers are hot, the sealing compound may be molten so that there is a slight possibility that a drop of cooling water could be pulled through the seam by the vacuum forming in the headspace. Although this undesirable occurrence is only a remote chance (usually between 1 in 10^4 and 1 in 10^5) 'leaker spoilage' is, by far, the commonest microbiologically-based reason for rejection of canned foods.

There are several available techniques for cooling water chlorination, ranging from direct injection of chlorine gas to addition of compounds that liberate chlorine by reaction with water, like sodium hypochlorite. Since chlorine is very reactive, especially with organic debris, which may be found on the containers and in the retort due to spillage during the filling operation, the dosage should only be fixed by the level of free residual chlorine found in the water draining from the retort. This should be monitored continually, and the dosage level adjusted accordingly. Residual free chlorine levels greater than 10 p.p.m. accelerate unsightly external corrosion of the containers, so dosages must be controlled within fairly strict limits.

Prolonged cold-water cooling of cans results in them leaving the retort wet and necessitates the use of a can drier before any handling can be permitted. A droplet of water on the seam of a can emerging from a retort may, very easily, be contaminated with micro-organisms on post-process line machinery or on workers' hands and may, subsequently, be drawn through the seam by the headspace vacuum to re-contaminate the contents.

Post-process hygiene measures, such as the prohibition of wet-can handling and the regular chlorinated cleaning of processed container conveyors, cannot be too highly stressed. Major disasters such as the Aberdeen typhoid epidemic (1965) and the Birmingham botulism cases (1979) have been traced back to failures in this area.

5.4 Cannery operations for specific canned-fish products

The setting up of process lines for the manufacturer of specific canned-fish products followed a scheme dictated by the nature of the raw material and the desired end-product within the constraints of speed and cost. The market for canned fish products is, mainly, long-established. As such, maximum output volume at minimum cost is the prime objective. Production lines have, therefore, become a series of linked, continuous processes that have become increasingly automated to reduce labour costs.

5.4.1 *Small pelagics*

5.4.1.1 *Anchovies* (Engraulis *spp.*). These are exceptional among canned foods in that they are not heat-processed but packed in dry salt and sealed in large cans. A degree of proteolytic ripening occurs through storage and the product is frequently sold in this form. Anchovy fillets, however, are removed from the whole fish after salting and packed in oil. These are semi-preserves with a 3–6 month shelf-life under refrigeration.

5.4.1.2 *Sardines and the sardine-like products.* Once called 'Norwegian Sardines' these are now labelled as Brisling (small sprats) or Sild (small

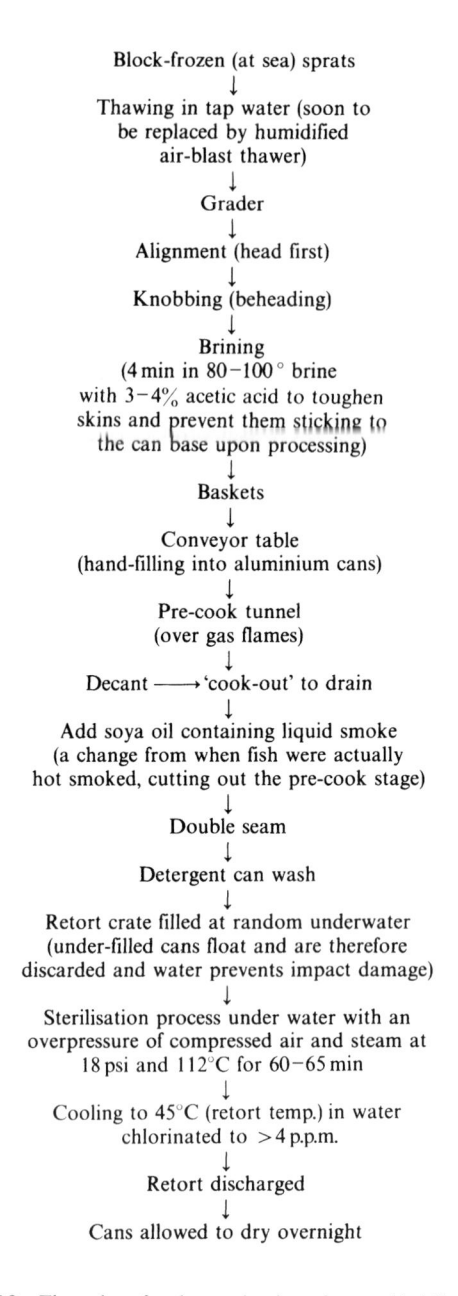

Block-frozen (at sea) sprats
↓
Thawing in tap water (soon to
be replaced by humidified
air-blast thawer)
↓
Grader
↓
Alignment (head first)
↓
Knobbing (beheading)
↓
Brining
(4 min in 80–100° brine
with 3–4% acetic acid to toughen
skins and prevent them sticking to
the can base upon processing)
↓
Baskets
↓
Conveyor table
(hand-filling into aluminium cans)
↓
Pre-cook tunnel
(over gas flames)
↓
Decant ⟶ 'cook-out' to drain
↓
Add soya oil containing liquid smoke
(a change from when fish were actually
hot smoked, cutting out the pre-cook stage)
↓
Double seam
↓
Detergent can wash
↓
Retort crate filled at random underwater
(under-filled cans float and are therefore
discarded and water prevents impact damage)
↓
Sterilisation process under water with an
overpressure of compressed air and steam at
18 psi and 112°C for 60–65 min
↓
Cooling to 45°C (retort temp.) in water
chlorinated to > 4 p.p.m.
↓
Retort discharged
↓
Cans allowed to dry overnight

Figure 5.9 Flow-chart for the production of canned brisling in oil.

herring) and are most frequently bought in block-frozen form. Seasonal catching and large hauls of such small fish favour such primary processing.

From thawing to packing cans of finished product into master cartons and palletising, almost the full sequence of operations has been automated. Hand packing of the fish into cans remains, on most lines, labour intensive. It is, however, useful to maintain one point on the line at which inspection and quality control of the contents can be practised. Figure 5.9 represents a flow-chart for the production of canned brisling in oil.

5.4.2 Tuna and mackerel

The pre-cooking of scombroid species is carried out in steam under atmospheric pressure for times that vary according to the size of the fish; 0.25 kg mackerel, for example, requires only 30 min, while large 20 kg tuna may require 4 h. In the former case, the mackerel are often cooked in the open can followed by decanting of the separated liquor, sauce or oil injection, sealing and retorting. Tuna, however, are cooked and allowed to cool for up to 24 h to allow the flesh to set and become manageable. They are then cleaned and trimmed, which comprises the removal of head, skin, spine and dark flesh underlying the lateral line. All skin must be painstakingly removed for the sake of pack presentation, and remaining flesh is packed as a solid steak, as chunks or as flakes, by hand or machine. In the latter two cases, this involves moulding the pieces into the correct shape, cutting and compressing them into the can. Subsequently, dry salt is added, followed by oil if an 'in oil' pack is required. McLay (1980) describes an automatic system where brined fillets are fed on to a conveyor so that they overlap each other by one third of their length and, after passing through a cooker, the continuous layer of fillets is cut into can-size lengths. The portions thus formed are discharged into cans already containing sauce. The sequence is completed by topping up the container with more sauce, sealing, washing and heat processing. The finished product contains about 60% fish and 40% sauce.

A typical flow-chart for the canning of tuna is shown in Figure 5.10. Mackerel may be similarly treated but have increasingly been packed as skinless fillets in a variety of sauces as shown by the flow-chart in Figure 5.11.

5.4.3 Crustacea and molluscs

Shrimp used for canning are normally in the medium (65–90 per kg) to very small (more than 140 per kg) size range. After removal of the head and shells, the meats are blanched for 6–10 min in boiling, 11–13% brine. This causes the meats to shrink and curl, and produces the characteristic pink colour. Larger shrimp must be de-veined (eviscerated) before blanching. After inspection and grading according to size, the shrimp meats are hand packed into cans, which are then topped up with 7–8% brine. The choice of small shrimp for canning is

156   FISH PROCESSING TECHNOLOGY

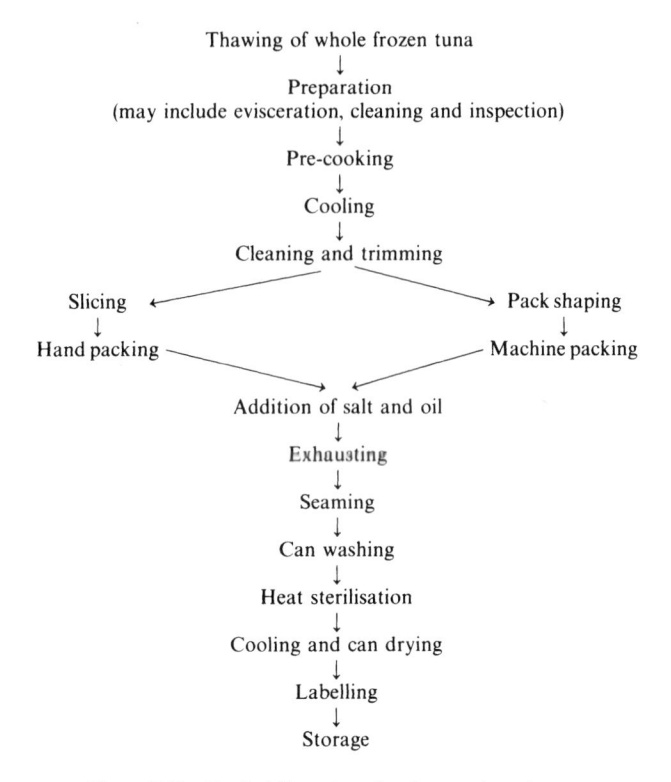

```
              Thawing of whole frozen tuna
                          ↓
                    Preparation
     (may include evisceration, cleaning and inspection)
                          ↓
                    Pre-cooking
                          ↓
                      Cooling
                          ↓
                Cleaning and trimming
              ╱                      ╲
         Slicing  ←                → Pack shaping
            ↓                             ↓
       Hand packing ╲              ╱  Machine packing
                    →            ←
               Addition of salt and oil
                          ↓
                     Exhausting
                          ↓
                      Seaming
                          ↓
                    Can washing
                          ↓
                  Heat sterilisation
                          ↓
                Cooling and can drying
                          ↓
                     Labelling
                          ↓
                      Storage
```

Figure 5.10 Typical flow-chart for the canning of tuna.

due, partly to the poorer market for this size and partly to the need to use as mild a sterilisation process as possible. Shrimp, being non-fatty, suffer adverse texture and flavour changes leaving the meat tough, dry and bland after the long heat-processing necessary to sterilise large pieces of meat.

Pre-cooking, however, is not always carried out in hot water or steam. In some cases, especially where a relatively low-fat raw material is to be preserved by heating and pickling, frying yields a more succulent end-product. An internationally popular example of such a product is 'mejillones en escabeche' (canned pickled mussels). A flow-chart for their production described by Lopez-Capont (1980) is shown in Figure 5.12. (A final acetic acid content of approximately 0.8% ensures a high acid product, which will not support the growth of pathogenic micro-organisms.)

5.4.4 *New canned-fish products*

Fish pastes and spreads have been preserved by heat-processing for many years, mainly in glass jars rather than cans. More recently, alternative comminuted fish products, like pâtés and terrines, have been introduced. These are

CANNING FISH AND FISH PRODUCTS

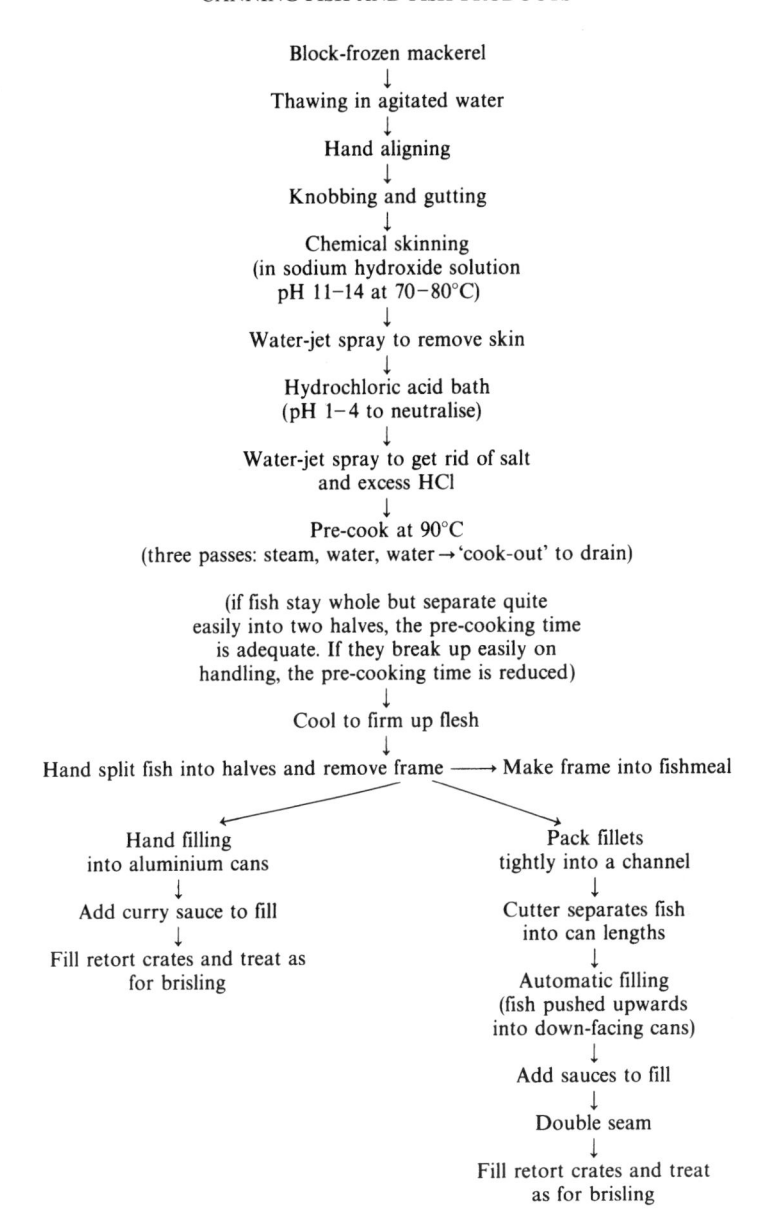

```
Block-frozen mackerel
        ↓
Thawing in agitated water
        ↓
Hand aligning
        ↓
Knobbing and gutting
        ↓
Chemical skinning
(in sodium hydroxide solution
    pH 11–14 at 70–80°C)
        ↓
Water-jet spray to remove skin
        ↓
Hydrochloric acid bath
(pH 1–4 to neutralise)
        ↓
Water-jet spray to get rid of salt
      and excess HCl
        ↓
Pre-cook at 90°C
(three passes: steam, water, water → 'cook-out' to drain)

(if fish stay whole but separate quite
easily into two halves, the pre-cooking time
is adequate. If they break up easily on
handling, the pre-cooking time is reduced)
        ↓
Cool to firm up flesh
        ↓
Hand split fish into halves and remove frame ——→ Make frame into fishmeal
```

Hand filling into aluminium cans	Pack fillets tightly into a channel
↓	↓
Add curry sauce to fill	Cutter separates fish into can lengths
↓	↓
Fill retort crates and treat as for brisling	Automatic filling (fish pushed upwards into down-facing cans)
	↓
	Add sauces to fill
	↓
	Double seam
	↓
	Fill retort crates and treat as for brisling

Figure 5.11 Flow-chart for the packing of mackerel as skinless fillets in a variety of sauces.

similar to spreads and pastes, in that the fish components need to be thoroughly cooked, before chopping and mixing with other ingredients, to achieve a spreadable consistency in the finished product. A pâté, to be sliceable and spreadable, requires the addition of a gelling agent that will withstand

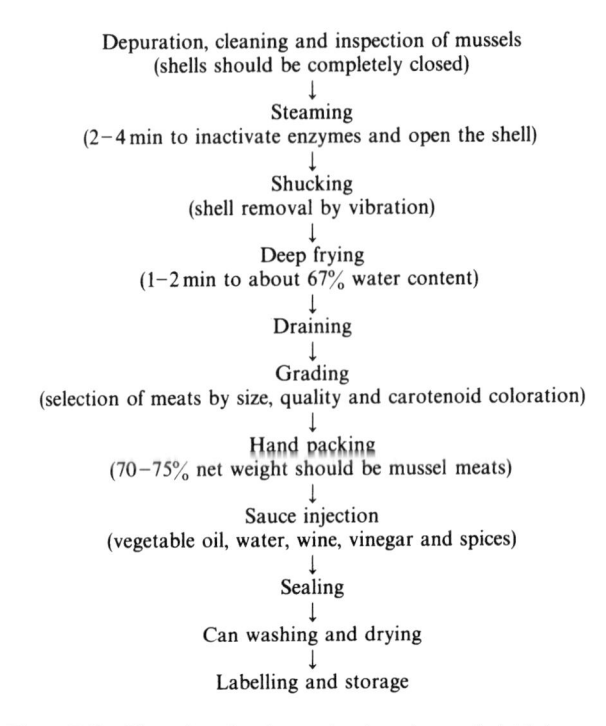

Depuration, cleaning and inspection of mussels
(shells should be completely closed)
↓
Steaming
(2–4 min to inactivate enzymes and open the shell)
↓
Shucking
(shell removal by vibration)
↓
Deep frying
(1–2 min to about 67% water content)
↓
Draining
↓
Grading
(selection of meats by size, quality and carotenoid coloration)
↓
Hand packing
(70–75% net weight should be mussel meats)
↓
Sauce injection
(vegetable oil, water, wine, vinegar and spices)
↓
Sealing
↓
Can washing and drying
↓
Labelling and storage

Figure 5.12 Flow-chart for the production of canned pickled mussels.

retort process temperatures that would largely destroy the gelling ability of the gelatine naturally associated with the fish flesh. Various seaweed-derived hydrocolloids can be used for this purpose.

More exotic (at least to the UK market!) canned-fish products, like octopus in its own ink, stuffed-squid mantles, bouillabaise and clam chowder, may increasingly find a market as consumers and delicatessen store managers become more adventurous.

It would be encouraging to see more development towards maximising the value of this high-quality protein resource, although some traditional combinations, like prawns in an acidic cocktail sauce, may remain impracticable, not because they cannot tolerate the heat sterilisation process, but because of the long-term incompatibility of the ingredients. It should always be remembered that many canned-fish merchandisers expect, at least, a 2-year shelf-life, at ambient temperature, from their products.

References

Adams, J.B. (1991), Review: Enzyme inactivation during heat processing of food-stuffs. *Int. J. Food Technol.* **26**, pp. 1–20.

Aitken, A. and Dunbar, E.M. (1990), Fish names in the European Community. *Torry Advisory Note No. 96*, MAFF.
Ali Khayat (1978), Colour changes in canned tuna during canning and colour improvement by chemical modification of haeme proteins. *J. Food Technol.* **13**, 117–127.
Ball, C.O. (1923), Thermal process time for canned food. Washington D.C.: *Bull. Nat. Res. Council* **7**, No 37.
Bender, A.E. (1972), Processing damage to protein food: A review. *J. Food Technol.* **7**, 239–250.
Berry, M.R. Jr and Bush, R.C. (1988), Thermal processing retortable plastic containers with metal lids in steam and water with comparisons to metal cans. *J. Food Sci.* **53**, 1877–1879, 1886.
Boone, P. (1966), Mode of action and application of nisin. *Food Manufacture* **41**, 49–51.
Cowell, N.D. (1968), Methods of thermal process evaluation. *J. Food Technol.* **3**, 304–305.
Durand, H. and Thibaud, Y. (1980), A study of canned fish behaviour during storage. In *Advances in Fish Science and Technology*, Connell, J.J. (Ed.), Fishing News Books, Farnham.
Gaze, J.F. and Brown, K.L. (1988), The heat resistance of spores of *Clostridium botulinum*, 213B over the temperature range 120 to 140°C. *Int. J. Food Technol.* **23**, 373–378.
Govaris, A.K. and Scholefield, J. (1988), Comparison of a computer evaluation with standard evaluation of retort thermal processing. *Int. J. Food Technol.* **23**, 601–606.
Jones, M.C. (1968), The temperature dependence of the lethal rate in sterilisation calculation. *J. Food Technol.* **3**, 31–38.
Lopez-Capont, F. (1980), The technology of Spanish canned mussels (*Mytilus edulis* L.) from raft cultivation. In *Advances in Fish Science and Technology*, Connell, J.J. (Ed.), Fishing News Books, Farnham, pp. 240–242.
McLay, R. (1980), Canned Fish. *Torry Advisor Note No. 41*, HMSO, Edinburgh.
Metal Box Company (1973), Double seam manual. Metal Box Company, London.
Nagaoka, C., Yamagata, M. and Horimoto, K. (1971), A method for predicting 'greening' tuna before cooking. In *Fish Inspection and Quality Control*, Kreuzer, R. (Ed.), Fishing News Books. Farnham, p. 96.
OECD (1978), *Multilingual Dictionary of Fish and Fish Products*. Fishing News Books, Farnham.
Ohlsson, T. (1979), Optimisation of heat sterilization using C-values. In *Food Process Engineering*. (V ol. I) Linko, P., Malkki, Y., Olkku, J. and Larinkari, J. (Eds), Applied Science, London, pp. 137–145.
Perigo, J.A. and Roberts, T.A. (1968), Inhibition of clostridia by nitrite. *J. Food Technol.* **3**, 91–94.
Pham, Q.T. (1990), Lethality calculation for thermal processes with different heating and cooling rates. *Int. J. Food Technol.* **25**, 148–156.
Richardson, P.S. (1987), Review: Flame sterilization. *Int. J. Food Technol.* **22**, 3–14.
Stumbo, C.R. (1953), New procedures for evaluation of thermal processes for foods in cylindrical containers. *Food Technol.* **7**, 309.
Thijssen, H.A.C., Kerkhof, P.T.A.N. and Liefkens A.A.A. (1978), A short-cut method for the calculation of sterilisation conditions yielding optimum quality retention for conduction-type heating of packaged food. *J. Food Sci.* **43**, 1096.
Tyndall, J. (1877), Further researches on the department and vital persistence of putrefactive and infective organisms from a physical point of view. *Phil. Trans. Royal Soc. Lond.* **167**, 149–206.

6 Methods of identifying species of raw and processed fish
I.M. MACKIE

6.1 Introduction

This review is concerned with methods of identifying the species of fish or shellfish when it is not possible to do so by the usual morphological criteria – shape and size, pigmentation of skin – characteristics which are the basis of zoological classification systems. This situation can arise when these species-distinguishing features have been removed on processing, as into fish fillets or fish fingers which are completely lacking in any features that would enable the identity of the fish to be established. When fish species are closely related, the flesh is often similar in appearance (and taste) and, although trained taste panellists and other connoisseurs of fish gastronomy may be able to identify the species some of the time, their pronouncements or judgements have on many occasions been shown to be faulty and little better than the rest of the apparently less discriminating sector of the population. If it is then necessary to establish the identity of the species unequivocally to check, for example, that it is as declared on the label of a product or that it is as specified in a commercial contract, a reliable means of identification other than by sensory criteria is required. For the purposes of this review, the term 'fish' will be assumed to include shellfish and crustacea.

There is general legislation protecting the consumer from misrepresentation (Trade Descriptions Act, 1968; Consumer Protection Act, 1987) but additionally, for fish, there are legal requirements for the use of specific names for commercial species (The Food Labelling Regulations, 1984; The Food Labelling (Scotland) Regulations, 1984). These regulations require that prescribed names for fish species are used; for example, that sole is *Solea solea* and not Canary sole *Solea canariensis*, that *Squalus* spp. are called dog fish and not rock salmon, that *Anarichas* spp. are catfish and not rock turbot. Scampi, otherwise known as Norway lobster or Dublin Bay prawn, is required to be the flesh of *Nephrops norvegicus*, but not that of its near relatives from the Pacific, the *Metanephrops* spp. Similarly, rainbow trout (*Oncorhynchus mykiss*), the widely farmed species of trout which was orginally introduced from North America, cannot be substituted for the native European trout (*Salmo trutta*), more commonly known as sea trout or brown trout, or salmon (*Salmo salar*).

The substitution of cheaper for more expensive foods and other commodities is, of course, not a new problem. It has no doubt been practised by unscrupulous operators ever since trading began but today, as more species of fish are processed into products almost completely devoid of the morphological characteristics of the animal, there are possibly more opportunities for substitution or adulteration than in earlier times.

Species of the Gadidae family such as cod (*Gadus morhua*), saithe (*Pollachius virens*), lythe (*Pollachius pollachius*), haddock (*Melanogrammus aeglefinus*) and whiting (*Merlangius merlangus*) for example, are all white-fleshed species of low fat content and as there are large differences in price between whiting and cod for example (UK Sea Fisheries Statistics 1991, 1992), there are incentives for the unscrupulous trader to substitute the cheaper for the more expensive species. Also, because of the general shortage of cod and haddock, in Europe there is the possibility of imported white fish such as Alaska pollack (*Theragra chalcogramma*) or hoki (*Macruronus novaezelandiae*) from New Zealand being substituted.

Substitution rather than adulteration is more likely to be practised as the fish processing industry, in the UK at least, is still based largely on processing whole fish into fillets. Opportunities for adulteration of more expensive species such as cod with cheaper species such as whiting or ling can, however, arise in the production of the frozen fish blocks which are subsequently used for the manufacture of fish fingers or fish portions. These products are usually made by filling rectangular trays or moulds with fillets of fish and if fish fillets of different species are of similar appearance, there are opportunities for adulteration. Products made from this raw material, although containing other 'white fish' of equivalent nutritional value, when labelled as 'cod fish fingers' would be contravening the Trade Descriptions Act. Similarly, for crustacean products such as breaded scampi, which are often prepared from comminuted tails of the Norway lobster, opportunities can arise for adulteration with cheaper crustacean species such as imported tropical shrimp or cheaper white fish species.

Although not a product of the UK fish processing industry, surimi, the raw material used for the manufacture of shellfish analogues or substitutes, for example 'crab sticks' (Piggot, 1986; Lanier and Lee, 1992; Mackie 1992, 1994a) has the potential for adulteration and substitution. This material has been used in Japan for many centuries to manufacture the traditional fish gel products of that country namely kamakoko, and since the early 1960s it has been the basis of a huge, largely sea-based, processing industry which has used Alaska pollack (*Theragra chalcogramma*) the white-fleshed species of the North Pacific, as the raw material for these and the fish sausage type of gel products. Alaska pollack is now used by Japan, USA and other countries, for the manufacture of shellfish substitutes of which 'crab sticks' is the most widely known product (Piggot, 1986; Anon, 1987a). Other products are mock lobster claws, crab claws and scallop substitutes, all of which could be substituted for authentic shellfish in a seafood platter. The surimi is essentially the minced flesh of Alaska pollack which has been washed free of all water-soluble components including the water-soluble proteins, and mixed with relatively high concentrations of mono- or disaccharides (*e.g.* sorbitol or sucrose), to protect the proteins from denaturation on frozen storage as a 'surimi block' (Mackie, 1992). When surimi is allowed to thaw and then blended with salt, it forms a sol which, on heating, sets to form a firm gel with textural properties similar to those of shellfish and crustacea.

As pressure on the Alaska pollack fishery has increased, the annual catch has decreased and, as so often happens in all fisheries throughout the world, what was once a cheap raw material is now no longer so. As a result, there have been moves to find alternative species for surimi manufacture and as the product requires specifications for quality and composition to comply with the laws of the countries in which the shellfish substitutes are sold, means of identifying the component species are required. For 'crab sticks' there are often national legal requirements that the fish component is declared and that the content of crab is also given (Anon, 1987a). Clearly, reliable methods for determining the species of fish used and its content in such products are required if standards are to be set and maintained.

Examples of substitution of cheaper for more expensive species of fish are not difficult to find. In the fresh fish trade the cheaper flat-fish species such as megrim and witch can be substituted for plaice or lemon sole and black halibut or Greenland halibut (*Reinhardtius hippoglossoides*) can be passed off as the true halibut *Hippoglossus hippoglossus*. For the last named species substitution can be unintentional as it is often assumed that a fish named Greenland halibut is indeed the true halibut (*Hippoglossus hippoglossus*) from Greenland. In actual fact it is a much smaller species with a black upper skin, often called such in other languages, for example *flétan noir* (French) or *hipogloso negro* (Spanish). The texture of the flesh is much softer than that of halibut and it is generally agreed that this fish is of inferior eating quality to halibut. Another fish which is sometimes found masquerading as a more expensive species is rainbow trout (*Oncorhynchus mykiss*). The species, when farmed in the sea, often develops a silvery appearance and this has encouraged the unscrupulous seller to substitute it for seatrout or salmon trout (*Salmo trutta*). It is not unknown for it to undergo a transformation into 'smoked salmon' (*i.e.* smoked *Salmo salar*). Again because of the similarity in appearance of all smoked salmon, the Pacific salmon (*Oncorhynchus* spp.) can also be found as a substitute for *Salmo salar*, the mainstay of the European fish-farming industries. It is worthwhile pointing out here that care must be taken in understanding the difference in meaning between 'Scottish smoked salmon' and 'smoked Scottish salmon' – the latter product is likely to have been farmed and smoked in Scotland while the former could mean smoked in the Scottish manner but farmed elsewhere.

Moving on to the other problem of differentiating wild from farmed salmon brings one into another controversial subject. Strong views are expressed on the superior eating quality of wild salmon over farmed salmon (and prices in restaurants reflect this perceived difference). However, the claimed superior eating quality is often attributed to its superior physical fitness acquired on its migration to and from Greenland waters and while this may often be the case, examples are also found of wild salmon in much poorer nutritional state than their farmed cousins. In general, farmed salmon is more likely to be of a more consistent eating quality but with a higher fat content than the wild counterpart (Torry Research Station, 1985). It is not possible to differentiate wild and farmed salmon consistently by sensory criteria and as the wild stocks have interbred to varying extents with escaped farmed fish, the authenticity of the wild population may itself be under question.

Canned fish can also present problems of identity when only steaks, fillets or minced flesh are available. The potential of substituting one species for another, already discussed in relation to smoked salmon, also applies to canned salmon as it is usually produced in the form of steaks which have few, if any, species-identifying features. The species used for canning have historically been those of the *Oncorhynchus* spp. (Pacific salmon), of which chum salmon (*O. keta*), coho salmon (*O. kisutch*), sockeye salmon (*O. nerka*) are possibly the best known. Atlantic salmon (*Salmo salar*) is not usually canned as it is seen as a high-value fish but with over-production of this species in fish farming, it might also be used for canning. The Labelling of Food Regulations (1984) require that each species be so labelled on canned and other products. Similarly, canned steaks or minced flesh of tuna and bonito cannot be identified reliably by sensory criteria and, as some of the species are much cheaper than others, for example bonito (*Sarda sarda*) and skipjack tuna (*Katsuwonus pelamis*), reliable means of differentiating the various tuna species are required. Bonito, in particular, has a lower import tariff than the tunas and not surprisingly it is believed that some of the canned tuna entering the EU is more likely to be bonito than tuna. For commercial reasons there is also a need to differentiate the true tuna species within the *Thunnus* genus – yellowfin (*Thunnus albacares*) and albacore (*Thunnus alalunga*), each of which is usually sold by name, when canned, in countries such as Spain and France.

There can also be problems over the name 'sardine', which in Europe is confined to the species *Sardina pilchardus* but in North America can be used for several species of the Clupeidae family. Other related species, for example *Sardinops* and *Sardinella* spp., are also called sardines in other parts of the world and, if imported into Europe, could raise issues of identity which might require unequivocal non-sensory methods to resolve them. Because of the close similarity of these various species, it is often difficult to rely on sensory criteria even if the whole fish (less the head) is canned, as the heat processing destroys the morphological features used to differentiate the raw fish.

Problems of identity can also arise with sardine and anchovy (*Engraulis encrasicolus*) as it is known that sardine fillets can be taken through the salted anchovy process to give a similar product to 'anchovy'. Again, differences in appearance are not usually sufficiently reliable for establishing the identity particularly when only broken fillets of fish are available, as on a pizza.

Other products that require objective methods of species identification are preserved fish eggs or roe. Of particular interest from a substitution point of view is the true caviar, most of which is produced from three species of sturgeon by the countries bordering the Caspian Sea – Russia and Iran. The familiar black eggs are obtained from the species sevruga, oscietra and beluga of which those from beluga are the most expensive (Anon, 1987b). Clearly, with an expensive and exclusive product such as caviar there must be pressure to substitute the cheaper for the more expensive species, particularly when differentiation is based on sensory criteria such as colour of the yolk and pigmentation of the 'skin' of the egg. Substitutes for caviar itself are also produced from other species of fish and are marketed as 'mock caviar', for example lumpfish eggs, but here the difference in egg size is likely to rule out any considerations of substitution for true caviar.

6.2 Requirements for non-sensory methods of fish species identification

For non-sensory methods to be of value in identifying the species of a portion of muscle tissue, they have to be based on the analysis of components of animal cells that are as unique to the species, as are the more obvious morphological features of the whole animal. As these very features are themselves a reflection of genetically determined differences in their constituent structural proteins, it is evident that analytical techniques which can differentiate the proteins of animal flesh are likely to be suitable for species identification. Physicochemical differences in size and net charge of proteins are often revealed as differences in mobility when proteins of animal tissues are subjected to electrophoresis and chromatographic techniques (Mackie, 1980, 1990, 1994b; Durand *et al.*, 1985; Rehbein, 1990; Yman, 1992; Sotelo *et al.*, 1993). Thus, when the sarcoplasmic or water-soluble proteins of a species of fish are separated by electrophoresis, they give profiles that are unique to the species and which, by reference to profiles of authentic samples, enable the species to be established. Similarly, other protein separation methods based on chromatography (Sotelo *et al.*, 1993) are also of potential use in providing species-specific separation profiles.

The specificity of interaction between antibodies and antigens – as utilised in immunoassay systems – can also be considered as it offers an alternative means of differentiating proteins and thereby of differentiating species (Patterson, 1985; Hitchcock, 1988). These methods depend upon the prior production in an animal of an antibody to a particular foreign protein or polysaccharide. When this foreign molecule (antigen) is introduced into the body of the animal, usually by injection, the animal responds by producing antibodies as part of its defence mechanism. Each antibody specifically interacts with the antigen that elicited its production in the first place and effectively removes it from the animal's body. Antibodies so raised against a particular protein or group of proteins of a species of fish can then be used to identify that species in a standard immunoassay analytical procedure. One major disadvantage of this method of species identification is its poor sensitivity to cooked products as the antibodies are normally raised against solutions of native proteins isolated from raw flesh. As the native structure is often destroyed on heating, the specificity of the antigen–antibody reaction is also lost. The extent of denaturation of proteins by heat varies considerably and if certain species have more stable proteins there will be a greater likelihood of success with this method as many of the characteristics of protein are lost on heating. In general, however, the immunoassay system is only suitable for the identification of raw species of fish.

Another disadvantage relates to the large number of species of fish available commercially compared with the three or four main 'meat' animals – beef, pork, sheep and possibly goat, likely to be involved in species substitution. This raises questions of the practicability of producing antibodies to so many likely species and, unlike the protein electrophoretic method, there is no way of comparing a profile of a sample with those of reference species to establish, for example, whether it is a mixture of two or more species or a single species. The assay gives no indication of the presence or absence of antigens of other species as it is specifically an analysis for a pre-selected species for which an antibody has been produced. Possibly, its greater value lies in detecting and determining the extent of adulteration of a product with a known adulterant, for example soya protein or horse meat in a beefburger. Immunoassay test kits for such adulterants are commercially available.

The third method of species identification is based on the analysis of deoxyribonucleic acid (DNA), the basic genetic material of all life. Techniques such as the widely known DNA finger-printing analysis can be applied equally well to the identification of fish species as to individual humans but it requires prior knowledge of the sequences of DNA of a large number of commercially important species of fish. Compared with the very extensive data bases available on humans and higher animals, relatively little information is available on DNA of fish although for certain families, such as tuna and salmon, extensive sequence data bases have been built up because of phylogenetic interest in the closely related species of these families (Foote et al., 1989; Thomas and Beckenbach, 1989; Bartlett and Davidson, 1991, 1992; Block et al., 1993; Carnegie, 1994; Russell and Carnegie, 1994; Unseld et al., 1995). For these studies, parts of the cytochrome b gene of mitochondrial DNA have been used and as a result extensive sequence data are now in the public domain and available for comparisons among the different species of these families. Mitochondrial DNA was selected for these studies in preference to nuclear DNA for a number of reasons, important among them being its high abundance in total cellular nucleic acid preparations and its high mutation rate which makes it more likely that sequence differences will be found between closely related species. Sequence studies have demonstrated that while function and order of the mitochondrial genome are highly conserved in vertebrates, considerable inter- and intraspecific variation of the nucleotide was observed. Additionally, sequence differentiation of mitochondrial DNA becomes more pronounced at the sub-species and species level (Thomas and Beckenbach, 1989). DNA technologies require specialised training in the particular techniques used and as equipment is still relatively expensive, the analytical systems are still far removed from routine use in an analytical laboratory. None the less, they are likely to be of value for the more difficult problems in fish species identification, such as the identification of heat-processed (canned) tuna, salmon or sardine species. DNA is less affected by heat than are proteins and the information provided is potentially greater as the analysis is of the basic genetic material itself rather than an expression of it – the proteins.

6.3 Principles of electrophoresis and isoelectric focusing

Electrophoresis as an analytical procedure was first introduced by the Swedish scientist, Tiselius, in 1930 (Kleparnik and Bocek, 1991). The detailed theory is very complicated but for the purposes of this review it is sufficient to say that it refers to the separation of charged molecules such as proteins or DNA in an aqueous medium under the influence of an electric field applied between positive and negative electrodes under constant temperature. The fundamental requirement for separation by electrophoresis is that the molecules under study must have either a net positive or net negative charge under the pH of the buffer selected for the analysis. Charged species with a high net charge will tend to move more quickly than those with a lower net charge and when molecules have the same net charge, the smaller ones will tend to move more quickly than the larger ones, that is, movement is dependent on both charge and size of the molecules being separated. The net result is that after a period of time, separation of the components of the mixture takes place depending upon differences in their respective mobilities.

If a protein is taken as an example of a molecule with both positive and negative charges, a zwitterion, it is evident that the net charge will be dependent on the pH of the solution, as it will be determined by the degree of ionisation of its constituent amino and carboxyl groups (Figure 6.1). At low pH values the free carboxyl group will be unionised and the amino groups will be protonated giving a net positive charge, the actual value of which will be dependent on the relative proportion of the two chargeable groups in that particular protein. Conversely, at high pH values the free carboxylic acid groups will be ionised but the amino groups will not be protonated, giving as

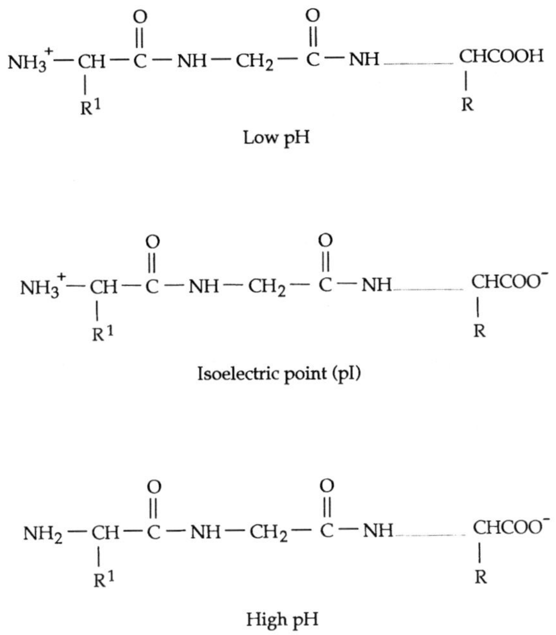

METHODS OF IDENTIFYING SPECIES 167

$$NH_3^+ - CH(R^1) - C(=O) - NH - CH_2 - C(=O) - NH \cdots CH(R)COOH$$

Low pH

$$NH_3^+ - CH(R^1) - C(=O) - NH - CH_2 - C(=O) - NH \cdots CH(R)COO^-$$

Isoelectric point (pI)

$$NH_2 - CH(R^1) - C(=O) - NH - CH_2 - C(=O) - NH \cdots CH(R)COO^-$$

High pH

Figure 6.1 Charges on zwitterion at different pH values.

a result a net negative charge overall. At intermediate values of pH some of the free side-chain groups will be protonated while others will be neutral depending upon their individual pK values. There will be one particular pH at which a protein molecule will have a net zero charge, that is, when there is a balance of positive and negative charges. This value is called the isoelectric point of the protein and it is designated pI. It is characteristic of each individual protein – at lower pH values than the pI, the protein will have a net positive charge, whereas at pH values higher than the pI the protein will have a net negative charge.

6.3.1 *Electrophoretic systems*

When first developed, electrophoresis was carried out in free solution in a moving boundary system (Andrews, 1986; Kleparnik and Bocek, 1991) and while giving very accurate values for fundamental mobility measurements of charged particles it was not, however, useful as a routine analytical tool. It has been largely superseded by systems in which the liquid medium is supported in a gel, to reduce diffusion. Usually polyacrylamide or agar gels are used, depending upon the nature of the molecules being separated. Earlier systems employed starch gel as the supporting medium but this has been largely replaced by polyacrylamide and agar gels because of their superior resolving power and more robust handling characteristics.

6.3.1.1 *Polyacrylamides gels.* Polyacrylamide gels are co-polymers formed from acrylamide monomer $CH_2\!=\!CH\ CONH_2$ and dimer, N,N'-methylene-bisacrylamide, the cross-linking agent. The reaction requires an initiator and a catalyst; ammonium persulphate is commonly used as an initiator and tetramethylethylenediamine (TEMED) as the catalyst. The reaction takes place via vinyl polymerisation and gives a randomly coiled gel structure. Since the polymerisation reaction is very sensitive to certain chemical impurities that may be present in the components of the mixture, it is important to use highly purified reagents.

Gels of widely varying pore size and hence sieving properties are obtained simply by altering the total concentration of the monomer and cross-linking agent. These compounds are usually present in a ratio 95:5 which in a 7.5% gel (total of monomer and dimer) gives a pore size of about 5 nm whereas in a 30% gel it is 2 nm. The pore size can be even more closely tailored to the dimensions of the proteins being separated by altering the ratio of monomer to dimer.

It is the flexibility in design of the characteristics of this support medium which makes polyacrylamide gel electrophoresis such a powerful tool in the analysis of proteins. These gels, unlike those used for chromatography, are not granular but are formed into continuous columns or slabs depending upon the mould used. The gels are, in effect, a network of polymer molecules surrounded by and penetrated by buffer, with the spaces within this framework being the 'pores'. When an external voltage is applied between the electrodes, protein molecules, together with the anions and cations of the buffer solution, migrate towards the appropriate electrode and as they move they experience a frictional resistance to their movement that is related to the relative sizes of the gel pores and the radii of the charged species; that is, a sieving effect, in addition to the net charge and mass effects also influences the separation of the proteins.

Once adequate separation of the proteins has taken place, the gel is removed from the system. It is then necessary to 'fix' the proteins by precipitating them with either a solution of perchloric acid or trichloroacetic acid (Mackie, 1980, 1990; Laird *et al.*, 1982). These 'fixed' proteins must then be visualised by staining the gel with a suitable dye (*e.g.* Coomassie Blue in a solution of ethanol, acetic acid and water typically in a ratio of 25:8:67). Subsequent washing of the stained gel with the same solvent removes unbound dye giving blue bands of proteins (*i.e.* protein separation profiles against a water-clear background). Fixing and staining procedures are required for all electrophoretic systems which utilise gel supports, as proteins which meet the requirements of electrophoresis must be in solution and for satisfactory fixing and visualisation, they have to be first of all removed from solution by a protein precipitant and then made visible by a suitable staining procedure (Rehbein *et al.*, 1995). Subsequent examination of the developed gels can be done visually or by using a densitometer or image analyser to measure positions and concentrations of the proteins in the profiles or separation patterns.

6.3.1.2 *Agarose gels.* Agarose is a linear polymer of D-galactose and 3,6-anhydro-D-galactose residues which is obtained from seaweed (*Agar agar*) (Serwer, 1983). It is readily dissolved in boiling water and on cooling to about 38°C forms a gel through the formation of hydrogen bonds. As the pores are relatively large compared with those those of acrylamide, agarose gels are more suitable as a support medium for the separation of larger molecules such as DNA and RNA, which can range in size from 10 000–60 000 base pairs (Sambrook *et al.*, 1989). Generally, flat-bed systems are used, as the support provided from below allows lower concentrations of agarose to be used. All polynucleotides such as DNA and RNA contain one phosphate group per base residue and as phosphate groups form a negative charge under the alkaline conditions required to dissolve DNA, they move towards the anode according to their molecular mass. In contrast to proteins which are ampholytes and can be in solution in both acidic and alkaline media, DNA is in effect a polyanion, soluble only under alkaline conditions. Detection of the separated DNA polymers is usually achieved by incorporating a fluorescent dye, ethidium bromide, into the gel. At the end of the electrophoretic run the separated DNA molecules are then identified by photographing under UV light. No separate staining or destaining steps are required. Once the photograph is developed, comparison with reference markers or other DNA residues can then be made as with protein separation profiles obtained by dyeing or silver staining.

6.3.2 *Separation systems*

6.3.2.1 *Zone electrophoresis.* Zone electrophoresis, in contrast to the original moving boundary electrophoresis of Tiselius (Kleparnick and Bocek, 1991), applies to a system in which the components of a mixture separate completely from one another to form discrete zones. Support media such as polyacrylamide or agarose are used to stabilise the system and by reducing diffusion, discrete separation zones of proteins are maintained throughout the electrophoretic run.

Gel rods. For the gel rod system, polyacrylamide gels are prepared in glass or plastic cylindrical tubes of internal diameter about 5 mm and length 70–100 mm. Typically six tubes are mounted through grommets in an upper electrolyte compartment with their bottom ends dipping into the buffer solution of the lower electrolyte compartment to give continuity of ionic strength and pH throughout the system. The same buffer solution (typically

TRIS-glycine pH 8.0) is used for both upper and lower electrolyte compartments. The sample is then applied by pipette to the top of the gel in a sucrose solution to minimise diffusion into the upper electrolyte compartment and reference samples are applied to the other gels. After the electrophoretic run has been completed, the gel is removed from the tubes by irrigation with water from a syringe, fixed with trichloroacetic acid and perchloric acid and stained with a suitable dye (*e.g.* Coomassie Blue). Comparison of the sample profile with those of reference samples is then made to identify the species (Mackie, 1969).

Slab gels. Polyacrylamide slab gels have now largely replaced the rods as supports for electrophoresis (Mackie, 1980). They have distinct advantages over the former system in being able to analyse a number of samples at the same time on the same gel and therefore under the same conditions of electrophoresis. As the gel is mounted on a cooling plate, the removal of heat produced during the electrophoresis is more uniform, giving superior resolving power and greater flexibility. Gels of a wide range of thicknesses can be prepared, mounted on plastic backing or glass. Gel dimensions can be readily altered to suit the requirements and samples can be applied to any position on the gel. For some analyses, for example, the quality of resolution is dependent on whether the sample is applied at the anodic or cathodic end.

The preparation of the gel is carried out in a similar manner to that for the rod system. After the glass mould is filled with the polymerising solution a slot former, or comb, is placed on top just before the gel polymerises. After polymerisation is complete the comb is removed, leaving sample wells separated from one another by continuous strips of gel. The samples are loaded into the wells as for gel rods in buffer containing glycerol or sucrose to increase their density and reduce diffusion into the upper electrolyte compartment. After the electrophorectic run has been completed the gels are removed, fixed, stained and destained as for gel rods to produce the developed gel (Hames and Rickwood, 1990). Although more expensive, slab gels are available commercially and offer substantial savings in operator time. It is appropriate to point out that the acrylamide monomer is a nerve toxin and that precautions should be taken to avoid uptake via skin absorption or by inhalation of acrylamide dust. Once polymerisation has taken place the resulting gels are relatively non-toxic and can be handled safely.

6.3.2.2 *SDS polyacrylamide gel electrophoresis.* When protein molecules are dissolved in solutions of the anionic detergent, sodium dodecyl sulphate (SDS), their individual charges are lost and a net negative charge is formed as a result of the anion dodecyl sulphate complexing with the protein (Weber and Osborn, 1969). The uptake of detergent is the same per unit mass (1.4 g SDS per g of protein) for all proteins and, in consequence, the mobility on electrophoresis is proportional to the molecular mass.

Separation on the basis of molecular mass of proteins provides an alternative method of analysis of flesh proteins and thereby of fish species identification. As will be discussed in Section 6.5.2, this method of electrophoresis is used to identify the species of samples of cooked fish as the proteins, although denatured, are extractable in 2% SDS solution to give species-specific profiles on electrophoresis in SDS solution.

Depending upon the size of proteins to be separated, a concentration of acrylamide is selected to optimise the sieving effect (Rehbein and Karl, 1985; Scobbie and Mackie, 1988). In addition, a stacking gel of lower acrylamide concentration is used to allow rapid movement of the sample to the top of the main separating gel, and thereby form a narrow zone of proteins prior to entry into the main separating gel. Other refinements such as concentration gradients of acrylamide can be used to provide increased resolution of closely separating proteins. Indeed, sophisticated gradient-forming devices now available make it possible to devise gradients of any shape to optimise protein separations (Dunn, 1987).

6.3.2.3 *Isoelectric focusing.* In contrast to SDS electrophoresis, isoelectric focusing (IEF) is a method which separates molecules on the basis of charge. In this system, a stable pH gradient is created between the anode and cathode by the focusing, or movement to positions of electrical neutrality, of a mixture of synthetic ampholytes, usually polyaminopolycarboxylic acids, selected to provide a range of pH values at which they are electrically neutral (*i.e.* a pH gradient). Once the pH gradient is formed, it is then in equilibrium since any displacement of ampholyte will result in the formation of a net charge on the molecule, causing it to migrate back to its isoelectric point under the influence of the applied electric field. Electrophoresis of the protein takes place in the pH gradient until the respective pI is attained, at which point the protein remains focused. Narrow range pH gradients, for example pH 4–6 or pH 5–8, can be used to amplify a particular part of a separation profile while wide range gradients (pH 3.5–9.5) are often used for more general separations. The resolving power of IEF is extremely high, as differences of as little as 0.01 between pIs of proteins can be separated. Once a steady state has been attained, the gel is removed, fixed, stained and washed as for other polyacrylamide gels. The synthetic ampholytes themselves are washed out during the initial fixing step and do not interfere with the uptake of dye by the proteins. By reference to standards of known pI, run at the same time, the pIs of the proteins of the species being examined can be established (Rehbein *et al.*, 1995).

6.3.2.4 *Capillary electrophoresis.* Capillary electrophoresis (CE) is a relatively recent development in electrophoresis (Ewing *et al.*, 1989; Novotny *et al.*, 1990; Kleparnik and Bocek, 1991). It is based on free zone electrophoresis in buffer-filled, fused silica capillary tubes of typical dimensions, 50 µm i.d. and 100 cm length which are used as the separation chambers.

Potentials of as high as 30 kV are applied and, under those conditions, rapid migration of proteins and other charged molecules takes place with minimal zone spreading. Because of the small sample volumes of solutions used, however, it is necessary to have highly sensitive detection systems. UV and fluorescence detectors have been developed for CE systems but as the signal to noise ratio is low for UV detection it is inappropriate for the small volumes of proteins applied. None the less, CE offers the prospect of a fully automated instrumental system with high separation efficiency and as the detection system is on line, continuous monitoring, both qualitative and quantitative, of the separated proteins can be obtained. Although promising results have been published within the last 2 years, it is still not at a stage where it can be considered as an alternative analytical procedure for fish species identification.

6.4 Fish flesh proteins

6.4.1 *Structure of muscle*

Before discussing the analysis of the proteins of fish flesh, from the point of view of fish species identification, some understanding of their nature and of the effects on them of treatments such as *post mortem* storage in ice, frozen storage and heat processing is required. Fish flesh itself is contractile muscle and, like that of other animals, it comprises specialised cells or fibres containing thick and thin filaments which interact with one another in response to nerve impulses and produce the contraction and relaxation required for movement of the animal. In all animals these muscle cells or fibres are arranged parallel to one another and are held together by connective tissue of collagen or elastin. In mammals and birds this membrane connects with the tendons, also of connective tissue, which anchor the muscle cells to the skeleton of the animal. In fish, however, a very different arrangement of muscle cells is found: instead of being connected to tendons the cells are bound together one cell deep to form segments or mytomes of muscle. These bundles of muscle cells are shaped like the letter 'W' and are oriented across the mid-plane of the fish with the central parts directed towards the head of the animal. The ends of the cells are attached to sheets of connective tissue called 'myocommata' which separate one block of mytomes from another. As mycommata are largely composed of collagen they break down on heating to release the mytomes, still in their original shape, more commonly recognised as the characteristic 'flakes' of cooked fish. Compared with mammalian and avian muscle cells, those of fish are short, generally less than 20 mm, even in large fish. In mammals and birds they can run the whole length of the muscle and may be as long as 300 mm (Love, 1970; Hultin, 1984, 1992).

Fish muscle consists of two types of cells, white and red, with the latter always present as the minor component. White muscle, as in the white fish species cod, *Gadus morhua*, has an anaerobic metabolism and being dependent upon glycogen for energy, is well supplied with glycolytic enzymes which permit intense contraction during periods of flight. Red muscle, on the other hand, has an aerobic metabolism and is rich in haemoglobin, myoglobin and mitochondria. In fatty species, particularly the surface-swimming pelagic fish of which tuna, mackerel and herring are typical examples, red muscle is well-developed to provide energy for sustained high swimming speeds. It is the white muscle, however, that is of interest as far as fish species identification is concerned and for the purposes of this review, it will be assumed to be what is implied by the term 'fish flesh'. As there are differences in the separation profiles of the two types of muscle, it is important that white muscle is free of red muscle and blood if reproducible results are to be obtained.

6.4.2 *Structure of myofibrils*

Within the muscle cells are the myofibrils, the long thin contractile elements which account for most of the volume and which have the characteristic striated pattern of alternate light and dark bands when viewed through a microscope. This is due to the repetitive arrangement of proteins in the myofibril. In polarised light, one set of bands, the A bands, is anisotropic or berefringent while the lighter bands, the I bands, are isotropic. At the centre of each I band is a dark line called the Z line or Z disc and at the centre of the A band is a light area called the H zone, down the centre of which is a darker M Line. The distance from one Z disc to another is called the sarcomere, the repeating unit of the myofibril (Figure 6.2).

Surrounding the myofibrils is the sarcoplasmic reticulum which, through tubules and vesicles, plays an essential role in the mechanism of muscle contraction and relaxation, releasing and recapturing Ca^{2+} in response to nerve signals. These myofibrillar structures run for the full length of the cell and are immersed in the enzymic pool, the sarcoplasm, within which are the various discrete cell compounds such as nuclei, mitochondria and glycogen granules (Squire, 1981; Bechtel, 1986).

6.4.3 *Muscle proteins*

As with all contractile muscle, fish flesh contains three main groups of proteins: the sarcoplasmic or water-soluble proteins, the myofibrillar proteins and the connective tissue proteins which are readily separated by fractional extraction techniques employing water and strong salt solutions (Goll *et al.*, 1977; Mackie, 1993).

6.4.3.1 *Sarcoplasmic proteins.*
The sarcoplasmic or water-soluble proteins make up 20–35% of the total protein content of muscle, depending upon the

174   FISH PROCESSING TECHNOLOGY

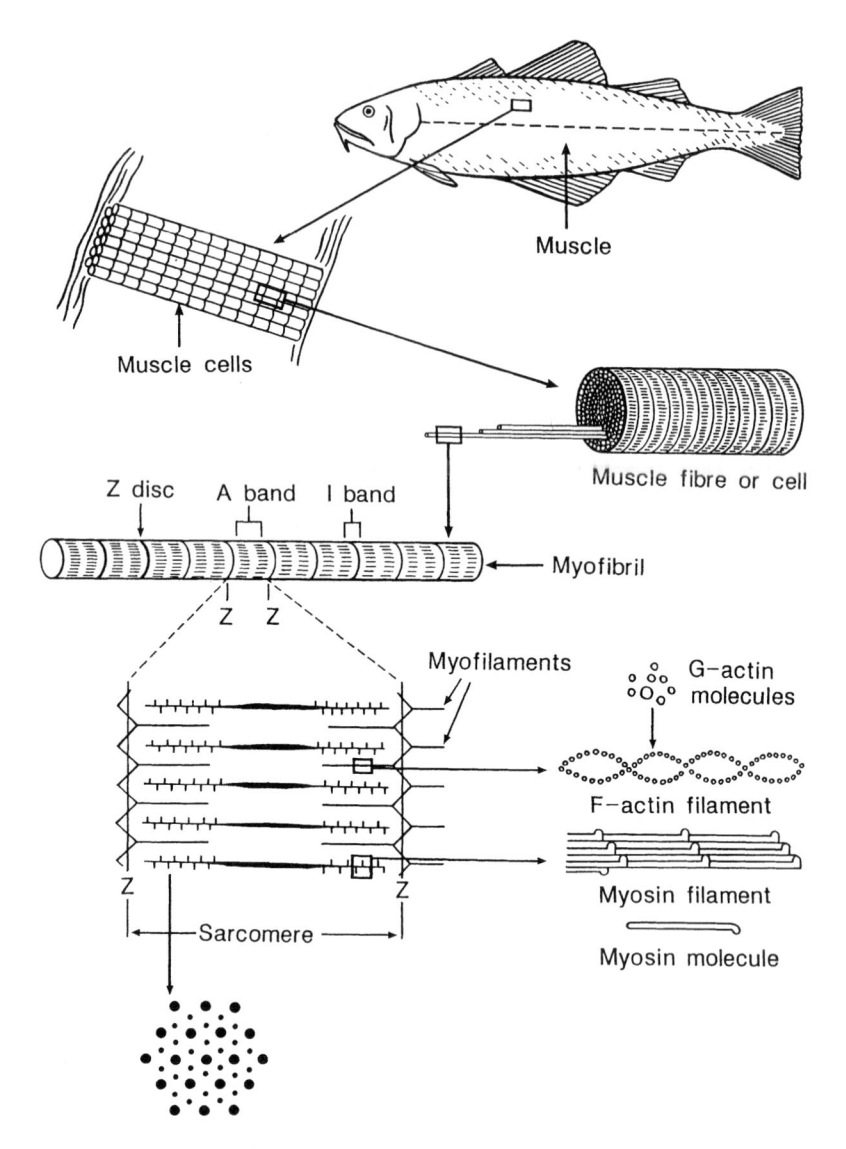

Figure 6.2 Muscle cells and components.

species (Mackie, 1993) (Table 6.1). They are present within the muscle cells usually at concentrations as high as 20% and are largely enzymes responsible for metabolism within the cell. The oxygen storage and transporting protein, myoglobin, is a major component of this fraction of red muscle but in white muscle it is present only in trace amounts. As the sarcoplasmic proteins are of low molecular weight (40–60 kDa), readily extractable with water and, of

Table 6.1 Main groups of proteins in the flesh of fish and mammals

Protein group	% Composition	
	Fish	Mammals
Sarcoplasmic	20–35	30–35
Myofibrillar	65–75	52–56
Connective tissue	3–10	10–15

course, electrically charged, they are ideally suited for separation by electrophoresis. Also, it would appear that genetic differences between species are more developed in this than in the other groups of proteins, responsible as they are for widely divergent enzymic transformations in the muscle cell. The unique nature of the sarcoplasmic proteins of each species is such that the separation patterns of profiles obtained on electrophoresis or IEF can be used for the unequivocal identification of the species.

6.4.3.2 *Myofibrillar proteins.* The A bands of the myofibrils are composed of thick filaments and the I bands of thin filaments. Each thick filament is formed from an ordered arrangement of about 400 myosin molecules. Myosin has a molecular weight of $\simeq 500$ kDa and is made up of two identical polypeptide chains. At the head end of the molecule are four light chains of molecular weights ranging from 16–25 kDa depending upon the species (Mackie, 1993).

Actin is the second most abundant protein, representing 15–30% of the total myofibrillar protein of muscle, and is the major constituent of the thin filaments. As the G or monomeric form, it has a molecular weight of 42 kDa but it exists in muscle as the polymeric form, F actin (Figure 6.2). Associated with F actin are the regulatory proteins, tropomyosin and the troponins TnT, TnI and TnC, which together account for 10% of the myofibrillar proteins. In addition, there are minor proteins such as C protein and α-actinin, also regulatory proteins concerned with regulation of the filamentous structure of myofibrils (Hultin, 1984; Asghar *et al.*, 1985; Bechtel, 1986).

Other proteins also present in the cell are those that comprise the 'gap filaments'. They maintain the structural framework within which the contractile mechanism operates (Mackie, 1993), otherwise called 'scaffold' or 'backbone' proteins. These gap filaments run longitudinally along the myofibril and appear in the 'gap' between the A and I bands of stretched muscle. Proteins of these filaments are 'connectin' or 'titin' and other related proteins, desmin and nebulin, which together comprise about 10% of the total contractile proteins of the cell (Robson *et al.*, 1984; Asghar *et al.*, 1985).

With the exception of the gap proteins, the myofibrillar proteins are extractable in strong salt solution while the former are only extracted in SDS solution often in the presence of high concentrations of urea. Like the sarcoplasmic proteins, the myofibrillar proteins are readily denatured by heat.

6.4.3.3 *Connective tissue proteins.* The connective tissue proteins of fish skeletal muscle consist mostly of collagen which exists in several polymorphic forms, the most common being Type I and Type III collagens. The collagen molecule is a long cylindrical protein comprising three polypeptide chains wound around each other in a supra helical coil and with inter- and intra-molecular cross-linking generally increasing with the age of the animal. In its native form, collagen is insoluble in low and high ionic strength salt solutions but on heating it breaks down to its denatured form, gelatin, which is soluble in water. Compared with those of mammals, fish connective tissue proteins are present in relatively low concentrations (Table 6.1) reflecting the different structural arrangement of contractible muscle, and in consequence they do not contribute much to the textural properties of fish muscle. As a general rule, fish collagens are less stable than those of mammals and they are denatured at lower temperatures (Montero and Mackie, 1992). These proteins in their denatured form, together with the myofibrillar proteins, are likely to contribute to the differences in the electrophoretic profiles of total muscle extracts such as those obtained on SDS electrophoresis of heat-denatured flesh.

6.5 Experimental procedures for electrophoretic methods

For as long as the sarcoplasmic proteins of flesh remain in the native state, they are readily extractable with water or dilute salt solutions and are then in a suitable form for separation by IEF or electrophoresis. However, this condition does not apply when the fish is cooked, as the sarcoplasmic proteins together with the structural proteins are denatured and precipitated by heating. They are no longer in their native forms and they cease to be extractable in water. For these and even more severely heat-denatured fish products, such as heat-sterilised canned fish, alternative methods of analysing for species-specific proteins, now denatured, have to be considered.

It is convenient to consider fish species identification under the following three categories of processing:

- raw fish flesh
- cooked but not autoclaved fish flesh
- heat-sterilised, autoclaved fish flesh.

6.5.1 *Raw fish flesh*

The species of raw fish is normally readily established by comparing the IEF profile of an aqueous extract with those of aqueous extracts of authentic species. Although the separation profiles of individual fish, even of the same species, show some degree of variation, they none the less have a sufficient number of major zones in common to enable a profile to be recognised unequivocally.

The procedure used varies to some extent between laboratories but in general a portion of fish muscle is extracted with water in the ratio 1:2 at low temperature (less than 5°C). The mixture is homogenised from 30 s to 1 min and the homogenate centrifuged at low centrifugal force (*e.g.* 8100 × *g*) for a few minutes. The supernatant solution can be stored in a refrigerator for no more than 2 days until it is used for IEF. For IEF, a large gel system (*e.g.* 245 × 110 × 1 mm) or a small gel slab system (*e.g.* Phast (43 × 40 × 035 mm)) is used. A wide range (*e.g.* 3.5–9.5) or narrow range pH gradient (*e.g.* pH 5.8, pH 4.6) is used depending upon the isoelectric points of the main proteins to be separated. This is illustrated in Table 6.2 in which are given the conditions used in a recent collaborative experiment among eight European laboratories (Rehbein *et al.*, 1995). In this study, frozen samples of reference materials and unknown samples were sent to seven laboratories by the coordinating laboratory as a preliminary exercise aimed at identifying those steps in the methodology, either procedures for preparation of samples, conditions of IEF or subsequent staining procedures, which were critical for the reproducibility of the results. Each laboratory used its own method of IEF (Table 6.2) to attempt identification of 10 unknown samples by reference to profiles obtained for authentic reference materials supplied at the same time and run on the same gel. The results obtained are given in Table 6.3 and show that four of the seven laboratories, including the one which used the Phast system, identified all the samples correctly. Two laboratories failed to identify one species and one laboratory failed to identify three species. Failure to identify all of the species was attributed to inexperience and lack of attention to critical details which were fully recognised by those other laboratories identifying the samples correctly. It was not an indication of lack of robustness in the IEF method of fish species identification.

Additional studies showed that water was satisfactory as the extractant and that the position of application of the sample solution to the IEF gel had a great influence on the band pattern. Other variables between laboratories such as type of ampholyte, protein content of sample, volt–hour product and staining procedure were found to affect the protein profiles obtained but not necessarily the quality of the results (Rehbein *et al.*, 1995). These results support the widely accepted view that, to have full confidence in the identification of a species, it is necessary to have samples of authentic species run on the same gel. It is not sufficient to rely on profiles obtained under apparently the same conditions of IEF. Continuing studies among the laboratories are aimed at optimising the procedure so that data on pIs of the proteins of a species can be measured from marker proteins of known pI and used as the basis of identification of species from a computer data base of pIs and the

Table 6.2 Comparison of isoelectric focusing (IEF) procedures (Rehbein et al., 1995)

Laboratory	Gel type	Run conditions	Sample	Staining procedure
1	Phast Gel IEF 3–9 (43 mm × 50 mm × 0.35 mm); Pharmalyte	Maximal voltage 2000 V, power × time 245 AVh, with prefocusing	Volume 1 µl, protein content adjusted to 10 mg ml^{-1}	The proteins were fixed by trichloroacetic acid (TCA) and stained by 'Phast Blue R'; the gels were destained by methanol/acetic acid and water. The protein bands were evaluated densitometrically by the 'Phast Image System'
2	(I) Servalyte^R Precote^R 3–10 (125 mm × 125 mm × 0.3 mm) (II) PAG Plate (Pharmacia-LKB), Ampholine 3.5–9.5, thickness 1 mm	(I) Maximal voltage 2000 V, voltage × time 3000 Vh, no prefocusing (II) Maximal voltage 1300 V, no prefocusing	(I) Volume 7.5 µl (II) Volume 10 µl, protein content adjusted to 10 mg ml^{-1}, the samples were applied with a SMI micropettor syringe	(I) Fixing solution 200 g TCA in 1000 ml ethanol 95% staining solution: 100 mg Coomassie R-250 in 250 ml destaining solution. Destaining solution: ethanol 95%/acetic acid/water, 4/5/1, v/v/v (II) Fixing and staining were carried out simultaneously in the following solution: Coomassie R-250 0.25 g methanol 75 ml water 155 ml sulphosalicyclic acid (SSA) 8 g TCA 25 g Destaining solution: ethanol 95%/acetic acid/water, 375/120/1000, v/v/v
3	PAG plate (Pharmacia-LKB), Ampholine 3.5–9.5, (245 mm × 110 mm × 1 mm)	Setting: 1500 V, 50 mA, 30 W, 1.5 h, no prefocusing	Volume 7.5 µl, application by means of pieces of filter paper	Fixing solution: 57.5 g of TCA + 17.25 g of SSA + 500 ml water. Staining solution: 0.46 g Coomassie Blue R in 400 ml destaining solution. Destaining solution: 500 ml ethanol + 160 ml acetic acid to 2 l of water
4	Pharmalyte 3–10 (7%) (245 mm × 110 mm × 0.3 mm)	Settings: 2000 V, 15 mA, 8 W; no prefocusing, voltage × time 3000 Vh	Application by means of pieces of filter paper	Fixing solution: 10% TCA, 5% SSA. Staining solution: 0.04% Coomassie Blue R-250 in destaining solution. Destaining solution: ethanol/acetic acid/water, 4/1/6, v/v/v
5	Ampholine 3.5–9.5, thickness 0.5 mm	Settings: 1500 V, 50 mA, 25 W. 1 h; no prefocusing	Volume 7.5 µl, application by means of pieces of filter paper	Fixing solution: 57.5 g of TCA + 17.25 g of SSA + 500 ml water. Staining solution: 0.46 g Coomassie Blue R-250 in 400 ml destaining solution. Destaining solution: 500 ml ethanol + 160 ml acetic acid + water to 2 l

6	PAG plate (Pharmacia-LKB), Ampholine 3.5–9.5, dimensions (245mm × 100mm × 1 mm)	Settings: 15 W for the first 30 min and 20 W for the rest of the run, no prefocusing	Volume 7.5 µl, application by means of pieces of filter paper	Fixing solution: 11.5% TCA, 3.5% SSA. Staining solution: 0.115% Coomassie Blue R-250 in destaining solution. Destaining solution: 25% ethanol, 8% acetic acid
7	Ampholine 3.5–10, thickness of the gel 0.5 mm	Prefocusing (if used): 500 V, 20 W, 50 mA; after application of samples: 500 V, 20 W, 50 mA; after removing of strips: 1200 V, 20 W, 50 mA, 4 h	Volumes 7.5, 10 or 18 µl; application by means of pieces of filter paper	Fixing solutions: (I) ethanol/acetic acid, 5/1, v/v (II) 20% TCA Staining solution: 290 mg of Coomassie Blue R-250 or G250 (Serva Blau R or G) in 250 ml of destaining solution. Destaining solution: ethanol/acetic acid, 25/8/67, v/v
8	Servalyte^R Precote^R 3–10, (245 mm × 125 mm × 0.3 mm)	Prefocusing with settings: 240 V, 30 mA, 8 W, 30 min; focusing, after application of samples, with settings: 2000 V, 30 mA, 8 W, voltage × time: 5000 or 7000 Vh	Volume 7.5 µl, applicator strip (slots 7 mm × 1 mm; silicone rubber)	Fixing solution: 20% TCA. Staining solution: 0.1 of SERVA Violet 49 in destaining solution. Destaining solution: methanol/acetic acid/water, 25/10/65, v/v/v

180   FISH PROCESSING TECHNOLOGY

Table 6.3 Summary of the results of the collaborative study on fish species identification using reference material and different isoelectric focusing (IEF)

Laboratory	Code of samples[a]									
	A	B	C	D	F	G	H	I	J	K
1	+	+	+	+	*	+	+	+	+	+
2	+	+	+	+	*	+	+	+	+	+
3	+	+	+	+	*	+	+	+	+	=
4	+	+	−	+	*	+	=	+	−	+
5	+	+	+	+	*	+	+	+	+	+
6	+	+	+	+	*	+	+	+	+	+
7	=	+	+	+	*	+	+	+	+	+

+, Fish species was correctly identified; *, fish species was designated as not included in the references; −, fish species was not identified, although it was included in the references; =, fish species was not correctly identified.
[a] A, halibut; B, North Atlantic hake; C, cod; D, herring; F, Alaska pollack; G and H, haddock; I, redfish; J, saithe; K, ling.

corresponding concentration of proteins specific to that species of fish. IEF profiles of raw fish of the same species have been shown to be remarkably constant, although minor differences attributable to polymorphism are sometimes found. Extensive studies on the effects of geographical location of fishing ground have shown that there is little variation in the profiles of a species. These results give confidence in the use of the method for fish species identification. Cod (*Gadus morhua*), for example, whether obtained off Aberdeen, the Baltic Sea or Northern Norway, has the typical cod profile which is different from that of Pacific cod (*Gadus macrocephalus*) (Mackie and Ritchie, 1981).

An example of a typical isoelectric focused profile obtained on the large gel system is given in Figure 6.3. In this case a narrow range pH gradient (pH 4.0–7.0) has been used as it gives greater resolution of the anodic proteins which are useful in differentiating the gadoid species. The gel shows the profiles of a wide range of commercially important species, each of which has been shown to be highly reproducible and unique to the species. The closer the genetic relationship, such as that which exists among cod (*Gadus morhua*), haddock (*Melanogrammus aeglefinus*), whiting (*Merlangius merlangus*) and pollack (*Pollachius pollachius*), the greater the similarity in the profiles, but even among these species the profiles are readily differentiated, enabling unequivocal identification of the species to be made. Similarly, scampi (*Nephrops norvegus*) can be differentiated from its relative *Metanephrops andamanicus* from the Pacific. Cold-smoked fish such as smoked salmon can be identified by this procedure also, as the sarcoplasmic proteins are still largely in their native state. However, because of the presence of salt, the aqueous extract must first be dialysed against water (at low temperatures) prior to IEF (Sotelo *et al.*, 1992).

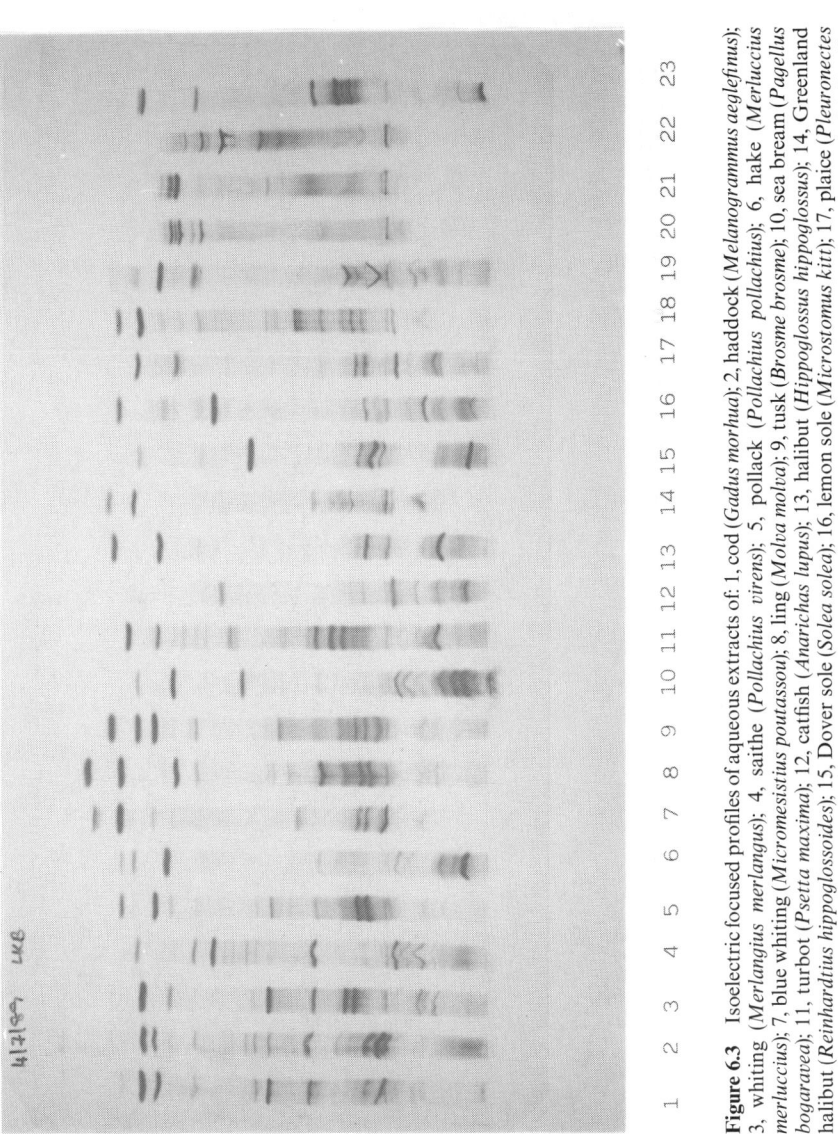

Figure 6.3 Isoelectric focused profiles of aqueous extracts of: 1, cod (*Gadus morhua*); 2, haddock (*Melanogrammus aeglefinus*); 3, whiting (*Merlangius merlangus*); 4, saithe (*Pollachius virens*); 5, pollack (*Pollachius pollachius*); 6, hake (*Merluccius merluccius*); 7, blue whiting (*Micromesistius poutassou*); 8, ling (*Molva molva*); 9, tusk (*Brosme brosme*); 10, sea bream (*Pagellus bogaravea*); 11, turbot (*Psetta maxima*); 12, catfish (*Anarichas lupus*); 13, halibut (*Hippoglossus hippoglossus*); 14, Greenland halibut (*Reinhardtius hippoglossoides*); 15, Dover sole (*Solea solea*); 16, lemon sole (*Microstomus kitt*); 17, plaice (*Pleuronectes platessa*); 18, witch (*Glyptocephalus cynoglossus*); 19, dab (*Limanda limanda*); 20, scampi (*Nephrops norvegicus*); 21, Pacific scampi (*Metanephrops andamanicus*); 22, shrimp (*Penaeus monodon*); 23, monkfish (*Lophius piscatorius*). (pH range 4–7).

182 FISH PROCESSING TECHNOLOGY

When adulteration rather than substitution takes place, the electrophoretic system can also be applied to the problem, provided that the profiles of the species of interest have species-specific zones which do not coincide with those of the other species in the mixture. Adulteration of scampi with white fish can be detected and to some extent quantified by either visual or densitometric measurements. In Figure 6.4a, for example, the IEF profiles of the white fish species, cod, haddock and whiting, have species-characteristic zones well separated from those of scampi, thereby enabling the presence of adulterants to be established. The reputed 'scampi' samples 5, 7 and 9 clearly contain white fish species (possibly haddock) as well as scampi.

In contrast to the IEF profiles of fish, those of crustacea have few protein zones, with the majority of them being found over the same narrow pH range. This is shown in Figure 6.4b where the IEF profiles of scampi, Pacific scampi

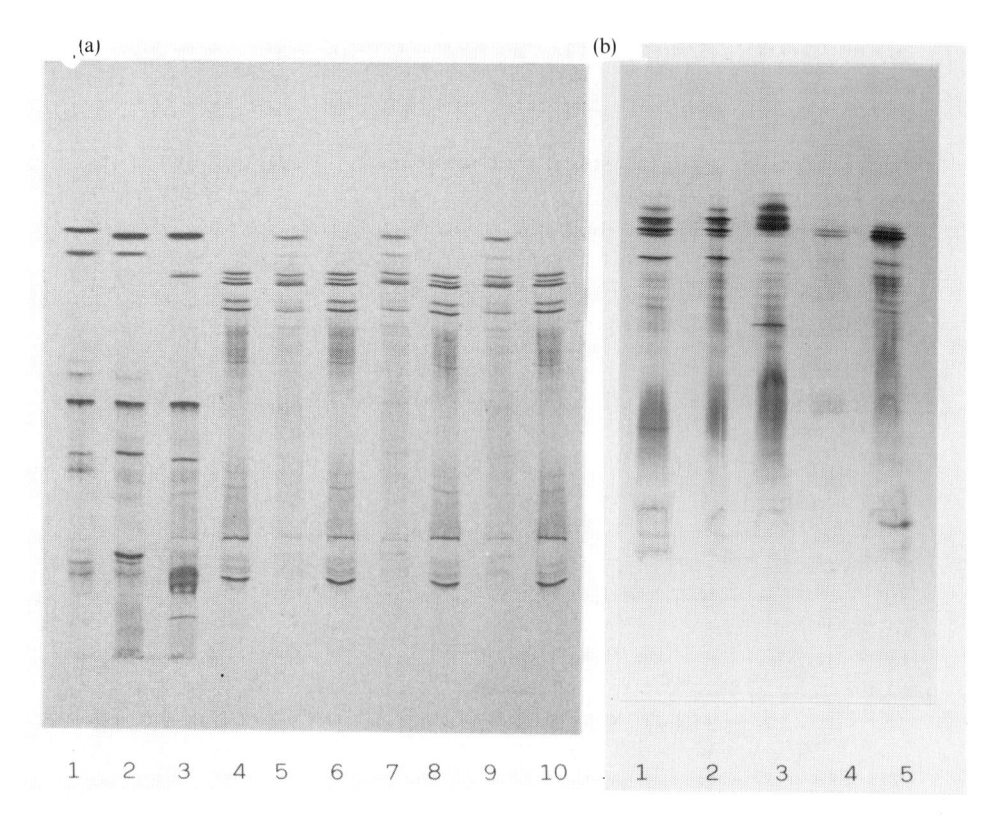

Figure 6.4 (a) Isoelectric focused profiles of aqueous extracts of: 1, cod (*Gadus morhua*); 2, haddock (*Melanogrammus aeglefinus*); 3, whiting (*Merlangius merlangus*); 4, scampi (*Nephrops norvegicus*); 5, unknown; 6, scampi; 7, unknown, 8, scampi; 9, unknown; 10, scampi. (b) 1, scampi (*Nephrops norvegius*); 2, sample; 3, Pacific scampi (*Metanephrops andamanicus*); 4, sample; 5, Pink Atlantic shrimp (*Pandalus borealis*); (pH range 4–7).

and pink shrimp (*Pandalus borealis*) are given for the pH range 4–7. When these and other crustacea species are present in a mixture the absence of species-specific zones, well separated from those of other species, makes identification of the components' species difficult and in many cases impossible. For this particular problem, the alternative SDS procedure, which separates on the basis of molecular weight, has been found to be more discriminating.

Separation profiles obtained with the Phast system are given in Figure 6.5. Although the gels are considerably smaller than the standard large gels, the definition obtained is more than adequate for differentiating even those closely related species. Automatic staining and destaining reduces operator input considerably and enables the result to be obtained in 2 h. One disadvantage of the Phast system is that it tends to lose some of the anodic proteins – the parvalbumins, which are useful in differentiating gadoid species – into the anodic solution. Consequently, differentiation of these species has to be based more on proteins of pIs between pH 5 and 7. In either system the identity of an unknown sample is established by reference to those of likely authentic species run on the same gel.

6.5.1.1 *Influence of iced storage.* During iced storage, changes in the composition of the sarcoplasmic proteins may occur as a consequence of proteolysis by endogenous and bacterial enzymes. In addition, proteins can be leached out by melting ice. Studies in Torry Research Station and elsewhere have shown that the profile remains remarkably constant even to the stage of spoilage when the fish has become unfit for human consumption.

6.5.1.2 *Frozen storage.* The effect of frozen storage has been shown to have a minimal effect on the profiles. This is not surprising as it is well established that sarcoplasmic proteins are more resistant to denaturation during frozen storage, compared with the myofibrillar proteins (Mackie, 1993). For this reason, it is considered that samples of flesh of authentic species can be stored at $-30°$ for at least 1 year without any significant quantitative loss in the sarcoplasmic proteins or any alteration in the IEF profile of a species. If the flesh is minced, however, and stored at relatively high temperature (*e.g.* $-10°C$) a reduction in the solubility of the sarcoplasmic proteins will occur. This is believed to be due to the effect of mincing which breaks cell walls and releases potential reactants (*e.g.* trimethylamine oxide (TMAO) and the enzyme trimethylamine oxide demethylase) thereby accelerating considerably the deteriorative changes which inevitably take place on frozen storage. As rates of reactions increase with a rise in temperature, storage at the higher temperature results in more rapid rates of reaction and consequently greater degrees of protein denaturation.

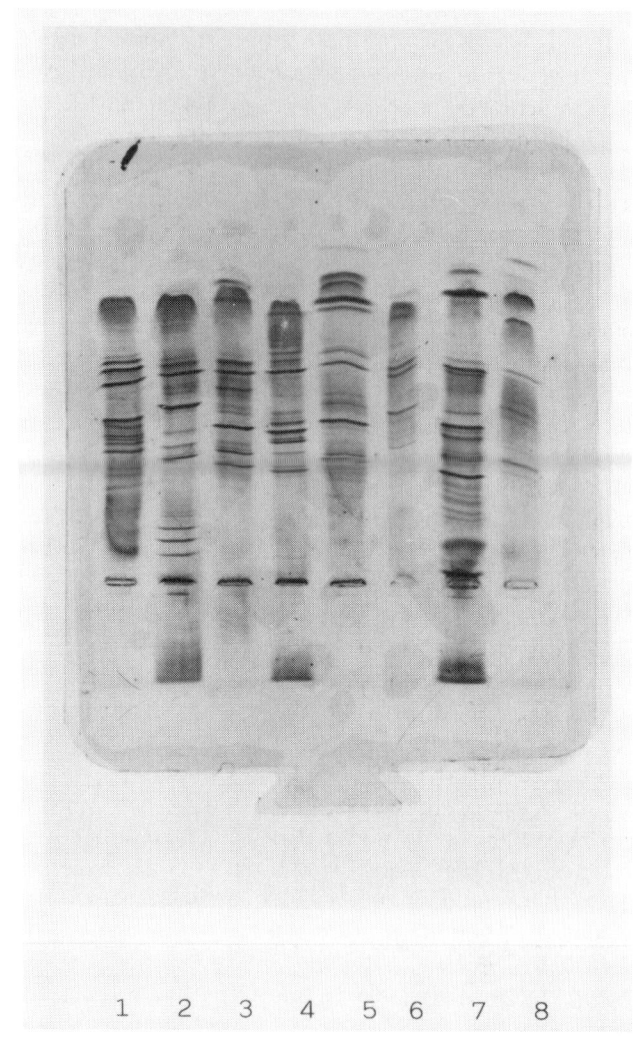

Figure 6.5 Isoelectric focused profiles of aqueous extracts of raw fish obtained on the Phast system. 1, cod (*Gadus morhua*); 2, haddock (*Melanogrammus aeglefinus*); 3, whiting (*Merlangius merlangus*); 4, saithe (*Pollachius virens*); 5, ling (*Molva molva*); 6, Alaska pollack (*Theragra chalcogramma*); 7, hake (*Merluccius merluccius*); 8, hoki (*Macruronus novaezelandiae*); (pH range 3.5–9.5).

6.5.2 Cooked but not autoclaved fish

When proteins are denatured on heating, an irreversible process of aggregation leading to the formation of precipitates usually takes place. These aggregates are inextractable in the normal protein solvents such as water or strong salt solutions as the native properties of the proteins have been destroyed. It is believed that the bonds that are formed are largely non-covalent, hydrophobic and hydrogen bonds, possibly with some degree of electrostatic interaction between the charged groups of the denatured proteins (Mackie, 1993, 1994a). Some —S—S— bonds may also be formed. It has been shown that heat-denatured proteins can be dissolved in a concentrated solution of SDS, a solvent which splits hydrogen bonds. A reducing agent such as dithiothreitol is often added to break any —S—S— bonds which may be present. The extracted protein residues can then be separated by SDS electrophoresis and, as for raw species, by comparing the protein profile of the unknown with those of authentic samples, which have been treated in the same way, identification of the species can be made. Instead of using SDS solutions as solvents, high concentrations of urea (8 M) have been used (An et al., 1988). Electrophoresis has also been carried out in urea (Mackie, 1980; Rehbein, 1990) but problems can arise due to the crystallisation of the urea. For these reasons, SDS electrophoresis is the preferred system.

While there are fewer differences between species largely because of the major contribution to the profile by the myofibrillar and connective tissue proteins, it is none the less possible to differentiate very closely-related species such as those of the cod family. This is illustrated in Figure 6.6 where an SDS profile of an unknown (reformed extruded scampi) is compared with those of authentic species of gadoid which could have been used to extend or adulterate the product. In this case the profiles of the gadoids can be differentiated from one another. The test sample corresponds to authentic scampi with no detectable presence of any other species. For these analyses silver staining (Blum et al., 1989; Craig et al., 1995) has been used as it gives more clearly defined profiles and higher sensitivity than Coomassie Blue. Although scampi is a raw product, the SDS procedure is preferred to IEF of aqueous extracts for its identification as more widely separated species-specific proteins are obtained than with the latter system (Craig et al., 1995). It also has an advantage in detecting adulteration of scampi with other crustacea such as 'warm water prawns' which tend to have IEF profiles similar to that obtained for scampi (Craig et al., 1995) with few if any species-specific zones that would be detected in a mixture. With SDS electrophoresis on the other hand, there are clear differences in mobility of one or two species-specific zones (Craig et al., 1995) making it possible to establish the presence of non-scampi proteins.

Typical conditions for the preparation of samples are to suspend 0.5 g of muscle in 2 ml 2% SDS. The mixture is homogenised and heated at 60°C for 30 min. It is then centrifuged at $12\,000 \times g$ for 10 min, diluted $\times$ 20 with sample buffer containing 2% SDS. The samples are applied to wells at the top of vertical slab gel and electrophoresis is carried out for 4–5 h. The gel is then fixed in 40% methanol, 10% acetic acid and stained with either Coomassie Blue (Craig et al., 1995) or silver (Blum et al., 1989); the latter is 10 times more sensitive and produces protein zones of different shades of brown which can assist comparison between species.

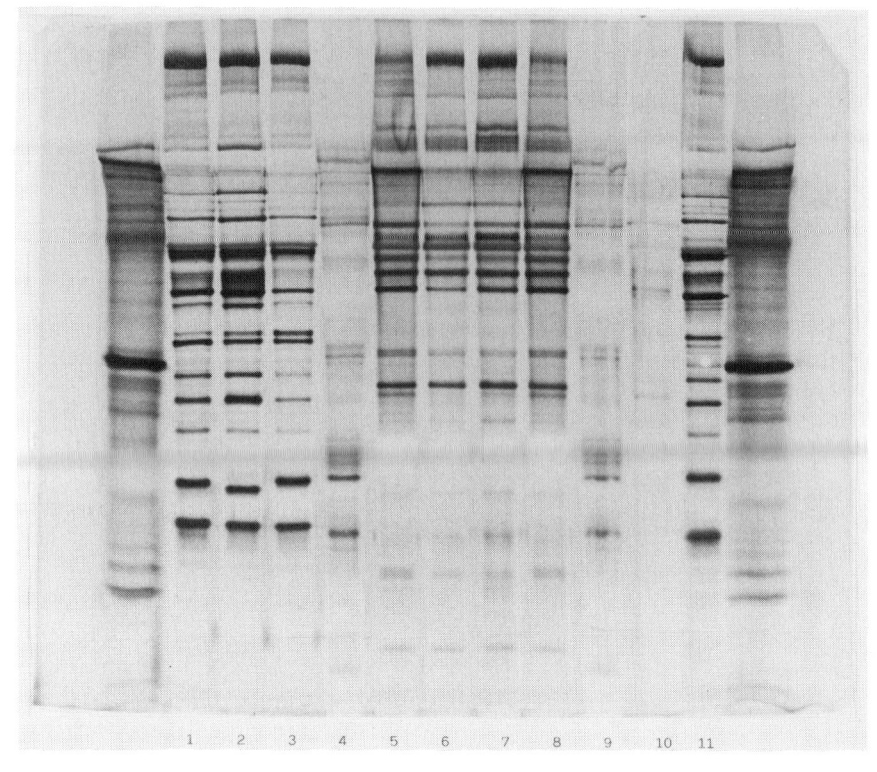

Figure 6.6 Separation profiles obtained on SDS electrophoresis of extracts of cooked: 1, cod (*Gadus morhua*); 2, haddock (*Melanogrammus aeglefinus*); 3, whiting (*Merlangius merlangus*); 4, soya protein; 5, scampi (*Nephrops norvegicus*); 6, unknown; 7, scampi; 8, unknown; 9, soya protein; 10, monkfish (*Lophius piscatorius*); 11, saithe (*Pollachius virens*); (standard molecular weight markers are on L and R).

Typical SDS separation profiles of some crustacea obtained with silver staining are given in Figure 6.7. Species differences are often more evident in the low molecular weight proteins as many of the higher molecular weight protein residues are common to all species. The profiles show that even with closely related species such as scampi (*Nephrops norvegicus*) and its counterpart from the Pacific (*Metanephrops andamanicus*) there are sufficient differences in profiles to differentiate the species. The profiles in turn are readily distinguishable from those of shrimps. The profiles of commercial samples of reputed 'scampi' products in lanes 2, 4, 6, 8 and 10, show to varying extents, the presence of non-scampi proteins. The main proteins in *Penaeus indicus* indicated by arrows are clearly present in samples 4, 8 and 10, showing that when species-specific proteins in a mixture are well resolved, adulteration can be established. The SDS technique is of value in differentiating cooked flesh in general and when salted smoked fish are being analysed it is often preferable to

METHODS OF IDENTIFYING SPECIES   187

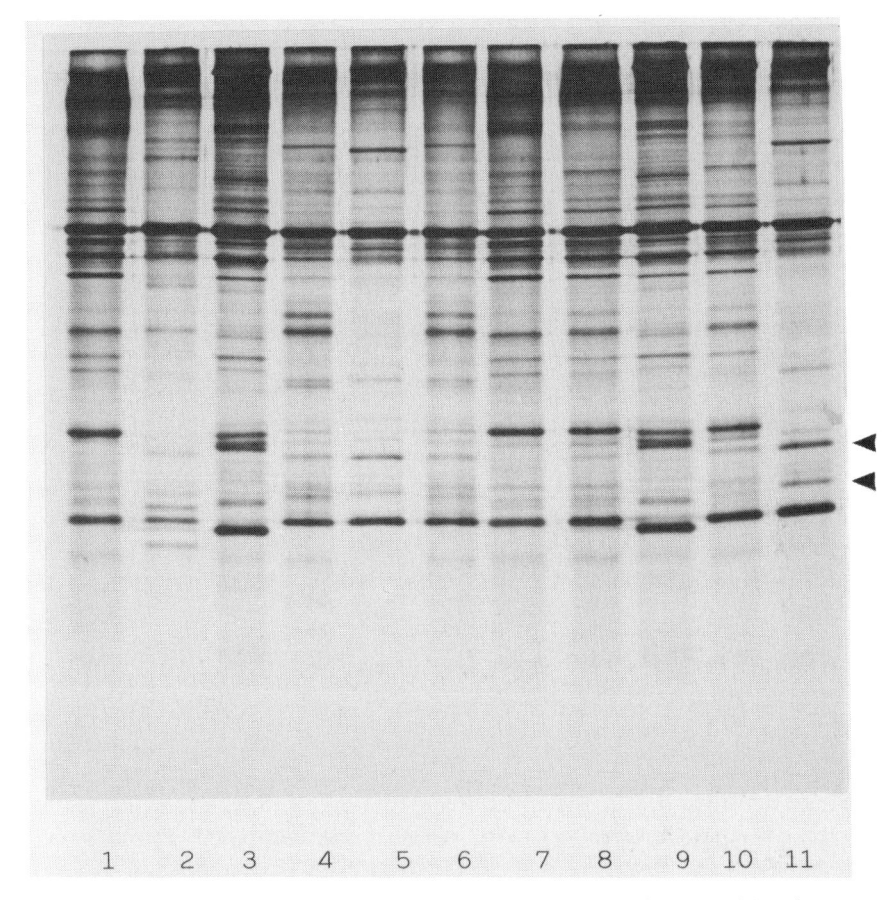

Figure 6.7 Separation profiles obtained on SDS electrophoresis of: 1, scampi (*Nephrops norvegicus*); 2, sample C; 3, Pacific scampi (*Metanephrops andamanious*); 4, sample D; 5, tropical shrimp (*Penaeus indicus*); 6, sample E; 7, scampi; 8, sample F; 9, Pacific scampi; 10, sample G; 11, tropical shrimp.

complement the IEF profiles of aqueous extracts, previously dialysed to remove salt, with those obtained on SDS electrophoresis to give greater confidence in the result.

SDS electrophoresis has been used successfully to analyse surimi-based gel products which are manufactured into shellfish substitutes. Sufficient differences in the SDS profiles of what are essentially sarcoplasmic protein-free muscle proteins have been obtained to enable the identity of the species of fish used in the production of surimi to be established (Torry Research Station, 1986; Rehbein, 1992). However, the success of the procedure in identifying the species is dependent on the extent of the relationship between the species. Distinct profiles can be obtained from cod, blue whiting (*Micromesistius poutassou*) and sardine (*Sardina pilichardus*) for example, but not between cod and Alaska pollack (*Theragra chalcogramma*), one of the main species used for the manufacture of surimi. The solution to the problem is to separate myosin heavy chains by SDS electrophoresis and, after limited proteolytic digestion, to separate the peptides obtained by SDS electrophoresis to give species-specific profiles.

6.5.3 Heat-sterilised and autoclaved products

When fish flesh is subjected to the high temperatures and pressure of the canning process (Horner, 1992), the constituent proteins undergo further denaturation, possibly in the form of increased covalent bond formation. The nature of these changes is not well understood. However, evidence for continued denaturation of proteins in the canning process is seen in the loss of protein zones on SDS electrophoresis and increased background staining (Mackie et al., 1992). These features of SDS electrophoresis of SDS extracts have meant that the method has little if any application to canned fish products; the few protein zones that are observed are often common to all the species examined and the strong background staining adds to the problems of differentiation.

A procedure using cyanogen bromide in formic acid was shown by the author (Mackie and Taylor, 1972) to give species-specific profiles on electrophoresis of the protein hydrolytic fragments. In this procedure, the protein chains are split specifically at the methionine residues on reaction with cyanogen bromide in formic acid (Gross, 1966). The protein fragments obtained can then be dissolved in water or 6 M urea and when separated by IEF they give species-specific profiles which can be compared with those of authentic species. Since the first demonstration of the application of this reaction, modifications have been made in the procedure and by using IEF, a suitable system has been developed for heat-sterilised canned fish products in general. However, when the species are closely related, such as members of the tuna or salmon families, the differences in profiles obtained are insufficient to enable differentiation to be made. Within the salmon species examined, two groupings have been made, namely Group 1: chum, sockeye and pink salmon and Group 2: spring, coho and Atlantic salmon (*Salmo salar*); differentiation below these levels was not possible. Similarly within the tuna family, skipjack tuna could be differentiated from the others (Mackie et al., 1992). Within the *Thunnus* genus itself, that is, yellowfin tuna (*Thunnus albacares*), big eye tuna (*Thunnus obesus*), bluefin tuna (*Thunnus thynnus*) and albacore (*Thunnus alalunga*), however, the profiles obtained were indistinguishable from one another.

In this procedure, 1.0 g of fish is suspended in 20.0 ml 70% formic acid to which is added 0.5 g cyanogen bromide to give a 200-fold excess over methionine. It is important for the specificity of the reaction that mild conditions are employed and that pH is on the acid side. After allowing the mixture to stand for 24 h at room temperature, the cyanogen bromide and formic acid are removed under reduced pressure in a rotary evaporator and the residue suspended in distilled water. After exhaustive dialysis against water, the solution of peptide residues is concentrated and then subjected to IEF using a wide range pH gradient (3.5–9.5). Fixing, staining and destaining are carried out as previously described for IEF acrylamide gel but because of the smaller size of the protein residues there is a greater risk of re-dissolving the precipated protein residues. It is therefore important to photograph the gel once developed as storage in the wash solvent will lead to a reduction in the intensity of the protein residue zones.

In Figure 6.8, the IEF profiles of the cyanogen bromide peptides of canned salmon, pilchard, bonito and herring are given to illustrate the overall similarity of the separation profiles of all species. In the middle pH range, however, differences between species are sufficient to allow identification to be established provided that the species are well separated zoologically (*e.g.* herring from salmon and bonito). The two profiles for the salmon groups illustrate the limitation of this method of species identification.

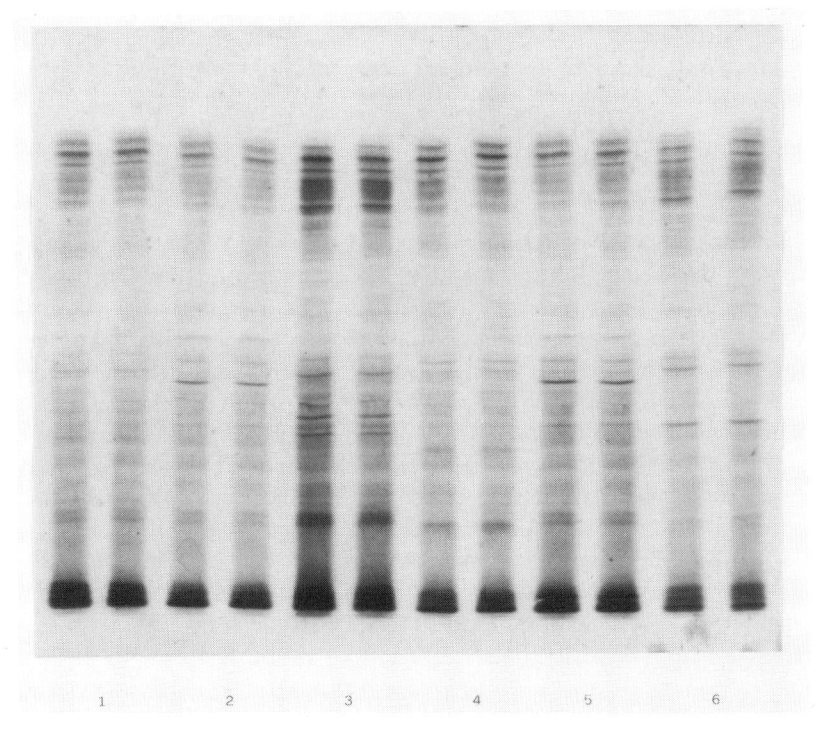

Figure 6.8 Isoelectric focused profiles of cyanogen bromide peptides of heat-sterilised canned: 1, Atlantic salmon (*Salmo salar*); 2, pink salmon (*Oncorhynchus gorbuscha*); 3, pilchard (*Sardina pilchardus*); 4, bonito (*Sarda sarda*); 5, chum salmon (*O. keta*); 6, herring (*Clupea harengus*). (pH range 3.5–9.5).

6.6 Alternative protein-based methods of fish species identification

6.6.1 *Immunoassay procedures*

Immunoassay systems have not been used to any great extent for fish species identification; however, two recent studies using the enzyme-linked immunosorbent assay (ELISA) procedure have been reported. An *et al.* (1990) developed a simple ELISA to verify the identity of rock shrimp (*Sicyonia brevirostris*) and to detect and quantify it in a mixture of raw and heat-processed seafood. In this study monoclonal antibodies were raised to a heat-denatured protein specific to rock shrimp. Another approach (Taylor and Jones, 1992) has been the development of an ELISA to identify canned sardine (*Sardina pilchardus*) in the presence of mackerel, herring and sild. In this study a polyclonal antiserum was raised against a crude extract of water-soluble proteins from authentic canned sardine but problems with cross-reaction with other species were not fully resolved. The robustness of these methods has not been fully determined, however, and further testing and modification would be required. The practicality of using the ELISA system for fish species identification has to be raised because of the large number of species involved and for which specific assay kits would have to be developed. With this method it is not possible to say that an unidentified species is present; it is limited to the detection and quantitation of species for which antibodies have been raised.

6.6.2 *Capillary electrophoresis*

While the potential of capillary electrophoresis has been recognised, its application to authenticity problems in fish and other foods is still very limited (Ewing *et al.*, 1989). A review of the methodology of capillary electrophoresis for protein separation has been published (Novotny *et al.*, 1990), and potential conditions for the analysis of model proteins, serum and milk protein have been established (Chen, 1991; Chen *et al.*, 1992). Recently Le Blanc *et al.* (1994) have shown that clear differences in profiles of species of the gadoid family can be obtained using high ionic strength phosphate buffer. The profiles, however, are very much subject to pH and unlike conventional electrophoresis, changes attributable to frozen storage were obtained. Recent papers by Gallardo *et al.* (1995) and by Mopper and Sciacchitano (1994) give encouragement to the view that capillary electrophoresis offers a realistic alternative to the conventional electrophoretic systems for species identification. The profiles obtained by Gallardo *et al.* (1995) for eight flat-fish species were shown to be highly reproducible indicating that no significant protein–silica surface interaction had occurred. The potential of this method for fish species identification is now established but further work is needed to determine the reliability of the system with long-term use, as protein adsorption on to the capillary columns is a recognised problem. At a later stage its performance in identifying species and in detecting adulteration would have to be compared with the present system. Mopper and Sciacchitano (1994) incidentally demonstrated high reproducibility of CE for identifying tuna and related species while using it for the determination of histamine.

6.6.3 HPLC

Like capillary electrophoresis, high-performance liquid chromatography (HPLC) uses a detection system of either UV absorbance or fluorescence. As these detectors also respond to non-protein substances which will be present in muscle extracts (*e.g.* nucleotides and amino acids) they lack the protein specificity of dye-binding as used for the detection of proteins after conventional gel electrophoresis. For this reason, caution is required in interpreting perceived differences in separation profiles. Two papers which have been published (Osman *et al.*, 1987; Armstrong *et al.*, 1992) emphasise the simplicity and speed of the analysis but they also point out that intraspecies variation is found and that there is loss of data through interference from the internal standard which is required to establish relative retention times of the peaks. Change in peak retention times, width and resolution arise with even slight variation in the chromatographic parameters such as mobile phase composition, gradient generation and temperature. Further work, including comparative studies with conventional electrophoresis, is required before any comment can be made on its reliability with unknown samples of fish. It has not yet been established to what extent factors such as physiological condition of the fish and *post mortem* changes during iced and frozen storage affect the separation profiles.

6.7 DNA techniques of fish species identification

There is interest in the application of DNA technologies to identify species of fish which cannot be identified by existing methods (Unseld *et al.*, 1995). As with proteins, DNA is also degraded on heating but procedures are now available which make it possible to isolate species-distinguishing fragments from such degraded material. Basic to all of the procedures is the polymerase chain reaction (PCR) which enables a species-distinguishing sequence of DNA to be targeted and amplified. Most of the studies have concentrated on the cytochrome b gene of mitochondrial DNA for which, even for fish, there are a number of sequences available for comparison, for example those of the tuna species which have been the subject of extensive phylogenetic studies (Block *et al.*, 1993). DNA probes rely on the specific interaction between the sequence of the probe and the complimentary sequence in the segment of the DNA of interest. On subsequent separation by electrophoresis on agar, the amplified DNA product can be identified and isolated for further analysis by an ever increasing range of technologies. The sequence of nucleotide residues can be determined and compared with those in a computer data base or the product can be digested with enzymes (endonucleases) which act on specific nucleotide sequences. The digested DNA can then be subjected to electrophoresis on agar gel and examined under UV light to reveal a series of fluorescent bands which, as for protein profiles, would be specific for the species assuming that the correct enzymes had been selected. The great advantage of the DNA approach is that although the DNA is considerably degraded as a result of the canning process, there is sufficient material to be recognised by the primers. However, it is well known that there is considerable variation in the degree of heating in commercial practice and as it is usually well in excess of what is required for sterilisation, any evaluation of the DNA technology for the identification of species of canned fish must take these variables into account. The *post mortem* history of the fish prior to canning should also be considered as degradation of DNA is known to take place to some extent on iced and frozen storage.

A typical electropherogram of DNAs isolated from canned and raw tuna is given in Figure 6.9. The DNA isolated from all the canned samples is severely degraded and shows only residues of the order of 100 base pairs. By selecting primers specific for a known sequence of the cytochrome b mitochondrial gene, it has been possible to isolate sufficient quantities of the PCR products of the seven commercially important species when canned (Figure 6.10) and to obtain their sequences. By reference to a data base, each of them has been identified from their specific sequences. However, because of the great variability of conditions of storage of tuna prior to canning and of the heat treatments of the canning proves itself, validation of the method is required on commercially canned products.

The use of restriction enzymes has also been investigated and separation profiles obtained on electrophoresis have established that this simpler approach can be used for the routine identification of species of canned tuna and bonito (Torry Research Station, unpublished results).

6.8 Fish eggs

The identification of species of fish egg or roe is unreliable if it is based on sensory criteria only, for example visual appearance. As with fish flesh, problems arise, particularly when the species are closely related (*e.g.* eggs of haddock, cod and saithe cannot be identified or differentiated by marine biologists when present in plankton trawls). Similarly, the roe of rainbow trout (*Oncorhynchus mykiss*) is indistinguishable from that of Atlantic salmon (*Salmo salar*). However, when individual eggs are analysed by SDS electrophoresis it has been shown that species-specific profiles are obtained (Scobbie and Mackie, 1990). Of particular interest is the use of the technique to

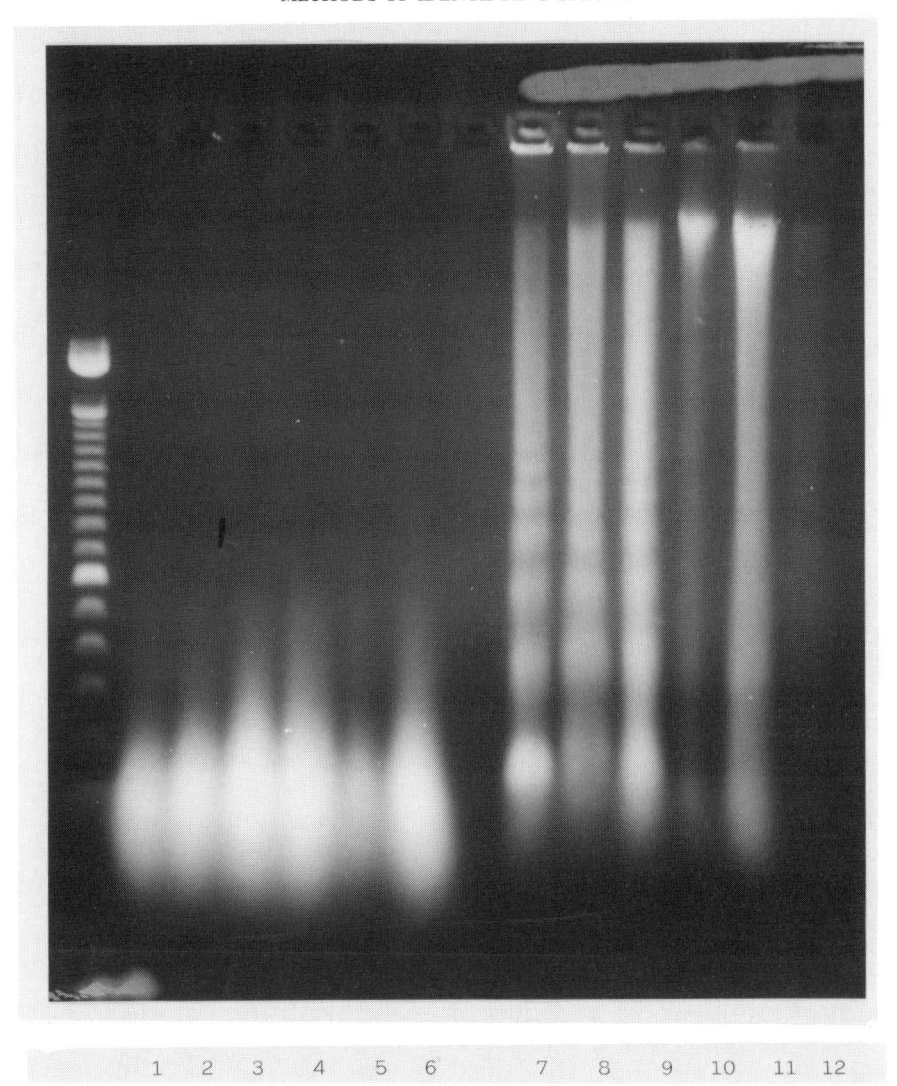

Figure 6.9 Agarose gel electrophoresis. DNA isolated from canned and raw tuna samples. 100 base pair DNA ladder: 1, bluefin tuna (*Thunnus thynnus*); 2, albacore (*Thunnus alalunga*); 3, skipjack tuna (*Katsuwonus pelamis*); 4, yellowfin tuna (*Thunnus albacares*); 5, bonito (*Sarda sarda*); 6, big-eye tuna (*Thunnus obestus*); 7, bluefin tuna; 8, albacore; 9, skipjack tuna; 10, yellowfin tuna; 11, bonito; 12, big-eye tuna (1–6 canned; 7–12 raw).

identify salmon eggs (Scobbie and Mackie, 1995), the possession of which is illegal in Scotland for historical reasons connected with the preservation of salmon fishing (Salmon Fisheries (Scotland) Act, 1868).

The SDS profiles of closely related species are given in Figure 6.11 and show similarities between salmon and rainbow trout and between cod and haddock.

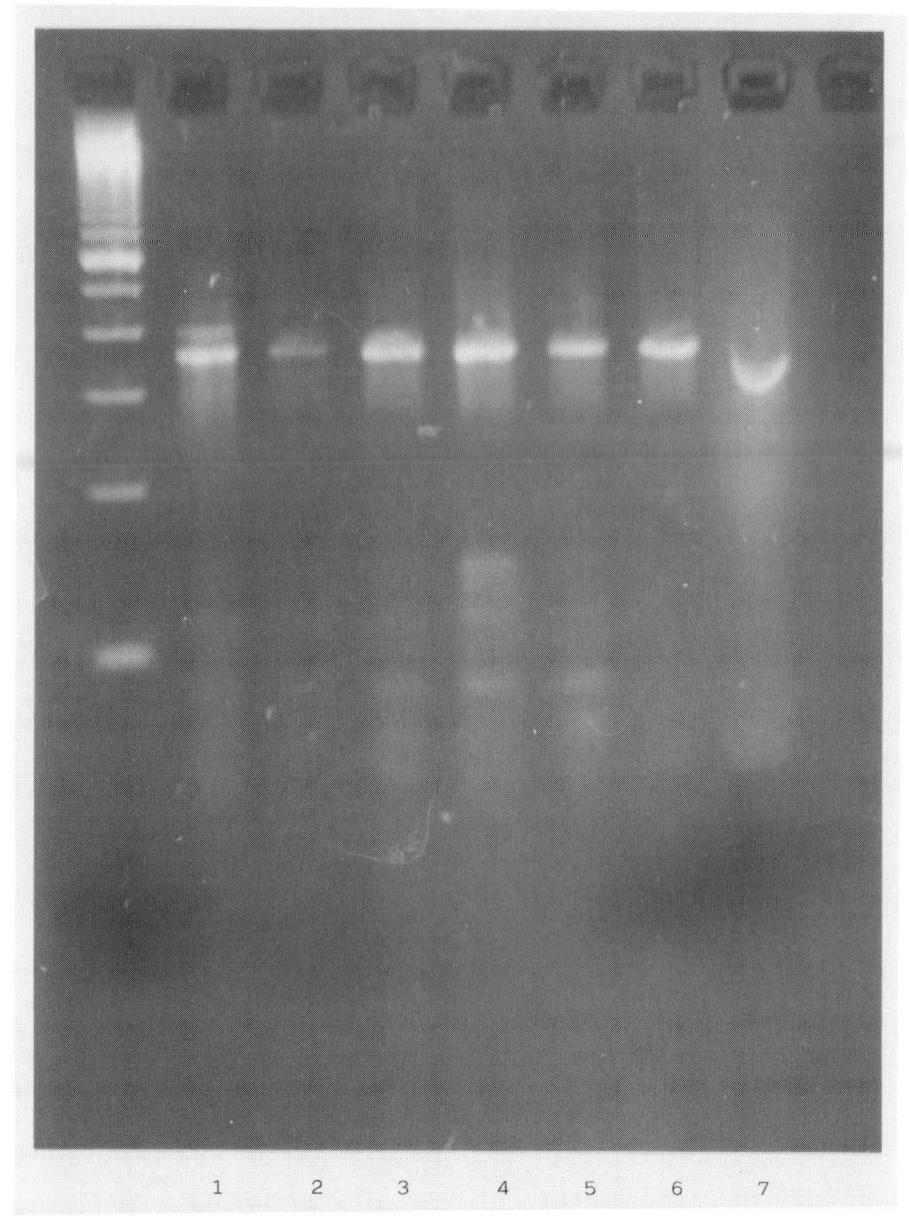

Figure 6.10 Agarose gel electrophoresis. 350 base pair PCR product of mitochondrial cytochrome b gene from canned tuna samples. 100 base pair DNA ladder: 1, skipjack tuna (*Katsuwonus pelamis*); 2, little tuna (*Euthynnus alleteratus*); 3, bluefin tuna (*Thunnus thynnus*); 4, albacore (*Thunnus alalunga*); 5, big-eye tuna (*Thunnus obesus*); 6, yellowfin tuna (*Thunnus albacares*); 7, bonito (*Sarda sarda*).

METHODS OF IDENTIFYING SPECIES 195

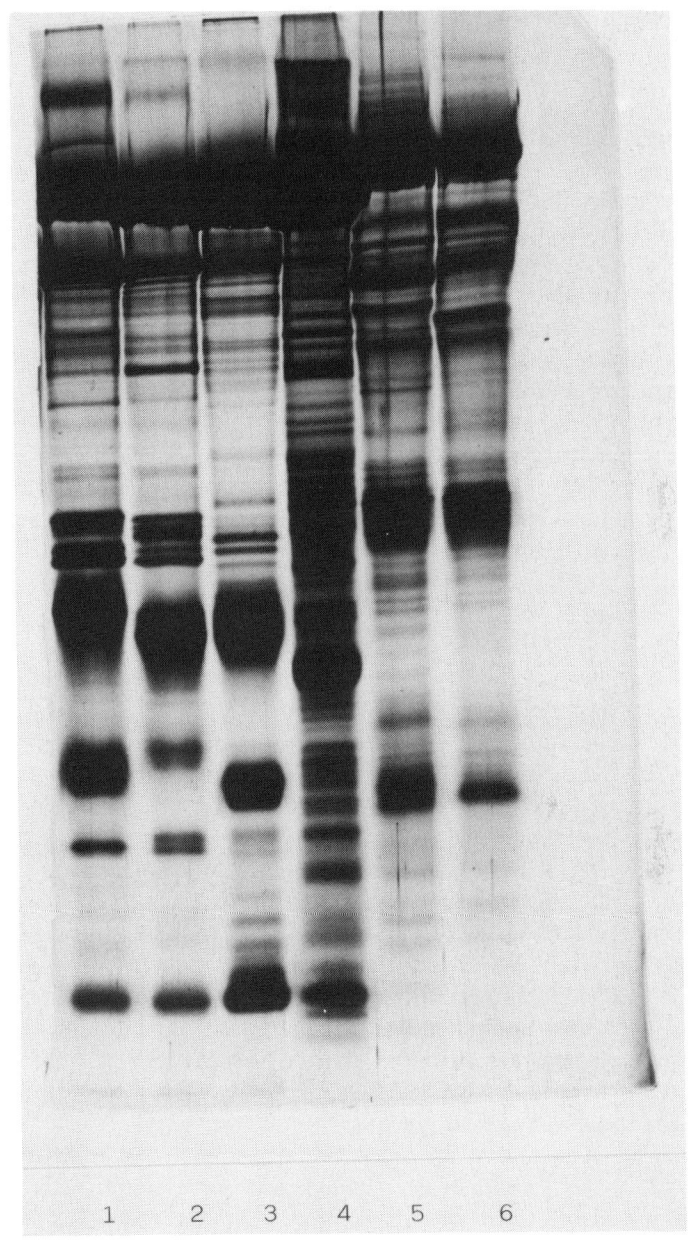

Figure 6.11 Separation profiles obtained on SDS electrophoresis of eggs of: 1, salmon (*Salmo salar*); 2, salmon; 3, rainbow trout (*Oncorhynchus mykiss*); 4, herring (*Clupea harengus*); 5, cod (*Gadus morhua*); 6, haddock (*Melanogrammus aeglefinus*).

This method of analysis is of potential value in identifying the species of egg in caviar and caviar substitutes.

6.9 General conclusions

Isoelectric focusing of aqueous extracts of fish muscle is the preferred procedure for the identification of species of raw fish. Wide experience of its use throughout the world has led to its general acceptance as the final arbiter in cases of dispute when the species requires to be established unequivocally. With recent advances in the use of densitometry and image analysis, data on pI values and peak height and areas can now be held in computer data bases. As this information continues to be built up it will have the potential of enabling species identification to move from simple visual comparison to a quantifiable computer-based analysis. These developments will make it more likely that the system will identify mixed species products and give an estimate of the quantitative composition of the respective species. Because of their sensitivity to heat, the sarcoplasmic proteins present in IEF profiles can be used to monitor the extent of heat processing. The parvalbumins, the proteins of gadoid species with low pIs, for example, remain soluble even after exposure to temperatures of 100°C. Although not fully evaluated as a system for species identification, CE appears to be a promising method. Further studies are required, however, before any conclusions regarding its use for routine analysis could be drawn.

SDS electrophoresis is likely to remain the preferred procedure for species identification of cooked fish products, surimi and raw fish products such as scampi where there is potential adulteration by other crustacea. Further developments in optimising the procedure will enable corresponding data on molecular weight and area of protein zone to be determined by image analysis and stored on a computer data base as for the IEF profiles of the sarcoplasmic proteins.

Although immunassay systems have not been applied to any significant extent to the identification of fish species, there may be particular problems such as the detection and estimation of crab-meat content of surimi products where they could have application; that is, an antibody to a specific crab protein is prepared and an ELISA technique developed for detecting and quantifying the specified component.

As far as DNA techniques are concerned, it is unlikely that they will replace the protein-based systems in the immediate future as they require a high degree of practical skill and often access to computer data bases of DNA sequences. The equipment required is also expensive. Their main use is likely to be in the identification of heat-sterilised canned fish products such as salmon, tuna and sardine. They are also likely to be used to complement data obtained by SDS electrophoresis in determining the species of fish eggs for either biological or food authenticity purposes.

References

Andrews, A.T. (1986), Monographs on physical biochemistry. In *Electrophoresis: Theory, Techniques and Biochemical and Clinical Applications*, 2nd edn., Peacock, A.R. and Harrington, W.F. (Eds), Clarendon Press, Oxford.
An, H., Marshall, M.R., Otwell, W.S. and Wei, C.I. (1988), Electrophoretic identification of raw and cooked shrimp using various protein extraction systems. *J. Food Science* **53**(2), 313–318.
An, H., Klein, P.A., Kao, K.J., Marshall, R.R., Otwell, W.S. and Wei, C.I. (1990), Development of monoclonal antibody for rock shrimp identification using enzyme-linked immunosorbent assay. *J. Agric. Food Chem.* **38**, 2094–2100.
Anon (1987a), *Surimi – It's American Now*. Alaska Fisheries Development Foundation Inc. The Alaska Writers Group, Anchorage, Alaska.
Anon (1987b), Caviar. *Seafood International*, 35–39.
Armstrong, S.G., Leach, D.N. and Wyllie, S.G. (1992), The use of HPLC protein profiles in fish species identification. *Food Chemistry* **44**, 147–155.
Asghar, A., Samejima, K. and Yasui, T. (1985), Functionality of muscle proteins in gelation mechanisms of structured meat products. *CRC Crit. Rev. Food Sci. Nutr.* **22**(1), 27–105.
Bartlett, S.E. and Davidson, W.S. (1991), Identification of *Thunnus* tuna species by the polymerase chain reaction and direct sequence analysis of their mitochondrial cytochrome *b* genes. *Can. J. Fish Aquat. Science* **48**, 309–319.
Bartlett, S.E. and Davidson, W.S. (1992), FINS (forensically informative nucleotide sequencing): a procedure for identifying the animal origin of biological specimens. *Biotechniques* **12**(3), 408–411.
Bechtel, P.J. (1986), *Muscle as Food*. Academic Press, New York.
Block, B.A., Finnerty, J.R., Stewart, A.F.R. and Kidd, J. (1993), Evolution of endothermy in fish: mapping physiological traits on a molecular phylogeny. *Science* **260**, 210–214.
Blum, H., Beier, H. and Gross, H.J. (1989), Improved silver staining of plant proteins, RNA and DNA in polyacrylamide gels. *Electrophoresis* **8**, 93–99.
Carnegie, P.R. (1994), Quality control in the food industries with DNA technologies. *Australian Biotechnology* **4**(3), 146–149.
Chen, F.T.A. (1991), Rapid protein analysis by capillary electrophoresis. *J. Chromatog.* **559**, 445–453.
Chen, F.T.A., Kelly, L., Polmieri, R., Bieler, R. and Schwartz, H. (1992), Use of high ionic strength buffers for the separation of proteins and peptides by capillary electrophoresis. *J. Liq. Chromatogr.* **15**, 1143–1161.
Consumer Protection Act (1987), HMSO, London.
Craig, A., Ritchie, A.H. and Mackie, I.M. (1995), Determining the authenticity of raw reformed breaded scampi (*Nephrops norvegicus*) by electrophoretic techniques. *Food Chem.* **52**, 451–454.
Dunn, M.J. (1987), Electrophoresis in polyacrylamide gels. In *New Directions in Electrophoretic Methods*, Jorgenson, J.W. and Phillips, M. (Eds), ACS Symposium Series 335, American Chemical Society, Washington, pp. 20–32.
Durand, P., Laudrein, A. and Quéro, J.-C. (1985), *Catalogue Electrophorétique des Poissons Commerciaux*, IFREMER, Centre de Nantes, Nantes.
Ewing, A.G., Wallingford, R.A. and Olefirowicz, T.M. (1989), Capillary electrophoresis. *Anal. Chemistry* **61**(4), 292A–303A.
Foote, C.J., Wood, C.C. and Withler, R.E. (1989), Biochemical genetic comparison of sockeye salmon and kokanee, the anadromous and non-anadromous forms of *Oncorhynchus nerka*. *Can. J. Fish Aquat. Sci.* **46**, 149–158.
Gallardo, J.M., Sotelo, C.G., Pineiro, C. and Perez Martin, R.I. (1995), Use of capillary zone electrophoresis for fish species identification. Differentiation of flat fish species. *J. Agric. Food Chem.* **43**, 1238–1244.
Goll, D.E., Robson, R.M. and Stromer, M.H. (1977), Muscle proteins. In *Food Proteins*, Whitaker, J.R. and Tannebaum, S.R. (Eds), Avi Publishing, Westford, CT, pp. 121–174.
Gross, E. (1966), The cyanogen bromide reaction. *Meth. Enzymol.* **II**, 238–255.
Hames, B.D. and Rickwood, D. (Eds) (1990), *Gel Electrophoresis of Proteins: a Practical Approach*, 2nd edn. IRL Press, Oxford.

Hitchcock, C.H.S. (1988), Opportunities and incentives for developing food immunoassays. In *Immunoassays in Veterinary and Food Analysis* I, Morris, B.A., Clifford, M.N. and Jackman, R. (Eds), Elsevier Applied Science, London, pp. 3–16.

Horner, W.F.A. (1992), Canning fish and fish products. In *Fish Processing Technology*, Hall, G.M. (Ed.), Blackie, London, pp. 114–153.

Hultin, H.O. (1984), Postmortem biochemistry of meat and fish. *J. Chem. Educ.* **61**(4), 289–298.

Hultin, H.O. (1992), Biochemical deterioration of fish muscle. In *Quality Assurance in the Fish Industry, Developments in Food Science*, 30, Huss, H.H., Jakobsen, M. and Liston, J. (Eds), Elsevier Science, Amsterdam, pp. 125–138.

Kleparnik, K. and Bocek, P. (1991), Theoretical background for clinical and biochemical applications of electromigration techniques. *J. Chromatog.* **569**, 3–42.

Laird, W.M., Mackie, I.M. and Ritchie, A.H. (1982), Differentiation of species of fish by isoelectric focusing on agarose and polyacrylamide gels–a comparison. *J. Ass. Public Analysts* **20**, 125–135.

Lanier, T.C. and Lee, C.M. (1992), *Surimi Technology*. Marcel Dekker, New York.

Le Blanc, E.L., Singh, S. and LeBlanc, R.J. (1994), Capillary zone electrophoresis of fish muscle sarcoplasmic proteins. *J. Food Science* **59**(16), 1267–1270.

Love, R.M. (1970), *The Chemical Biology of Fishes*. Academic Press, London.

Mackie, I.M. (1969), Identification of fish species by a modified polyacrylamide disc electrophoresis technique. *J. Assoc. Public Analysts* **7**, 83–87.

Mackie, I.M. (1980), A review of some recent applications of electrophoresis and isoelectric focusing in the identification of species of fish and fish products. In *Advances in Fish Science and Technology*, Connell, J.J. (Ed.), Fishing News Books, Farnham, pp. 444–451.

Mackie, I.M. (1983), New approaches in the use of fish proteins. In *Developments in Food Proteins–2*, Hudson, B.J.F. (Ed.), Applied Science, Barking, Essex, pp. 215–262.

Mackie, I.M. (1990), Identifying species of fish. *Analytical Proceedings, Royal Society of Chemistry* **27**, 89–92.

Mackie, I.M. (1992), Surimi from fish. In *The Chemistry of Muscle-based Foods*, Johnston, D.E., Knight, M.K. and Ledward, D.A. (Eds), Royal Society of Chemistry, Cambridge, pp. 200–207.

Mackie, I.M. (1993), The effects of freezing on flesh proteins. *Food Reviews International* **9**(4), 575–610.

Mackie, I.M. (1994a), Fish protein. In *New and Developing Sources of Food Proteins*, Hudson, B.J.F. (Ed.), Chapman & Hall, London, pp. 95–143.

Mackie, I.M. (1994b), Fish speciation. *Food Technology International Europe*, 177–180.

Mackie, I.M. and Ritchie, A.H. (1981), Differentiation of Atlantic cod (*Gadus morhua*) and Pacific cod (*Gadus macrocephalus*) by electrophoresis and isoelectric focusing of water-soluble proteins of muscle tissue. *Comp. Biochem. Physiol.* **68B**, 173–175.

Mackie, I.M. and Taylor, T. (1972), Identification of species of heat-sterilised canned fish by polyacrylamide disc electrophoresis. *Analyst* **99**, 609–611.

Mackie, I.M. Chalmers, M., Reece, P., Scobbie, A.E. and Ritchie, A.H. (1992), The application of electrophoretic techniques to the identification of species of canned tuna. In *Pelagic Fish, the Resource and its Exploitation*, Burt, J.R., Hardy, R. and Whittle, K.W. (Eds), Fishing News Book, London, pp. 200–207.

Montero, P. and Mackie, I.M. (1992), Changes in intra-muscular collagen of cod (*Gadus morhua*) during post mortem storage in ice. *J. Sci. Food Agric.* **59**, 89–96.

Mopper, B. and Sciacchitano, C.G. (1994), Capillary zone electrophoretic determination of histamine in fish. *J. AOAC Intl* **77**(44), 881–884.

Novotny, M.Y., Cobb, K.A. and Liu, J. (1990), Recent advances in capillary electrophoresis of proteins peptides and amino acids. *Electrophoresis* **11**, 735–749.

Osman, M.A., Ashoor, S.H. and Worsch, P.C. (1987), Liquid chromatographic identification of common fish species. *J. Assoc. Off. Anal. Chem.* **70**(4), 618–625.

Patterson, R.L.S. (1985), *Biochemical Identification of Meat Species*. Elsevier Applied Science, London.

Piggot, G.M. (1986), Surimi–the 'high-tech' raw material from minced fish flesh. *Food Res. International* **2**, 213–246.

Rehbein, H. (1990), Electrophoretic techniques for species identification of fishery products. *Z. Lebensum, Unters-Forsch* **191**, 1–10.

Rehbein, H. (1992), Fish species identification by peptide mapping of the myosin heavy chain. *Electrophoresis* **13**, 805–806.

Rehbein, H. and Karl, H. (1985), Identification of fish muscle proteins with buffers containing sodium dodecyl sulphate. *Z. Lebensm Unters-Forsch* **180**, 373–383.

Rehbein, H., Etienne, M., Jerome, M., Hattula, T., Knudsen, L.B., Jessen, F., Luten, J.B., Bouquet, W., Mackie, I.M., Ritchie, A.H., Martin, R. and Mendes, R. (1995), Influence of variation in methodology on the reliability of the isoelectric focusing method of fish species identification. *Food Chem.* **52**, 193–197.

Robson, R.M., O'Shea, J.M. Hartzer, M.K., Rathbun, W.E., La Salle, F., Schreiner, P.J., Kasang, L.E., Stromer, M.H., Lusby, M.L., Ridpath, J.F., Pang, Y.Y., Evans, R.R., Zeece, M.G., Parrish, F.C. and Huiatt, T.W. (1984), Role of new cytoskeletal elements in maintenance of muscle integrity. *J. Food Chem.* **8**, 1–24.

Russell, A.R. and Carnegie, P.R. (1994), Identification of gourmet meat using FINS (Forensically informative nucleotide sequencing). *Biotechniques* **17**(1), 24–26.

Salmon Fisheries (Scotland) Act 1868. Revised to 31 October 1978. HMSO, London.

Sambrook, J., Fritsch, E.F. and Maniatis, T. (1989), *Molecular Cloning. A Laboratory Manual*, 2nd edn. Cold Spring Harbor Laboratory Press, NY, USA.

Scobbie, A.E. and Mackie, I.M. (1988), The use of sodium dodecyl sulphate polyacrylamide gel electrophoresis in fish species identification – a procedure suitable for cooked and raw fish. *J. Sci. Food Agric.* **44**, 343–357.

Scobbie, A.E. and Mackie, I.M. (1990), The use of sodium dodecyl sulphate-polyacrylamide gel electrophoresis in species identification of fish eggs. *Comp. Biochem. Physiol.* **96B**(4), 743–746.

Scobbie, A.E. and Mackie, I.M. (1995), The use of sodium dodecyl sulphate-polyacrylamide gel electrophoresis to confirm the presence of salmon (*Salmo salar*) eggs in illegal fish bait. *Electrophoresis* **16**, 306–307.

Serwer, P. (1983), Agarose gels: Properties and use for electrophoresis. *Electrophoresis* **4**, 375–382.

Sotelo, C.G., Pineiro, C., Gallardo, J.M. and Perez-Martin, R.I. (1992), Identification of fish species in smoked fish products by electrophoresis and isoelectric focusing. *Z. Lebensm. Unters. Forsch.* **195**, 224–227.

Sotelo, C.G., Pineiro, C., Gallardo, J.M. and Perez-Martin, R.I. (1993), Fish species identification in seafood products. *Trends Food Sci. Tech.* **4**, 395–401.

Squire, J.M. (1981), *The Structural Basics of Molecular Contraction*. Plenum Press, London.

Taylor, W.J. and Jones, J.L. (1992), An immunoassay for verifying the identity of canned sardines. *Food Agric. Immunol.* **4**, 169–174.

The Food Labelling Regulations (1984), S.I. No. 1305. HMSO, London.

The Food Labelling (Scotland) Regulations (1984), S.I. No. 1519 (S. 128), HMSO, London.

Thomas, W. and Beckenbach, A.T. (1989), Variation in salmonid mitochondrial DNA: Evolutionary constraint on mechanisms of substitution. *J. Mol. Evol.* **29**, 233–245.

Torry Research Station (1985), Internal Report.

Torry Research Station (1986), Internal Report.

Trade Descriptions Act, 1968. HMSO, London.

UK Sea Fisheries Statistics 1991 & 1992. HMSO, London.

Unseld, M., Beyermann, B., Brandt, P. and Heisel, R. (1995), Identification of the species origin of highly processed meat products by mitochondrial DNA sequences. *PCR Meth. Applic.* **4**, 241–243.

Weber, K. and Osborn, M. (1969), The reliability of molecular weight determination by dodecyl sulphate–polyacrylamide gel electrophoresis. *J. Biol. Chem.* **244**, 4406–4412.

Yman, I.M. (1992), Meat and fish species identification. *European Food and Drink Review*, Autumn, 67–72.

7 Modified-atmosphere packaging of fish and fish products
A.R. DAVIES

7.1 Introduction

The ability of modified atmospheres to extend the shelf-life of foods has been recognised for many years. Reports of the use of modified atmospheres date back to the 1920s whilst reports for fish begin in the 1930s (Davis, 1993). Recent years have seen a marked expansion in the use and market share of modified-atmosphere packaging (MAP) generally. This is partly a result of the increasing consumer demand for fresh and chilled convenience foods containing fewer preservatives, which has led to a diversification in the range of products packaged in modified atmospheres. Currently, as well as fish and fish products, foods packaged in MAP include raw and cooked red meats and poultry, fruit and vegetables, fresh pasta, cheese, bakery products, crisps, coffee and tea.

In the past 60 years, many workers have studied and continue to study the use of MAP for fish and fish products. Indeed, despite the recent resurgence of interest in MAP and the concurrent diversification in the range of foods packaged in MAP, studies on fish and fish products still continue to account for a large proportion of the literature published on MAP. The use of MAP for fish has been excellently reviewed by Statham (1984), Stammen *et al.* (1990), Farber (1991), Reddy *et al.* (1992), Davis (1993) and Gibson and Davis (1995).

There are several techniques by which the atmosphere surrounding a product can be modified and often the terminology is confusing. The three techniques most relevant to fish and fish products are:

- Modified-atmosphere packaging (MAP); the replacement of air in a pack by a gas mixture, the composition of which differs from that of air, where the proportion of each component is fixed when the mixture is introduced, but no further control is exercised during storage.
- Controlled-atmosphere packaging (CAP); packaging in an atmosphere where the composition of gases is continuously controlled throughout storage. This technique is used primarily for the bulk storage of products and requires constant monitoring and control of the gas composition.
- Vacuum packaging (VP); the product is placed in a pack of low oxygen permeability, air is evacuated and the package sealed. The small remaining gaseous atmosphere of the vacuum package is likely to change during

storage because of metabolism of the product and/or micro-organisms and, therefore, the atmosphere becomes modified indirectly.

The three major gases used commercially in MAP are carbon dioxide, nitrogen and oxygen.

Carbon dioxide (CO_2) is both water- and lipid-soluble and is mainly responsible for the bacteriostatic effect seen on micro-organisms in modified atmospheres (Farber, 1991). The overall effect on micro-organisms is an extension of the lag phase of growth and a decrease in the growth rate during the logarithmic phase (Farber, 1991). This bacteriostatic effect is influenced by the concentration of CO_2, the age and load of the initial bacterial population, storage temperature and type of product to be packaged (Reddy et al., 1992). Although this bacteriostatic effect of CO_2 has been known for many years, the precise mechanism of its action is still not clearly understood. Farber (1991) summarised the theories regarding the influence of CO_2 on the bacterial cell as:

- alteration of cell membrane function including effects on nutrient uptake and absorption;
- direct inhibition of enzymes or decrease in the rate of enzyme reactions;
- penetration of bacterial membranes, leading to intracellular pH changes;
- direct changes to the physicochemical properties of proteins.

With high-moisture/high-fat foods such as seafood, meat and poultry, excessive absorption of CO_2 can lead to the phenomenon known as 'pack collapse' (Parry, 1993). Increased in-pack drip is also caused by dissolution of the gas into the surface of fresh muscle foods which reduces their pH sufficiently to reduce the water-holding capacity of the proteins (Parry, 1993). As a consequence of these factors and the fact that high CO_2 concentrations are reported to produce organoleptic changes (e.g. coarsening of the texture: Davis, 1993; alteration of the colour of the belly flaps, cornea and skin: Haard, 1992) the use of high CO_2 concentrations for fish is generally avoided.

Nitrogen (N_2) is an inert tasteless gas with low solubility in both water and lipid. It is used to displace oxygen in packs so as to delay oxidative rancidity and inhibit the growth of aerobic micro-organisms (Farber, 1991). Because of its low solubility, it is used as a filler gas to prevent pack collapse which may occur when products are packed in high CO_2-containing atmospheres.

Oxygen (O_2) will generally stimulate the growth of aerobic bacteria and can inhibit the growth of strictly anaerobic bacteria, although there is a very wide variation in the sensitivity of anaerobes to O_2 (Farber, 1991). The presence of O_2 can cause problems of oxidative rancidity in fatty fish and hence it is usually excluded from atmospheres used for these species in order to minimise these effects. The rationale for the inclusion of O_2 for the MAP of fish is questioned by Davis (1995). He argues that the evidence that the inclusion of

O_2 reduces the amount of exudate released from fish during storage is inconclusive, and suggests that O_2 may be included with lean fish because of the concerns about anaerobic packaging. The potential for the production of botulinum toxin in O_2-containing atmospheres is, however, well documented (Reddy et al., 1992) and Lindsay (1982) comments that the belief that the inclusion of some O_2 with N_2 or CO_2 will prevent the hazards from botulism in fresh packaged fish is a misconception and may offer a false sense of security.

A wide range of atmospheres has been examined for use with fish. Examples of some of these atmospheres can be found in the extensive tables presented by Stammen et al. (1990), Reddy et al. (1992) and Gibson and Davis (1995). Gas mixtures of 40% CO_2/30% N_2/30% O_2 for white fish and of 40–60% CO_2 balance N_2 for fatty species have been recommended (Anon, 1985; Tiffney and Mills, 1982) and are probably those most widely used.

One, if not the major, reason for the use of MAP is the extension of shelf-life that can be obtained. However, the reported increases in shelf-life for fish and fish products vary markedly and are small in comparison with those reported for several other products. Gibson and Davis (1995) discuss the reason for these variations in the reported shelf-life and suggest that it is a result of the definition and measurement of shelf-life which is both difficult and debatable for fish products. They suggest that if the end of shelf-life is taken at an early stage (high quality shelf-life), little practical extension of the shelf-life is observed, whilst if the end of shelf-life is taken at a point corresponding to that used by public health inspectors, then the shelf-life extension can be considerable. Reddy et al. (1992) have tabulated the reported extension in shelf-life obtained by several workers. They found the percentage increase in shelf-life (compared with aerobic storage) ranged from 0% (no increase) to 280%.

7.2 Microbial flora of fresh fish

The wide range of fish species, the vastly different environments from which they are harvested and the variety of microbiological sampling techniques used has resulted in widely varying reports of the numbers of organisms on fish. For example, Ward (1994) reports the range of counts to be: skin, 10^2–10^6 cm^{-2}; gills, 10^3–10^5 g^{-1}, intestine, from very few to 10^7 g^{-1} or greater. The lower counts for the skin and gills are commonly associated with clean, cold waters and the higher counts with tropical or sub-tropical waters and polluted areas. The count in the intestine relates directly to feeding activity, being high in feeding fish and low in non-feeding fish.

The bacterial flora of cold-water fish is dominated by the psychrotrophic Gram-negative genera (Shewan, 1977). Organisms involved belong to the genera *Acinetobacter*, *Flavobacterium*, *Moraxella*, *Shewanella* and *Pseudomonas*. Members of the Vibrionaceae (*Vibrio* and *Photobacterium*) and the

Aeromonadaceae (*Aeromonas* spp.) are also common aquatic bacteria and typical of the fish flora whilst Gram-positive organisms such as *Bacillus, Micrococcus, Clostridium, Lactobacillus* and coryneforms can also be found in varying proportions (Huss, 1995).

Shewan (1977) concluded that the Gram-positive *Bacillus*, and *Micrococcus* dominate the flora of fish from tropical waters; however, other workers have found that the microflora on tropical fish species is very similar to that on temperate species but with a slightly higher load of Gram-positive and enteric bacteria (Huss, 1995).

It is apparent from the above that there is potentially a very diverse range and number of organisms present on fish. The significance of these organisms in terms of spoilage of the fish and how this can be controlled by packing in modified atmospheres is of the greatest interest. As many of the organisms present on spoiled fish may not have played any direct role in the spoilage (Huss, 1995), it is important to understand which are the specific spoilage bacteria of most significance.

Huss (1994) presented an overview (Table 7.1) of the dominating and specific spoilage bacteria of fresh white fish when stored aerobically, under vacuum and in MAP at 0 and 5°C, and when stored aerobically at ambient temperature. *Shewanella putrefaciens* is the primary specific spoilage organism of marine temperate-water fish stored aerobically in ice.

In general, aerobic micro-organisms are sensitive to CO_2 and it is this, along with their requirement for O_2, that is utilised in MAP to control the spoilage of foods. Gram-negative bacteria are generally more sensitive to CO_2 than are Gram-positive bacteria (Lambert *et al.*, 1991). In chill-stored proteinaceous foods, such as meat and fish, this generally results in the inhibition of the Gram-negative *Pseudomonas*, Enterobacteriaceae and *Acinetobacter/Moraxella*, whilst the Gram-positive lactic-acid bacteria and *Brochothrix thermosphacta* become the dominant organisms. As expected from the above, *Shewanella putrefaciens* is strongly inhibited by CO_2; however, *Photobacterium phosphoreum* has been found to be highly resistant to CO_2 (Huss, 1995) and has an increased growth rate under anaerobic conditions. It is these factors that explain the importance of this organism in the spoilage of vacuum packaged/MAP cod and it is thought that this explains the reduced extension in shelf-life obtained with modified-atmosphere packaging of cod, in comparison with those obtained for other products, notably meats.

7.3 Pathogenic flora of fresh fish

Whilst the ability of MAP to extend the shelf-life of many products is well recognised, concern has been expressed by regulatory authorities, food industry groups and others that MAP may introduce undue safety hazards. These concerns are that suppression of the normal spoilage flora may result in an

Table 7.1 Dominating microflora and specific spoilage bacteria at spoilage of fresh, white fish (cod). From Huss (1995) by Courtesy of FAO

Storage temperature	Packaging atmosphere	Dominating microflora	Specific spoilage organisms
0°C	Aerobic	Gram-negative psychrotrophic, non-fermentative rods (*Pseudomonas* spp., *S. putrefaciens*, *Moraxella*, *Acinetobacter*)	*S. putrefaciens*, *Pseudomonas*[3]
	Vacuum	Gram-negative rods; psychrotrophic or with psychrophilic character (*S. putrefaciens*, *Photobacterium*)	*S. putrefaciens*, *P. phosphoreum*
	MAP[1]	Gram-negative fermentative rods with psychrophilic character (*Photobacterium*) Gram-negative non-fermentative psychrotrophic rods (1–10% of flora; *Pseudomonas*, *S. putrefaciens*) Gram-positive rods (LAB[2])	*P. phosphoreum*
5°C	Aerobic	Gram-negative psychrotrophic rods (Vibrionaceae, *S. putrefaciens*)	*Aeromonas* spp., *S. putrefaciens*
	Vacuum	Gram-negative psychrotrophic rods (Vibrionaceae, *S. putrefaciens*)	*Aeromonas* spp., *S. putrefaciens*
	MAP	Gram-negative psychrotrophic rods (Vibrionaceae)	*Aeromonas* spp.
20–30°C	Aerobic	Gram-negative mesophilic fermentative rods (Vibrionaceae, Enterobacteriaceae)	Motile *Aeromonas* spp. (*A. hydrophila*)

[1] Modified atmosphere packaging (CO_2-containing).
[2] LAB: Lactic acid bacteria.
[3] Fish caught in tropical waters or freshwaters tend to have a spoilage dominated by *Pseudomonas* spp.

organoleptically acceptable product, whilst either allowing or enhancing the growth of pathogenic organisms. MAP of fish and fish products has received the widest attention with respect to these concerns notably, with respect to the non-proteolytic, psychrotrophic strains of *Clostridium botulinum*. More recently, concern has also been expressed with respect to the other psychrotrophic pathogens, for example *Aeromonas, Listeria* and *Yersinia*.

7.3.1 Clostridium botulinum

Undoubtedly the single most important concern with respect to the use of MAP for many products (but particularly fish) is the potential for outgrowth and toxin production by the non-proteolytic, psychrotrophic strains of *Clostridium botulinum*. These strains can grow and produce toxin without producing overt signs of spoilage, which may also be absent as a result of an inhibition of the normal spoilage flora. Strains of type E and the non-proteolytic types B and F are the major concern in MAP, as they are able to grow at temperatures as low as 3.3°C, albeit slowly, and, as they do not putrefy proteins, they may not show obvious signs of spoilage.

To determine the potential botulism hazard from any food, it is necessary to have information as to the possibility of contamination of that food and the potential for outgrowth and toxin production in the food. Dodds (1993a) has reviewed the literature on the incidence of *C. botulinum* in the environment (soils, sediment, water) and on fish and other aquatic organisms obtained *in situ* and on fish in the market place (Dodds, 1993b). She states that, as the aquatic environment is frequently contaminated with spores, it is to be expected that fish will be contaminated as well. This contamination of fish may be as a result of exposure to spores before harvest or due to contamination during processing and handling. Data presented by Dodds (1993a) show an incidence of positive samples of fish and aquatic animals taken *in situ* ranging from 0 to 100% with a mean incidence calculated from her data (assuming a value of 0 for data presented as < 1) of 19.6%. Similarly, in fish taken from the market place in North America the incidence of positive samples ranged from 0 to 100% with a mean of 21.6%, and for fish from the market place in Europe and Asia a range from 0 to 63% with a mean of 11.7% (Dodds, 1993b). In summary, the isolation of all types of *C. botulinum* from marine environments, although highly variable by geographic location and season, is frequent enough that processors must assume its presence. This is substantiated by the variable but often high incidence of fish from throughout the World found to contain *C. botulinum* spores.

Numerous studies have examined the relationship between time to toxin production and signs of organoleptic spoilage for MAP fish, and these have been reviewed (Stammen *et al.*, 1990; Reddy *et al.*, 1992; Gibson and Davis, 1995). Unfortunately, because of the many variables between studies, for example fish type, size and site of inoculum, temperature, season, atmosphere, etc., direct comparisons between the studies cannot be made. Stammen *et al.* (1990) in their review concluded that

> with few exceptions, at temperatures above 20°C, organoleptic spoilage coincided with toxin production in many fresh fishery products, regardless of the modified atmosphere used. However, at lower temperatures, organoleptic spoilage preceded toxin development in all fresh fish products except cod and whiting fillet held in either an air, vacuum or CO_2 atmosphere. This trend was seen at storage temperatures from 4–12°C. The time interval between toxin development and organoleptic spoilage of MAP fish products generally decreased as storage temperatures increased. In contrast toxin development preceded organoleptic spoilage in cod and whiting fillets packaged in 100% CO_2 and held at refrigeration temperatures. These products were still acceptable for consumption even though botulinal toxin was found in them.

Baker and Genigeorgis (1990) developed a predictive model from over 18 700 samples analysed over a 5-year period. The utility of the model was demonstrated by its ability to predict the time before toxigenesis in inoculated fish stored under different modified atmospheres as reported in the international literature (Baker and Genigeorgis, 1990). Temperature explained 74.6% of experimental variation in the final multiple linear regression model ($r^2 = 0.883$) but, surprisingly, the gaseous atmosphere was of little importance (Baker and Genigeorgis, 1990).

One approach that may provide the safety required for MAP of fish with respect to *C. botulinum* is the use of pre-treatment in combination with MAP. Potassium sorbate, sodium chloride, nisin and irradiation in combination with MAP have all been shown to be effective (Stammen *et al.*, 1990). A more detailed discussion of the use of combination treatments to improve both safety and quality is presented below.

7.3.2 Other pathogens

As well as *C. botulinum*, other pathogens regarded as intrinsic and natural and found on aquatic products include *Vibrio parahaemolyticus*, *Vibrio vulnificus*, *Listeria monocytogenes* and *Aeromonas hydrophila*, whilst other emerging pathogens such as *Yersinia enterocolitica* have also been reported (Gibson and Davis, 1995). Whilst there have been many studies on the effects of MAP on *C. botulinum*, few studies have addressed the effect of MAP on these other foodborne pathogens. We have recently examined the effect on these other pathogens and these studies are reported in section 7.5.

7.4 Present applications of MAP to fish and fish products

The current use and market status of MAP in general is difficult to research because it is a relatively new sector of the packaging market and there is considerable confusion over definitions (Cakebread, 1994). Davies (1995) has reviewed what limited information is available and presents some of the factors deemed to be responsible for the differences between the American and European markets and also for the highly uneven spatial development across Europe. The United Kingdom and France dominate the European market as a result of their concentrated multiple retailer market shares. This is reflected in the fact that the most rapid growth in sales of fish and seafood in MAP has occurred in Europe, mostly the UK and France. Estimates of the total numbers of modified atmosphere fish packs produced in Europe between 1986 and 1990 rose five-fold to 250×10^6, and by 1992 the estimate for the UK alone had risen to 193×10^6, with predicted sales for 1996 of 340×10^6 (Davis, 1995). Day (1993a) reports that, in 1992, fish and seafood was the fourth largest sector of the UK Retail MAP market, accounting for 10%, with red meat (30%), snack and dried foods (14%) and cooked meat (13%) being the three largest sectors.

Cakebread (1994) reviewed the future prospects for the whole European MAP market and concluded that, despite a theoretically large potential, MAP will not continue to grow rapidly in Europe during the 1990s unless concrete solutions can be found to a number of problems that have restricted commercial development outside France and the UK. These problems include:

- the lack of a concentrated retail food and chill chain distribution structure in many European countries;
- entrenched environmental prejudice against PVC, plastics and 'unnecessary' packaging generally;
- competition from high-quality unpacked fresh produce;
- competition from vacuum-skin packaging in certain key sectors such as fresh and cooked meats and fish;
- concern over the safety of some chilled food products.

Many of these factors are obviously relevant for fish and fish products and also for the MAP market outside Europe. Probably the two most significant factors that negatively affect the MAP fish market are the concerns over the safety of retail packs with respect to *Clostridium botulinum* and the less dramatic increases in shelf-life that are obtained in comparison with those for some other commodities. Against this, the increasing demands on the world's fish stocks and the development of improved chill chains are likely to have a positive influence on the future market.

7.5 Experimental approach

7.5.1 *Introduction*

Recently, concerns have been expressed about the ability of the emerging psychrotrophic pathogens, for example *Listeria*, *Aeromonas* and *Yersinia*, to grow on MAP products (Beuchat, 1991; Farber, 1991; Reddy et al., 1992). Golden et al. (1989) found that the growth of A. hydrophila was enhanced when cells were incubated under nitrogen in comparison with air and Enfors et al. (1979) concluded that vacuum packaging (VP) or gas packaging with low CO_2 concentrations might involve certain health hazards following the observation of A. hydrophila growth on pork packaged under nitrogen. However, Gill and Reichel (1989) observed an inhibition of both A. hydrophila and Y. enterocolitica by CO_2 at 0, 2 and 5°C on high-pH beef. Studies by Eklund and Jarmund (1983) also noted CO_2 inhibition of Y. enterocolitica. They found, at 20°C, a 40% decrease in numbers when packaged in CO_2 in comparison with air and at 6°C a 98% reduction in numbers. Likewise, Avery et al. (1994) showed that saturated CO_2 packaging inhibited the growth of L. monocytogenes at both 5 and 10°C on beef striploin steaks, whilst Wimpfheimer et al. (1990) observed an inhibition of L. monocytogenes on raw chicken packaged in an anaerobic modified atmosphere.

The objectives of this study were to compare the growth/survival of Aeromonas, Listeria, Salmonella and Yersinia on MAP cod (Gadus morhua) and rainbow trout (Oncorhyncus mykiss) with that on aerobically stored fish.

7.5.2 Materials and methods

7.5.2.1 Fish supply

Cod. Cod (Gadus morhua) of 6–18 months (weighing c. 6–8 kg each) were landed off Rye, UK, eviscerated and stored on ice in insulated containers prior to delivery to the laboratory. Once in the laboratory, the fish were portioned into 100 g steaks within 36 h of capture.

Trout. Trout (Oncorhyncus mykiss) of 7–12 months (weighing c. 0.5 kg each) were purchased from a local fish farm freshly gutted and cleaned. The fish were transported on ice to the laboratory and, following inoculation, were cut in half and a head or tail section packaged as appropriate within 4 h *post mortem*.

7.5.2.2 Inoculation.
Bacterial strains and agar media used are shown in Table 7.2. Antibiotic-resistant (nalidixic acid and streptomycin) strains of Y. enterocolitica, Aeromonas spp. and S. typhimurium were produced using the method of Blackburn and Davies (1994). The required inoculum was prepared by sub-culturing each organism twice in Tryptic Soy Broth (Gibco) incubated at 30°C for 24 h. For experiments with either Aeromonas spp. or S. typhimurium, strains were pooled to produce a cocktail containing approximately equal numbers of each strain. Organisms were diluted 1 in 100 in sterile distilled water resulting in c. 10^{6-7} cfu ml^{-1}. Individual fish (trout) or steaks

Table 7.2 Strains and culture media used

Organism	Reference	Source	Media	Incubation
Aeromonas spp.	CPHL 66252 317 A 467 435	Dr B. Rowe[a] Dr J. Sheridan[b]	Aeromonas Agar Base* (Ryan; ASA; Unipath)	30°C, 48 h
S. typhimurium	7M-5522 S-5698 77-7628 7M-4987	Dr T. Roberts[c]	Xylose Lysine Agar* (XL; Difco)	30°C, 48 h
Y. enterocolitica	GER 0:3 P-	Dr J. Sheridan[b]	Yersinia-selective agar* (Unipath) excluding sodium desoxycholate, cefsulodin, irgason and novobiocin (YSA-SEL)	30°C, 48 h
L. monocytogenes	NCTC 11994	NCTC[d]	Listeria-selective agar (Oxford formulation; Unipath)	30°C, 48 h

* Media contained 50 µg ml^{-1} nalidixic acid and 1000 µg ml^{-1} streptomycin.
[a] Kindly supplied by Dr B. Rowe, Division of Enteric Pathogens, Central Public Health Laboratory, Colindale, UK.
[b] Kindly supplied by Dr J. Sheridan, National Food Centre, Dublin, Ireland.
[c] Kindly supplied by Dr T. Roberts, formerly the Institute of Food Research, Reading, UK.
[d] National Collection of Type Culture, Central Public Health Laboratory, Colindale, UK.

(cod) were fully immersed in the inoculum for 5 s (with belly flap exposed for trout), drained for 5 s and stored at 0°C for 1–2 h prior to packaging.

7.5.2.3 Packaging

Cod. Individual steaks were placed into high-density polyethylene trays (Dynopack Ltd) and were loosely overwrapped in a stomacher bag (Seward laboratories) for the aerobic samples, or packed in an atmosphere of 60% CO_2/40% N_2 or 40% CO_2/30% N_2/30% O_2. The modified-atmosphere packs were produced using food-grade gases (BOC) and a Mecapac (M500; ECM Mecapac) modified-atmosphere packer with a gas to product ratio of c. 2.75 l of gas per kg of fish. The top web used was Suprovac 90 (Kempner Ltd) with a gas permeability of O_2 c. 25, CO_2 c. 90 and N_2 c. 6 ml^{-3} m^{-2} d^{-1} bar^{-1} at 20°C, 50% RH. Once packed, cod were stored at 0 or 5°C.

Uninoculated (control) samples were treated in the same manner throughout.

Trout. Trout halves were placed into high-density polyethylene trays (Dynopack) and were loosely overwrapped in a stomacher bag (Seward) for the aerobic samples or packed in an atmosphere of 60% CO_2/40% N_2 or 80% CO_2/20% N_2. The modified-atmosphere packs were produced as for the cod. Once packed, trout were stored at 0 or 5°C.

Uninoculated (control) samples were treated in the same manner throughout.

7.5.2.4 Microbiological methods.
On each sampling occasion, duplicate packs were examined. Samples were examined for the inoculated pathogen and the total aerobic colony count at 25°C. Each sample was aseptically cut into small pieces and a 20 g sample added to 180 ml Maximum Recovery Dilutent (MRD; Unipath). The sample was stomached for 60 s (lab blender 400, Seward) and a decimal dilution series prepared in MRD and used to inoculate the appropriate media (Table 7.2) using the spiral plate (Don Whitley Scientific) procedure.

7.5.3 Results

The growth/survival of *L. monocytogenes* and *S. typhimurium* on cod and trout at 0 and 5°C is presented in Figs. 7.1 and 7.2, respectively. At 0°C, on cod, numbers of *L. monocytogenes* remained static during the early stage of the 20-day storage period but increased slightly (c. 0.5–1 $\log_{10}$) towards the end, whilst on trout, numbers declined slightly over the 14-day storage. At 5°C, on cod, in all atmospheres the maximum increase was <0.5 $\log_{10}$, whilst on trout, numbers increased by c. 3 $\log_{10}$ in the aerobically stored trout, increased slightly (c. 0.5 $\log_{10}$) in the 60% CO_2/40% N_2 atmosphere and remained static

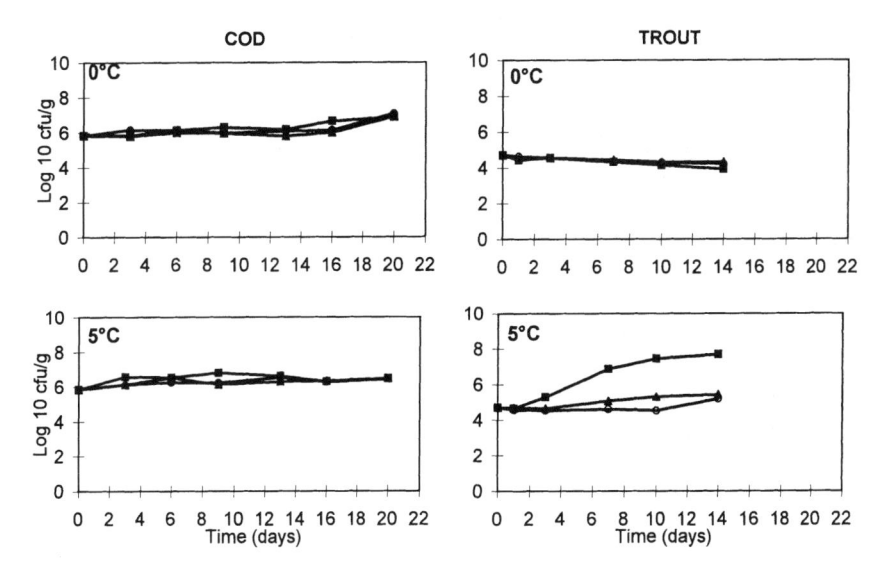

Figure 7.1 Growth\survival of *Listeria monocytogenes* on cod stored aerobically (■), in 60% CO_2/40% N_2 (▲) and in 40% CO_2/30% N_2/30% O_2 (●), and on trout stored aerobically (■), in 60% CO_2/40% N_2 (▲) and in 80% CO_2/20% N_2 (○) at 0 and 5°C.

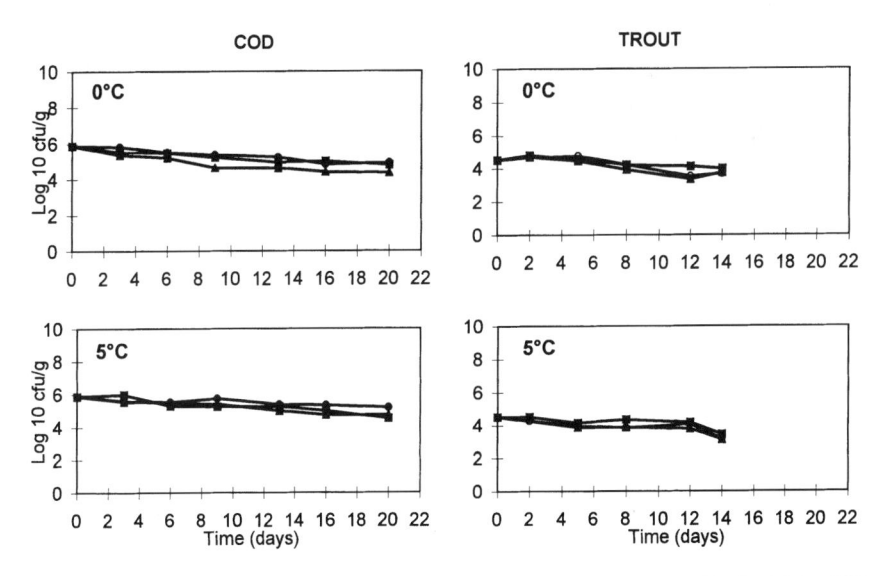

Figure 7.2 Growth\survival of *Salmonella typhimurium* on cod stored aerobically (■), in 60% CO_2/40% N_2 (▲) and in 40% CO_2/30% N_2/30% O_2 (●), and on trout stored aerobically (■), in 60% CO_2/40% N_2 (▲) and in 80% CO_2/20% N_2 (○) at 0 and 5°C.

212 FISH PROCESSING TECHNOLOGY

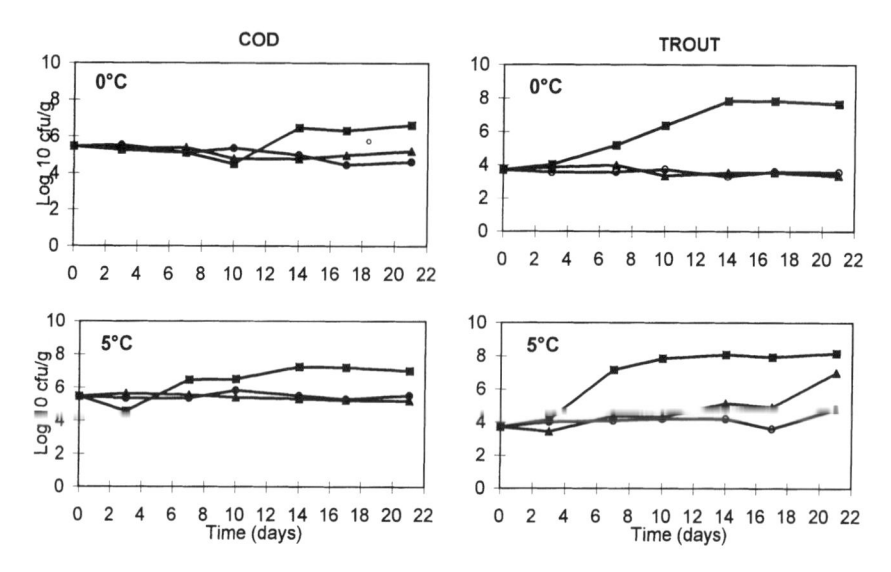

Figure 7.3 Growth\survival of *Aeromonas* spp. on cod stored aerobically (■), in 60% CO_2/40% N_2 (▲) and in 40% CO_2/30% N_2/30% O_2 (●), and on trout stored aerobically (■), in 60% CO_2/40% N_2 (▲) and in 80% CO_2/20% N_2 (○) at 0 and 5°C.

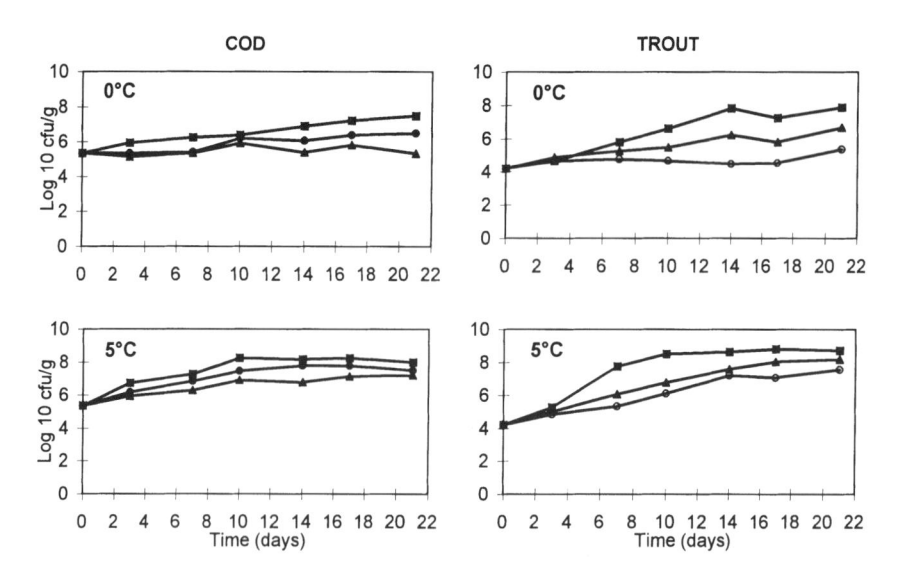

Figure 7.4 Growth\survival of *Yersinia enterocolitica* on cod stored aerobically (■), in 60% CO_2/40% N_2 (▲) and in 40% CO_2/30% N_2/30% O_2 (●), and on trout stored aerobically (■), in 60% CO_2/40% N_2 (▲) and in 80% CO_2/20% N_2 (○) at 0 and 5°C.

in the 80% CO_2/20% N_2 atmosphere. On both cod and trout, numbers of *S. typhimurium* remained static or showed a slight decline in all atmospheres at both temperatures over the storage period.

The growth/survival of *Aeromonas* spp. at 0 and 5°C on cod and trout is presented in Fig. 7.3. At both 0 and 5°C, on cod, numbers remained static in both modified atmospheres but increased (1–1.5 $\log_{10}$) in the aerobically stored packs. On trout, numbers of *Aeromonas* spp. remained static in both modified atmospheres at 0°C whilst at 5°C numbers increased in both modified atmospheres towards the end of the storage period, with the increase being greater in the lower CO_2-containing atmosphere. At both temperatures (0 and 5°C) numbers increased by c. 4 $\log_{10}$ on the aerobically stored trout.

The growth/survival of *Y. enterocolitica* at 0 and 5°C is presented in Fig. 7.4. On both cod and trout, numbers increased in all three atmospheres at both temperatures, with the greatest increase observed in the aerobically stored fish (c. 3.5–4.0 $\log_{10}$). The highest CO_2-containing atmosphere (60% CO_2/40% N_2 for cod and 80% CO_2/20% N_2 for trout) was the most inhibitory.

Inoculation of the fish with the pathogens was not found to affect the developing flora in any of these studies (results not presented).

7.5.4 Discussion

We observed a carbon dioxide-dependent bacteriostasis of *Salmonella* at chill temperatures on cod and trout, which concurs with the studies of Luiten *et al.* (1982) and Baker *et al.* (1986). Luiten *et al.* (1982) observed an increase in numbers of *S. typhimurium* at 10°C on beefsteaks overwrapped with an oxygen-permeable film, whereas in an atmosphere of 60% CO_2/40% O_2 numbers remained static over the 9-day storage period. Baker *et al.* (1986) observed an inhibitory effect of 80% CO_2 on *S. typhimurium* at 2, 7 and 13°C, with numbers remaining stable at 2°C over 5 days.

In the present study, both *Aeromonas* and *Yersinia* were able to grow on aerobically stored fish, with growth generally being somewhat greater on trout than on cod. Growth of both organisms was reduced when the cod and trout were packaged in the two modified atmospheres (60% CO_2/40% N_2 and 40% CO_2/30% N_2/30% O_2 for cod; 60% CO_2/40% N_2 and 80% CO_2/20% N_2 for trout). This inhibition of these organisms by CO_2 agrees with the studies of Gill and Reichel (1989) and Eklund and Jarmund (1983). With both fish, the higher CO_2-containing atmosphere was more inhibitory to both these organisms and, as found by others (Gill and Tan, 1980; Finne, 1982) the effect of CO_2 was diminished at the higher temperature.

In our studies with *L. monocytogenes*, at 5°C, differences were observed between the cod and the trout. On cod, numbers remained generally static in all atmospheres whilst on the trout, numbers increased by c. 3 $\log_{10}$ cfu g^{-1} in the aerobically stored pack and by c. $0.5\log_{10}$ in the 60% CO_2/40% N_2 atmosphere. Conflicting results regarding the growth/survival of *Listeria* on fish have also been reported by other workers. Harrison *et al.* (1991) observed no increase in the *L. monocytogenes* population of film-overwrapped and vacuum-packed fish and shrimp when stored on ice, whereas Lovett *et al.* (1988) obtained extensive growth on whitefish stored at 7°C. We observed no difference in the inhibition of *L. monocytogenes* between the aerobic (40% CO_2/30% N_2/30% O_2) and anaerobic (60% CO_2/40% N_2) atmospheres examined for cod. This differs from Wimpfheimer *et al.* (1990), who observed an increase of *L. monocytogenes* numbers of nearly $6\log_{10}$ cfu g^{-1} at 4°C on raw chicken packaged in an aerobic modified atmosphere (72.5% CO_2/22.5% N_2/5% O_2) whilst no growth was observed in an anaerobic modified atmosphere (75% CO_2/25% N_2).

There is little information on the incidence of *Yersinia* in fish, but Simard and Villemure (1989) found a high proportion of Atlantic cod naturally contaminated with *Y. enterocolitica*. Surveys of fish for motile aeromonads in England, New Zealand and Switzerland have found 32, 34 and 94% contaminated, respectively (Fricker and Tompsett 1989; Hudson and de Lacy, 1991; Gobat and Jemmi, 1993), whilst a survey of frozen seafood from several countries found 61% to be contaminated with *Listeria* spp., with 26% of these positive for *L. monocytogenes* (Weagent *et al.*, 1988). A survey in New Zealand found 35% of processed seafood contaminated with *L. monocytogenes* (Fletcher *et al.*, 1994). This high incidence of contaminated fish and the ability of these organisms to grow at refrigerated temperatures indicate the potential food safety issues from such foods. However, in this study pathogen growth on MAP fish was never greater and was generally markedly less than that on aerobically stored product.

To conclude, growth of the pathogens on the MAP fish was never greater and was generally markedly less than that observed on the aerobically stored fish. This reduction in growth was greater in the higher CO_2-containing atmosphere and at the lower temperature.

7.6 Future developments

Davies (1995) has reviewed several recent developments in MAP in general which offer the potential to improve further the safety of MAP and to extend the technology available to a wider range of products. These include the development of intelligent packaging, the use of predictive modelling, the use of combination treatments and developments in packaging films and equipment. All of these developments, along with a significant move towards the use of risk assessment based systems to control food safety, are likely to have some influence on the future use of MAP for fish and fish products. These are discussed below.

7.6.1 Combination treatments

The use of combined treatments (or hurdles) to control microbial safety and quality, whilst not being a new approach, is currently receiving renewed attention. The principles and uses of such applications have been reviewed by Leistner (1995).

Several workers (Fey and Regenstein, 1982; Regenstein, 1982; Statham et al., 1985; Daniels et al., 1986; Sharp et al., 1986; Barnett et al., 1987; Taylor et al., 1990) have examined the effect of chemical pre-treatments in combination with MAP for fish, whilst others (Eklund, 1982; Licciardello et al., 1984; Przybiski et al., 1989) have examined the potential of irradiation or irradiation and chemical pre-treatment in combination with MAP.

Chemical pre-treatments that have been examined include potassium sorbate, polyphosphates, sodium chloride, linoleic acid and nisin; these studies have been reviewed by Davis (1993) and Stammen et al. (1990). All these combination treatments (chemical and irradiation) have been shown, on occasions, to provide a benefit in shelf-life, safety, quality or a combination of these factors for various products. Occasionally, a synergistic effect of the treatments and MAP has been observed.

As a consequence of the recent interest in the use of natural antimicrobial systems, Tassou et al. (1996) have recently examined the effect of an olive oil mixture with oregano and lemon juice, in combination with modified atmosphere storage, on populations of inoculated *Salmonella enteritidis* and *Staphylococcus aureus* as well as on the autochthonous flora of fresh Greek fish. The combined treatment had bacteriostatic and bactericidal effects on both inoculated pathogens, as well as on the autochthonous flora. They found that *Brochothrix thermosphacta* and pseudomonads dominated the spoilage flora, under modified atmosphere and under air, respectively, whilst *Shewanella putrefaciens* was clearly inhibited.

Because of the paucity of information on the effect of these combination treatments, however, it is currently necessary to evaluate any proposed treatments for individual foods. Further studies are required to gain a fuller understanding of the true potential of these combined treatments and this is currently being addressed as part of an EU programme (Improving the quality and safety of whole fresh fish, AIR2-CT94-1496). It is suggested by Huss (1995) that, in the future, it is likely that MAP will be used in combination with preservation techniques that have been developed specifically to inhibit growth of CO_2-resistant trimethylamine oxide (TMAO)-reducing marine spoilage bacteria such as *P. phosphoreum*.

7.6.2 Predictive/mathematical modelling

Mathematical modelling affords the potential to make significant improvements in at least two areas of MAP. The first and most advanced of these is in the prediction of the microbial safety and shelf-life of MAP products. The second aspect is in the optimisation (gas to product ratio, film permeability, etc.) of MAP for specific products.

Predictive microbiology uses mathematical equations to estimate the growth, survival or death of micro-organisms as affected by extrinsic (processing and storage conditions) and intrinsic parameters of the food (*e.g.* salt concentration, pH or a_w; Willcox *et al.*, 1993). To date, few people have developed models that include gaseous atmosphere that would be relevant to MAP fish and fish products. However, some models have been developed and efforts are continuing in this area. For example, Einarsson (1992) has developed a model that could be used to predict the shelf-life of MAP cod fillets, whilst Willcox *et al.* (1993) modelled the influence of temperature and carbon dioxide upon the growth of *Pseudomonas fluorescens*. Dalgaard *et al.* (1994) have examined the potential of turbidimetric methods in determining the specific growth rates of several bacterial strains (some isolated from fish) for use in predictive modelling. Probably the most extensive predictive model developed for MAP fish is the *C. botulinum* model of Baker and Genigeorgis (1990) discussed in section 7.3.1.

In the UK, the Ministry of Agriculture, Fisheries and Food (MAFF) initiated a nationally coordinated 5-year programme of research into the growth and survival of micro-organisms in foods, with the aim of developing a computerised predictive microbiology data base in the UK (McClure *et al.*, 1994). The data base developed was launched as a commercial service in 1992 and a personal computer (PC) based version (Food MicroModel) is now available. Currently, the effect of gaseous atmosphere is only available for two organisms (verocytotoxigenic *Escherichia coli* and mesophilic *Bacillus cereus*); however, other organisms may be included in the future.

The major difficulty in modelling the effect of the gaseous environment is the very dynamic nature of the atmosphere, particularly with products such as fish. It is therefore necessary to have a good understanding of the changes in the gaseous composition with time (which in itself is dependent on numerous other variables), and a sufficiently sophisticated model to be able to predict the effect of this on microbial growth/survival.

At present, the selection of the appropriate modified atmosphere and the pack to product ratio is often a largely empirical trial-and-error exercise. As mentioned above, many variables interact within a modified atmosphere pack and because many of these variables feed back to alter related factors, it is difficult to manipulate all of these variables simultaneously in a quantitative manner (Day, 1993b). Computer models have been developed that aid in elucidating the salient variables active in a particular package design and in optimising the design while minimising development time and costly laboratory testing (Zagory, 1990). However, none of the mathematical models to date has been comprehensive enough to include all of the salient intrinsic and extrinsic parameters (Day, 1993b).

7.6.3 Intelligent packaging

The development of 'smart' films (Sneller, 1986), more recently described as 'active' packaging (Labuza, 1989) or intelligent packaging (Summers, 1992), is probably the aspect of MAP that has received the greatest attention in recent years. Intelligent packaging is defined by Summers (1992) as 'an integral component or inherent property of a pack, product or pack/product configuration which confers intelligence appropriate to function and use of the product itself and has the ability to sense or be sensed and to communicate'. Church (1994) suggests that intelligent packaging techniques could now be categorised under 12 headings:

- O_2 removal
- O_2 barrier
- water removal
- gas indicator
- ethylene removal
- CO_2 release
- antimicrobial action
- preservative release (*e.g.* ethanol generators)
- aroma release
- taint removal
- time–temperature indicators
- edible films.

These techniques have been described by Labuza and Breene (1989), Labuza (1990), Robertson (1991), Church (1994) and Davies (1995) amongst others. Of the techniques, O_2 scavengers have probably received the widest attention and greatest commercial acceptance. Two approaches to the development of O_2 scavenging systems have been examined. The most successful commercially have been the use of sachets which are included in the pack while the second (and more exciting area) has been the development of films containing immobilised enzymes. Abe (1990) discussed the attributes of oxygen absorbers that led to their widespread use in Japan, whilst Idol (1991) has critically reviewed their use, specifically addressing the resistance encountered to their adoption in the United States. This resistance largely results from a fear of litigation arising from the accidental ingestion of the scavenger. Recent developments in which the O_2 scavenger is presented in a separate sealed compartment, or is incorporated on the lidstock sealed under the paper label, may overcome these fears and further open the market (Church, 1994).

New types of O_2 absorbers have recently been developed. These include ones for use with frozen foods, which overcome the slow absorption speed of the standard types at low temperatures, some that are microwaveable, and some that emit the same volume of CO_2 as that of O_2 absorbed (Abe, 1990).

Whilst the development of O_2 scavengers is not new it is an area receiving greater attention for those involved with MAP. The same is also true for the use of time–temperature indicators (TTIs). Despite commercial systems being available and there being over 150 patents on the technology, market penetration has been low (Church, 1994). One of the problems with the use of TTIs for MAP is that often the indicator is some distance from the product and so does not truly represent the product temperature.

One further aspect of intelligent packaging that could benefit the fish sector is the development and commercialisation of gas indicators. CO_2- and O_2-sensitive labels, which change colour at set concentrations of gas, are now available. These labels may have considerable potential as non-destructive pre-spoilage indicators and could also be used to detect faulty packaging and product tampering (Church, 1994).

7.6.4 *Developments in packaging films and equipment*

As well as the developments in the intelligent film area, advances have been made notably in the area of in-line, non-destructive leak detection and in the development of retail and bulk MAP systems.

Much of the development work for MAP packaging in general has centred on retail packs and includes the development of microwaveable (Roberts, 1990) and resealable (Cakebread, 1994) MAP packs, and the development of triple-web packs which combine the advantages of both MAP and vacuum-skin packaging (VSP). Whilst they are currently not being extensively used for fish and fish products, all these developments could be applied to these products.

Recently, several companies have developed new bulk MAP systems and the New Zealand Captech system has entered the European market (Cakebread, 1994). These bulk MAP systems could be applied to fish and would circumvent the potential concerns regarding temperature abuse that may occur with retail packs. A current EU-funded project (AIR 1 CT92 0273, assuring the quality of fresh fish onboard and on shore by means of modified atmosphere, development of equipments, packaging and methods) is currently addressing the potential for applying modified atmospheres onboard ship. Initial results indicate the following:

- In the case of vessels on long trips, the medium or poor quality of fish landed at present can be considerably improved. The quality on landing is better in the case of treated fish, although the texture is softer than that of untreated fish. A prolongation of the total time at sea by about 1 week is possible.
- For these types of vessel, the preparation of the hold in order to achieve this aim is simple and easily adaptable according to size. In the case of new vessels, a solution has been studied to ensure a highly efficient and simple way of loading and unloading the fish hold.

- Fish which have been treated onboard retain a residual effect when they are put on display in air, thus considerably increasing their commercial shelf-life, even when distributed in the traditional way. Trays of medium permeability, which are therefore cheaper, may be used for distribution on land.

Such an approach obviously holds considerable potential as it enables application of the modified atmosphere at the earliest possible opportunity.

7.6.5 Quality Assurance of MAP

As with other food sectors, the hazard analysis critical control point (HACCP) system has now been widely adopted in the fisheries sector (see Chapter 8). Indeed, in the European Union, both the General and Fisheries Directives (93/43/EEC and 91/493/EEC) have a requirement for the use of HACCP or HACCP-based principles and, in the USA, HACCP will be mandatory for the fisheries sector by December 1997 (Federal Register 60 [242], 18/12/95 pp 65096-202).

Huss (1995) reviews the role of HACCP in the assurance of fresh fish quality and states that:

> the great advantage of the HACCP system is that it constitutes a scientific and systematic, structural, rational, multi-disciplined, adaptable and cost-effective approach of preventative quality assurance. Properly applied, there is no other system or method which can provide the same degree of safety and assurance of quality, and the daily running cost of a HACCP system is small compared with a large sampling programme.

The regulations and guidelines regarding the manufacture and sale of MAP and *sous vide* products have been reviewed by Farber (1995). With specific reference to fish and fish products in the United States, the National Advisory Committee on Microbiological Criteria for Foods (NACMCF) has developed recommendations on the safety of vacuum packaging or MAP for refrigerated raw fishery products (NACMCF, 1992). In the United Kingdom, the Sea Fish Industry Authority (Anon, 1985) has published guidelines for the handling of fish packaged in a controlled atmosphere and, internationally, the Codex Committee on Fish and Fishery Products has prepared a proposed draft code of hygienic practice for fish and fishery products in controlled and modified-atmosphere packaging (Codex Alimentarius Commission, 1994).

Of great significance are the overall comments of the National Advisory Committee (NACMCF, 1992) that temperature is the primary preventative measure against the possible hazard of toxin production by *Clostridium botulinum*, leading to the recommendation that the sale of vacuum-pack/MAP raw fishery products only be allowed when certain conditions are met. These conditions include storage of the product at $\leqslant 3.3°C$ at all points from packaging onwards, the use of high quality raw fish, adequate product labelling in respect of storage temperature, adequate shelf-life and cooking requirements and the use of a HACCP plan. It was also noted that, at all times, organoleptic spoilage and rejection by the consumer should come before the possibility of toxin production.

Acknowledgements

The experimental section is reproduced from a paper submitted for publication in the proceedings of the International Seafood Conference, Noordwijkerhout, The Netherlands, 13–16 November and is reproduced with their kind permission. The financial support of the European Union (Fisheries and Aquaculture Research [FAR] programme; Contract No. UP-2-515) and the Science and Technology Policy Committee of the Leatherhead Food Research Association, for the experimental work, is gratefully acknowledged.

References

Abe, Y. (1990), *Ageless – Oxygen Absorber*. Food Packaging Technology International. Interpack 90. The World of Packaging, Dusseldorf, 7–13.6.90, pp. 183–184.
Anon. (1985), *Guidelines for the Handling of Fish Packed in a Controlled Atmosphere*. Sea Fish Industry Authority, Edinburgh.
Avery, S.M., Hudson, J.A. and Penney, N. (1994), Inhibition of *Listeria monocytogenes* on normal ultimate pH beef (pH 5.3–5.5) at abusive storage temperatures by saturated carbon dioxide controlled atmosphere packaging. *J. Food Prot.* **57**, 331–333, 336.
Baker, D.A. and Genigeorgis, C. (1990), Predicting the safe storage of fresh fish under modified atmospheres with respect to *Clostridium botulinum* toxigenesis by modelling length of the lag phase of growth. *J. Food Prot.* **53**, 131–140.
Baker, R.C., Quereshi, R.A. and Hotchkiss, J.H. (1986), Effect of an elevated level of carbon dioxide containing atmosphere on the growth of spoilage and pathogenic bacteria at 2, 7 and 13 °C. *Poult. Sci.* **65**, 729–737.
Barnett, H.J. Conrad, J.W. and Nelson, R.W. (1987), Use of laminated high- and low-density polyethylene flexible packaging to store trout (*Salmo gairdneri*) in modified atmosphere. *J. Food Prot.* **50**(8), 645–651.
Beuchat, L.R. (1991), Behaviour of *Aeromonas* species at refrigeration temperatures. *Int. J. Food Microbiol.* **13**, 217–224.
Blackburn, C. de W. and Davies, A.R. (1994), Development of antibiotic-resistant strains for the enumeration of foodborne pathogenic bacteria in stored foods. *Int. J. Food Microbiol.* **24**, 125–136.
Brody, A.L. (1993), The market. In *Principles and Applications of Modified Atmosphere Packaging of Food*, Parry, R.T. (Ed.), Blackie, London, pp. 19–40.
Cakebread, D. (1994), *MAP Market Developments and Opportunities*. Leatherhead Food RA, Modified-atmosphere Packaging Course, T072.
Church, P.N. (1994), Developments in modified-atmosphere packaging and related technologies. *Trends Food Sci. Technol.* **5**, 345–352.
Codex Alimentarius Commission (1994), Proposed draft code of hygienic practice for fish and fishery products in controlled and modified atmosphere packaging. Codex Committee on Fish and Fishery Products. Joint FAO/WHO Food Standards program. May 2–6, 1994, Bergen, Norway.

Dalgaard P., Ross, T., Kamperman, L., Neumyer, K. and McMeekin, T.A. (1994), Estimation of bacterial growth rates from turbidimetric and viable count data. *Int. J. Food Microbiol.* **23**, 391–404.

Daniels, J.A., Krishnamurthi, R. and Rizvi, S.S.H. (1986), Effects of carbonic acid dips and packaging films on the shelflife of fresh fish fillets. *J. Food Sci.* **51**(4), 929–931.

Davies, A.R. (1995), Advances in modified-atmosphere packaging. In *New Methods of Food Preservation*, G.W. Gould (Ed.), Blackie, London, pp. 304–318.

Davis, H.K. (1993), Fish. In *Principles and Applications of Modified Atmosphere Packaging of Food*, Parry, R.T. (Ed.), Blackie, London, pp. 189–228.

Davis, H.K. (1995), Modified atmosphere packaging (MAP) of fish and seafood products. In *Proceedings Modified Atmosphere Packaging (MAP) and Related Technologies*, Campden and Chorleywood Food Research Association, Gloucestershire, UK, pp. 1–13.

Day, B.P.F. (1993a), Recent developments in MAP. *Eur. Food Drink Rev.* **Summer**, pp. 87–95.

Day, B.P.F. (1993b), Fruit and vegetables. In *Principles and Applications of Modified Atmosphere Packaging of Food*, Parry, R.T. (Ed.), Blackie, London, pp. 114–133.

Dodds, K.L. (1993a), *Clostridium botulinum* in the environment. In Clostridium botulinum, *Ecology and Control in Foods*, Hauschild, A.H.W. and Dodds, K.L. (Eds), Marcel Dekker, New York, pp. 21–51.

Dodds, K.L. (1993b), *Clostridium botulinum* in foods. In Clostridium botulinum, *Ecology and Control in Foods*, Hauschild, A.H.W. and Dodds, K.L. (Eds), Marcel Dekker, New York, pp. 53–68.

Einarsson, H. (1992), Predicting the shelf life of cod (*Gadhus morhua*) fillets stored in air and modified atmosphere at temperatures between $-4°C$ and $+16°C$. In *Quality Assurance in the Fish Industry*, Huss, H.H. et al. (Eds), Elsevier Science Publishers, Amsterdam.

Eklund, M.W. (1982), Significance of *Clostridium botulinum* in fishery products preserved short of sterilization. *Food Technol. Aust.* **36**, 12.

Eklund, T. and Jarmund, T. (1983), Microculture model studies on the effect of various gas atmospheres on microbial growth at different temperatures. *J. Appl. Bacteriol.* **55**, 119–125.

Enfors, S.-O., Molin, G. and Ternstrom, A. (1979), Effect of packaging under carbon dioxide, nitrogen or air on the microbial flora of pork at 4°C. *J. Appl. Bacteriol.* **47**, 197–208.

Farber, J.M. (1991), Microbiological aspects of modified-atmosphere packaging technology – a review. *J. Food Prot.* **54**, 58–70.

Farber, J.M. (1995), Regulations and guidelines regarding the manufacture and sale of MAP and *sous vide* products. In *Principals of Modified-atmosphere and* sous vide *Product Packaging*, Farber, J.M. and Dodds, K.L. (Eds), Technomic, Pennsylvania, pp. 425–548.

Fey, M.S. and Regenstein, J.M. (1982), Extending shelf-life of fresh wet red hake and salmon using CO_2-O_2 modified atmosphere and potassium sorbate ice at 1°C. *J. Food Sci.* **47**, 1048–1054.

Finne, G. (1982), Modified- and controlled-atmosphere storage of muscle foods. *Food Technol.* **36**, 128–133.

Fletcher, G.C., Rogers, M.L. and Wong, R.J. (1994), Survey of *Listeria monocytogenes* in New Zealand seafood. *J. Aqua. Food Prod. Technol.* **32**, 13–24.

Fricker, C.R. and Tompsett, S. (1989), *Aeromonas* spp. in foods: A significant cause of food poisoning? *Int. J. Food Microbiol.* **9**, 17–23.

Gibson, D.M. and Davis, H.K. (1995), Fish and shellfish products in *sous vide* and modified atmosphere packs. In *Principles of Modified-atmosphere and* sous vide *Product Packaging*, Farber, J.M. and Dodds, K.L. (Eds), Technomic, Pennsylvania, pp. 153–174.

Gill, C.O. and Reichel, M.P. (1989), Growth of the cold-tolerant pathogens *Yersinia enterocolitica, Aeromonas hydrophila* and *Listeria monocytogenes* on high-pH beef packaged under vacuum or carbon dioxide. *Food Microbiol.* **6**, 223–230.

Gill, C.O. and Tan, K.H. (1980), Effect of carbon dioxide on growth of meat spoilage bacteria. *Appl. Environ. Microbiol.* **39**, 317–319.

Gobat, P. and Jemmi, T. (1993), Distribution of mesophilic *Aeromonas* species in raw and ready-to-eat fish and meat products in Switzerland. *Int. J. Food Microbiol.* **20**, 117–120.

Golden, D.A., Eyles, M.J. and Beuchat, L.R. (1989), Influence of modified-atmosphere storage on the growth of uninjured and heat-injured *Aeromonas hydrophila*. *Appl. Environ. Microbiol.* **55**, 3012–3015.

Haard, N.F. (1992), Technological aspects of extending prime quality of seafood: A Review. *J. Aqua. Food Prod. Technol.* **1**, 9–27.

Harrison, M.A. Huang, Y., Chao, C. and Shineman, T. (1991), Fate of *Listeria monocytogenes* of packaged, refrigerated, and frozen seafood. *J. Food Prot.* **54**(7), 524–527.

Hudson, J.A. and De Lacy, K.M. (1991), Incidence of motile aeromonads in New Zealand retail foods. *J. Food Prot.* **54**, 696–699.

Huss, H.H. (1994), Assurance of seafood quality. FAO Fisheries Technical Paper No. 334. FAO, Rome.

Huss, H.H. (1995), Quality and quality changes in fresh fish. FAO Fisheries Technical Paper No. 348. FAO, Rome.

Idol, R.C. (1991), A critical review of in-package oxygen scavengers and moisture absorbers. In *Proceedings of CAP 91*, Sixth International Conference on Controlled/Modified Atmosphere/Vacuum Packaging, San Diego, California, Schotland Business Research Inc. pp. 181–190.

Labuza, T.P. (1989), Active food packaging technologies. In *Engineering and Food Vol. 2. Preservation Processes and Related Techniques*, Spiess, W.E.L. and Schubert, H. (Eds), Elsevier Applied Science, London, pp. 304–311.

Labuza, T.P. (1990), *Action Packs for Longer Life*. Food Packaging Technology International. Interpack 90. The World of Packaging, Dusseldorf, 7–13.6.90, pp. 190–194.

Labuza, T.P. and Breene, W.M. (1989), Application of 'active packaging' for improvement of shelf-life and nutritional quality of fresh and extended shelf-life foods. *J. Food Process. Preserv.* **13**, 1–69.

Lambert, A.D., Smith, J.P. and Dodds, K.L. (1991), Shelf-life extension and microbiological safety of fresh meat – a review. *Food Microbiol.* **8**, 267–297.

Leistner, L. (1995), Principles and applications of hurdle technology. In *New Methods of Food Preservation*, Gould, G.W. (Ed.), Blackie, London, pp. 1–21.

Licciardello, J.J., Revesi, E.M., Tuhkunen, B.E. and Racicot, L.D. (1984), Effect of some potentially synergistic treatments in combination with 100 krad irradiation on the iced shelf life of cod fillets. *J. Food Sci.* **49**, 1341–1346, 1375.

Lindsay, R.C. (1982), Controlled atmosphere packaging. In: *Proceedings of the Fourth Annual International Seafood Conference*, pp. 80–91, Washington, DC.

Lovett, J., Francis, D.W. and Bradshaw, J.G. (1988), Outgrowth of *Listeria monocytogenes* in foods. Society for Industrial Microbiology – Comprehensive Conference on *Listeria monocytogenes*, Rohnert Park, CA, Oct. 2–5.

Luiten, L.S., Marchello, J.A. and Dryden, F.D. (1982), Growth of *Salmonella typhimurium* and mesophilic organisms on beef steaks as influenced by type of packaging. *J. Food Prot.* **45**, 263–267.

McClure, P.J., Blackburn, C de W., Cole, M.B., Curtis, P.S., Jones, J.E., Legan, J.D., Ogden, I.D., Peck, M.W., Roberts, T.A., Sutherland, J.P. and Walker, S.J. (1994), Modelling the growth, survival and death of microorganisms in foods: the UK approach. *Int. J. Food Microbiol.* **23**, 265–275.

NACMCF (National Advisory Committee on Microbiological Criteria for Foods) (1992), Vacuum or modified atmosphere packaging for refrigerated raw fishery products. Adopted, March 20, 1992. Washington DC.

Parry, R.T. (1993), Introduction. In *Principles and Applications of Modified Atmosphere Packaging of Food*, Parry, R.T. (Ed.), Blackie, London, pp. 1–18.

Przybiski, L.A., Finerty, M.W., Grodner, R.M. and Gerdes, D.L. (1989), Extension of shelf-life of iced fresh channel catfish fillets using modified atmospheric packaging and low dose irradiation. *J. Food Sci.* **54**, 269–273.

Reddy, N.R., Armstrong, D.J., Rhodehamel, E.J. and Kauter, D.A. (1992), Shelf-life extension and safety concerns about fresh fishery products packaged under modified atmospheres: a review. *J. Food Saf.* **12**, 87–118.

Regenstein, J.M. (1982), The shelf-life extension of handdock in carbondioxide–oxygen atmospheres with and without potassium sorbate. *J. Food Qual.* **5**, 285–300.

Roberts, R. (1990), An overview of packaging materials for MAP. In *Proceedings of International Conference on Modified Atmosphere Packaging*, Campden Food and Drink Research Association, Chipping Campden, Glos, UK.

Robertson, G.L. (1991), The really new techniques for extending food shelf life. In *Proceedings of CAP 91*, Sixth International Conference on Controlled/Modified Atmosphere/Vacuum Packaging, San Diego, California, Schotland Business Research Inc., pp. 163–180.

Sharp, W.F. Jr, Norback, J.P. and Stuiber, D.A. (1986), Using a new measure to define shelf life of fresh whitefish. *J. Food Sci.* **51**, 936–939, 959.

Shewan, J.M. (1977), The bacteriology of fresh and spoiling fish and biochemical changes induced by bacterial action. In *Proceedings of the Conference on Handling, Processing and Marketing of Tropical Fish.*, Tropical Products Institute, London, 51–66.

Simard, R.E. and Villemure, G. (1989), Detection of pathogenic bacteria in Atlantic cod under carbon dioxide atmosphere. *Sci. Aliments.* **9**, 155–160.

Sneller, J.A. (1986), Smart films give a big lift to controlled atmosphere packaging. *Paper Film Converter* **12**, 58–59.

Stammen, K., Gerdes, D. and Caporaso, F. (1990), Modified atmosphere packaging of seafood. *CRC Crit. Rev. Food Sci. Nutr.* **29**, 301–316.

Statham, J.A. (1984), Modified atmosphere storage of fisheries products: the state of the art. *Food Technol. Aust.* **36**(5), 233–239.

Statham, J.A., Bremner, H.A. and Quarmby, A.R. (1985), Storage of Morwong (*Nedadactylus macropterus* Bloch and Schneider) in combinations of polyphosphate, potassium sorbate and carbon dioxide at 4°C. *J. Food Sci.* **50**, 1580–1584, 1587.

Summers, L. (1992), *Intelligent Packaging.* Centre for Exploitation of Science and Technology, London.

Tassou, C.C., Drosinos, E.H. and Nychas, G.J.E. (1996), Inhibition of resident microbial flora and pathogen inocula on cold fresh fish fillets in olive oil, oregano and lemon juice under modified atmosphere or air. *J. Food Prot.* **59**(1), 31–34.

Taylor, L.Y., Cann, D.D. and Welch, B.J. (1990), Antibotulinal properties of nisin in fresh fish packaged in an atmosphere of carbon dioxide. *J. Food Prot.* **53**(11), 953–957.

Tiffney, P. and Mills, A. (1982), Storage trials of controlled atmosphere packaged fish products. Sea Fish Industry Authority, Technical Report No. 191.

Ward, D.R. (1994), Microbiological quality of fishery products. In *Fisheries Processing, Biotechnological Application*, Martin, A.M. (Ed.), Chapman & Hall, London, pp. 1–17.

Weagent, S.D., Sado, P.N., Colburn, K.G., Torkelson, J.D., Stanley, F.A., Krane, M.H., Shields, S.C. and Thayer, C.F. (1988), The incidence of *Listeria* species in frozen seafood products. *J. Food Prot.* **51**, 655–657.

Willcox, F., Mercier, M., Hendrickx, M. and Tobback, P. (1993), Modelling the influence of temperature and carbon dioxide upon the growth of *Pseudomonas fluorescens*. *Food Microbiol.* **10**, 159–173.

Wimpfheimer, L., Altman, N.S. and Hotchkiss, J.H. (1990), Growth of *Listeria monocytogenes* Scott A, serotype 4 and competitive spoilage organisms in raw chicken packaged under modified atmospheres and in air. *Int. J. Food Microbiol.* **11**, 205–214.

Zagory, D. (1990), Application of computers in the design of modified atmosphere packages for fresh produce. In *Proceedings of International Conference on Modified Atmosphere Packaging*, Campden Food and Drink Research Association, Chipping Campden, Glos, UK.

8 HACCP and quality assurance of seafood
M. DILLON and V. McEACHERN

8.1 Introduction

The objective of this chapter is to review the approaches adopted in Europe and North America to overcome the practical problems in the development and verification of seafood quality systems. The provision of safe, quality seafood at the end of an increasingly complex food chain requires an effective system. Seafood businesses range in size from multinational organisations to sole traders – but the responsibility to provide the consumer with safe wholesome and quality food is the same. Safety is not an option but an essential and integral part of the planning, preparation and production of food products. Lack of management consideration to food safety at the product planning and factory development stage may represent serious threats to public health. The requirement for management to implement effective safety systems has been recognised within the new product-specific European Union (EU) legislation (*e.g.* meat and fish) and also within the broader horizontal directives. The emphasis within the General Directive on the Hygiene of Foodstuffs has moved from prescriptive structural and hygiene issues to the development of management systems based upon hazard analysis critical control point (HACCP). HACCP is a food safety management system which concentrates prevention strategies on known hazards and the risks of them occurring at specific points in the food chain (Dillon and Griffith, 1995). It is the specificity which makes HACCP so effective and the approach easily integrates into total quality management or accredited quality management systems (*e.g.* ISO 9000) which is increasingly required by the market or through legislation.

An increasing number of organisations from all sectors (20 000 by 1993 from within the UK), are developing quality management systems which meet the requirements of ISO 9000 (BS 5750, EN 2900). This is a model quality management system standard which outlines the key areas a business should address to ensure the effective operation of their quality management system. There has been an increasing number of companies wishing to develop appropriate quality systems as seen by the production of the food sector interpretation of ISO 9000 by the British Standards Institute (BSI) (1989, 1991). Robertson *et al.* (1992) described the development of ISO 9000 as the basis for effective quality systems in the shrimp and fish processing factories in Norway. The authors predicted that as 60% of the fish products were taken by the EU then ISO 9002 would become a purchaser requirement.

A detailed examination of the approach adopted by Canada in developing, implementing and verifying HACCP-based seafood quality systems provides an insight to both the problems facing the processor and the regulators of HACCP-based quality systems. The Canadian quality management programme (QMP) was the first mandatory food inspection programme based on HACCP principles. From February 1992, all federally registered fish processing plants operating in Canada were required by law to develop and implement an in-plant quality management programme. In early 1987 an industry/government working group began developing the fundamental principles of the quality assurance systems which was applied to all types of fish processing plants, large and small, from coast to coast (McEachern, 1992). The programme must be specific to the product and process as a condition of plant registration and a written QMP must be prepared by the company describing the processor control measures that ensure safe and wholesome food production and comply with regulatory requirements. QMP is based upon compliance with the Canadian Fish Inspection Regulations and is verified by the Department of Fisheries and Oceans (DFO).

The working group identified 12 generic elements of fish processing and their related hazards that must be considered when developing a QMP, which resulted in the DFO creating the QMP submission guide. Table 8.1 summarises the potential critical control points identified.

Denmark was the first country in the world to register fishing vessels in line with the requirements of another formal quality management system – ISO 9000. The increasing demands of the customer for safe food and the severe limitations of end-product microbial standards reported by Corlett (1991) have also resulted in the adoption by the United States of HACCP systems as the primary means to assure food safety. The recent changes in the technology of the manufacture, transport, distribution and consumer preparation of foods necessitated a more scientific basis for food safety management systems resulting in this increased application of HACCP.

8.2 Defining HACCP

HACCP is primarily a preventative control system rather than a cure and thus is well suited to achieving the cost-effective delivery of quality products at agreed prices. A critical examination of the organisation must occur by a suitably balanced team to identify hazards, assess risk and implement control, monitoring and corrective action procedures. The introduction of the Food Safety (General Food Hygiene) Regulations in 1995 ensured that common food hygiene rules were in place across the EU as set out in the directive (93/43/EEC). Industry guides to a good hygienic practice are voluntary guides providing more detailed advice on complying with the regulations as they relate to specific sectors. This horizontal EU legislation is linked to the

HACCP concept and the importance of effective hygiene management in controlling and identifying the potential hazards in a specific chain of production. HACCP has emphasised the benefits of preventative control in preference to end-product microbial testing which is time consuming and may be

Table 8.1 Potential critical control points (CCPs). From McEachern (1992) by courtesy of *Infofish International*

Items	Hazards	Critical control points
1 Fish	Health and safety risks	*Prior to processing* Unloading dock Receiving room – cool room
2 Other ingredients	Contamination of products with unapproved, unsafe compounds Use of compounds not meeting specifications; misapplication	*Prior to use* When received Before application or use Application area
3 Packaging material	Use of unapproved, damaged or unclean containers	*Prior to use* Packing area When received Immediately before use
4 Labels	Information not consistent with regulations	*Prior to use* Before application When received
5 Cleaning agents, sanitisers, lubricants	Contamination of product with unapproved, unsafe chemicals; misapplication	*Prior to use* When received Before application During application at application area
6 Construction/ maintenance of production facilities/processing equipment	Contamination of product due to faulty construction of plant or equipment Insufficient processing of product	*Prior to start-up/during operation* Twice per operating season Weekly evaluations at CCPs
7 Operation and sanitation	Contamination of product due to poor operation and sanitation practices	*Prior to start-up/during operation* Once per 3 months of operation Daily sanitation checks
8 Process control	Production of product that does not comply with safety, quality, wholesomeness and/or fair trade requirements	*During operation* Fish washing, can seaming, retort process, can cooling Cooling Freezing Fish washing, fish freezing
9 Storage	Decomposition or contamination of product due to poor storage conditions	*During operation of cold store*
10 Final product	Production of product that does not comply with safety, quality, wholesomeness and/or fair trade requirements	*Prior to packaging* On-line inspection Before packaging During storage
11 Recall procedures	Inability to trace product to customer	*During coding prior to shipping*
12 Employee qualifications	Production of product posing health and safety risks	*Prior to start-up* Retort operators

Table 8.2 The hazard analysis critical control point (HACCP) requirements of recent EU legislation. From Fish Processing, HACCP Workshop, Ireland (1994)

Regulation	Sector	Key phrases
90/667/EEC	Animal waste/feedstuffs	Tractability, risk category, identify and control CCP, representative samples, microbiological standards, corrective action
91/493/EEC	Marine foods	Identify CCP, monitoring, methodology, established/implemented, analysis, record results, appropriate measures taken
92/5/EEC	Meat products	Identify CCP, establishment/implementation, methods of monitoring, analysis, records, appropriate measures taken
92/46/EEC	Dairy products	Identify CCP, monitoring, appropriate methods, analysis, record results, appropriate measures taken
93/43/EEC	Horizontal (food)	Identify any step critical to food safety, adequate procedures identified, maintained and reviewed on HACCP principles including periodic monitoring

unreliable. HACCP utilises management resources more effectively by targeting the control or prevention of hazards.

The EEC directive 91/493 as seen in Table 8.2 shows the HACCP requirements of the seafood specific directive provided in more detail within 94/356. Dillon (1993) outlined the implications of the 1990 UK Food Safety Act with particular reference to the need for recognised appropriate quality systems based upon hazard analysis principles for seafood processors. The defence of due diligence has continued to be a central issue within food manufacturers and processors today. It was suggested by Pallet (1992) that HACCP would assist in attaining the defence of due diligence, due to its logical approach to the identification of potential hazards and implementation of controls within a given food processor's plant.

Guidelines on the statutory defence of due diligence were published in February 1991 and recommended that all reasonable precautions equated to the development of a control system and due diligence related to the effective operation of the control system (Lacots, 1991). Dillon (1991, 1993) reviewed the defence and emphasised that 'all reasonable precautions' and 'all due diligence' are related and cannot be separated. Case law has proved that every case is unique but factors such as the size and resources of a defendant's business and nature of circumstances of the specific offence will be considered. The divisional court distinguished between what was reasonable for a large retailer and a small village shop (Howells *et al.*, 1990), more will be expected of the former. 'Reasonable' will also be linked to the more stringent precautions required for the manufacture and distribution of high risk food (*e.g.* cook–chill seafood) and the recommendation to focus a quality system (ISO 9000) by

Table 8.3 Comparison of ISO 9000 and quality management programme (QMP)

ISO 9002	QMP
1 Management responsibility	Partially meets – QMP does not require management to perform internal audits
2 Quality systems	Meets – QMP submission includes system
3 Contract review	Does not address this aspect of the process
4 Design control	Does not address
5 Document and data control	Met through the approval of the processor's QMP submission and DFO verification
6 Purchasing	Covered through CCPs for input material
7 Control of customer – supplied product	Covered through CCP incoming fish
8 Product identification and traceability	Covered CCP recall
9 Process control	Covered
10 Inspection and test status	Covered
11 Control – inspection, measuring, equipment	Partially meets – only instruments used in controlling measuring critical processes
12 Inspection and test status	Covered
13 Control – non-conforming	Covered
14 Corrective action	Covered
15 Handling, packaging, store, transportation	Partially meets – does not cover delivery
16 Control – quality records	Covered
17 Internal quality audits	Third party audits by DFO
18 Training	Partially – employees controlling critical processes
19 Servicing	Does not address
20 Statistical techniques	Partially – processor must apply statistically valid product sampling plans

DFO, Department of Fisheries and Oceans; QMP, quality management programme.

employing HACCP was made. McEachern has compared ISO 9000 with the Canadian QMP programme in Table 8.3.

The author goes on to discuss the omissions and the differences within the QMP system in relation to the key elements of ISO 9002 (1994).

8.3 Application of QMP

8.3.1 *ISO 9002 elements not addressed by QMP*

The ISO 9002 elements of *3 (Contract Review)*, *4 (Design Control)*, and *19 (Servicing)* are either not applicable to the fish processing industry or are outside the mandate of a government food inspection agency and are therefore not addressed under the QMP.

8.3.2 *ISO 9002 elements partially addressed by QMP*

1 *Management responsibility.* QMP requires that the processor illustrates how QMP fits in to the plant organisation and how the programme principles, directives and strategies are realised in the day-to-day operation of the company. The processor must also identify the individual responsible for the in-plant QMP and describe the reporting relationship within the management structure. QMP does not specifically address the issue of the company performing internal audit, although it is understood that the processor must review the in-plant programme on an ongoing basis.

11 *Control of inspection, measuring and testing equipment.* QMP requires the processor to ensure the accuracy of equipment used to control, measure or test critical safety factors in a process. The processor is also required to maintain records of calibration.

15 *Handling, packaging, storage and delivery.* QMP requires controls to be in place for handling, packaging and storage of all incoming fish and input materials. QMP does not extend to controls over the delivery of the product.

18 *Training.* QMP requires the processor to ensure that employees operating retort equipment have received a recognised training course.

19 *Statistical techniques.* All sampling plans used by the processor must be statistically valid.

8.4 Practical aspects of planning and implementing HACCP systems

The seven principles of HACCP that must be addressed by any team involved in the design and implementation of a HACCP system are as follows:

1. Conduct analysis – identify hazards and specify preventative measures.
2. Identify critical control points (CCPs).
3. Establish target levels and critical limits for specified control measures.
4. Establish monitoring system.
5. Establish corrective action procedures.
6. Establish verification procedures.
7. Establish documentation for relevant system areas.

These principles have been expanded further within *Technical Manual 38* issued by the Chipping Campden Food Research Association (1992) who suggested dividing the principles into 14 stages during the completion of the HACCP study.

Many of the practical difficulties faced by seafood processors in developing a system are related to a lack of planning at the initial stages. Dillon and Horner (1994), Mortimer and Wallace (1995) and Dillon and Griffith (1995) have emphasised the contribution that simple project planning can make to effective HACCP system development. During the analysis stage it is vital that the company has a plan for the HACCP project. The initial phase of the project involves defining the terms of reference, agreeing the scope and the necessary skills needed within the team to achieve the effective development of the HACCP plan in an agreed time scale and within budget. The HACCP plan once developed must be verified against the agreed terms of reference and reviewed to ensure that it is still achieving the level of control necessary, which will be dealt with more specifically later in the chapter.

The effective implementation of the plan considered at the outset will now involve staff who may not have been primarily involved in deciding the significance of particular hazards. This group should have been made aware of the changes resulting from the new system and involved in modifying procedures or working practices relevant to their area. The training and awareness raising of staff is crucial to effective implementation. The ownership of the HACCP system must reside with the operatives as well as the management and their involvement at relevant stages should be planned to ensure effective change. The objective and scope of the HACCP plan should be defined at the outset. A project leader should be appointed and be involved in the development of the project plan and thus its effective management. Lack of information is another major problem area – the identification and compilation of objective data for the selected HACCP team to assess is another action which will reduce the time scale of implementation and focus the HACCP team discussions. The expected scope will define whether product safety and/or quality is to be considered and may also further define the initial hazard type to be considered (*i.e.* microbial, chemical or physical). Although it is recommended that the team develop an effective safety-based HACCP plan, market pressure has led to the practical necessity to develop quality-based systems focused on utilising the HACCP principles (QACCP – Dillon and Horner, 1995).

The team should begin by selecting the hazard type which is of major importance to the product (e.g. microbial) and complete the HACCP analysis – other hazard types can then be reviewed and added to this HACCP plan. Once the expected scope of the plan has been defined it is necessary for the HACCP team to collect the product details; this helps to ensure that the appropriate team members attend the preliminary meetings. Gathering the factual data on complaints, flow diagrams, factory layouts, customer base and shelf-life are also important. However, before any data are collected it is important that the purpose and the data collection method are established. It is important during verification to ensure that these data are accurate. Sufficient data are required to perform thorough analysis and need to be carefully recorded through simple checklists in order to avoid rework (Hudak-Roos and Spencer-Garret 1992). The project leader should chair the meetings utilising proforma with action minutes and detailed minutes maintained to allow proper management.

It is vital that the necessary support and commitment from the senior management team exists as this will be necessary to gain the required funding and aid in the removal of a number of potential barriers to completion, for example deployment of staff, release for training courses. Once this has been achieved and the budgets agreed, the HACCP team needs to agree the scope of the HACCP system it wishes to organise. The defined scope for the fictional company (MD Fisheries) was as follows: to manufacture to stringent food safety guidelines, to highlight and control specific risk areas and to design methods to monitor the risk areas identified. With this specification, the implementation of a HACCP system becomes a goal-orientated project and so therefore should be planned as such. The next step is for the team to decide upon the necessary tasks that they must undertake in order to achieve their goal. When a list of tasks has been drawn up, along with the corresponding duration times, a GANNT chart can be constructed. This is a simple way of illustrating the interdependencies of the tasks that must be completed and can allow those involved to see clearly what needs to be done. The GANNT chart for MD Fisheries is shown in Figure 8.1. Another technique that often accompanies GANNT charts is the milestone plan. This highlights the main points of the project and assigns them to a specific function within an organisation, for example, managerial, financial or technical. Each of the milestones has against it a completion date which allows those involved to see accurately if the project is on schedule. These two project management tools are excellent ways of methodically planning a HACCP system.

The actual 'hazard analysis' forms a large portion of the work involved in designing and implementing a HACCP system. Identifying contamination points and determining severity and probability of hazard occurrence are the main components of hazard analysis. Dillon and Horner (1994) proposed a risk matrix that allows for a balanced analysis of hazards including an estimate of risk and severity. This matrix may be used to assess CCP decisions and ensure that CCPs are necessary. An alternative American system for categorising hazards based on risk was explained by Bryan (1992) and further by Corlett and Pearson (1992). The risk assessment scheme is shown in Figure 8.2.

The team can employ the codex decision tree (Figure 8.3) to assist in identifying CCPs or utilise judgement and experience. After the CCPs are identified then the factors which support the associated control measures will require target values, management through monitoring and agreed corrective action for deviation. The documented system, needed to both prove that the system exists and is working, will then be finalised and eventually verified. An example of a completed HACCP plan is given in Figures 8.4–8.6.

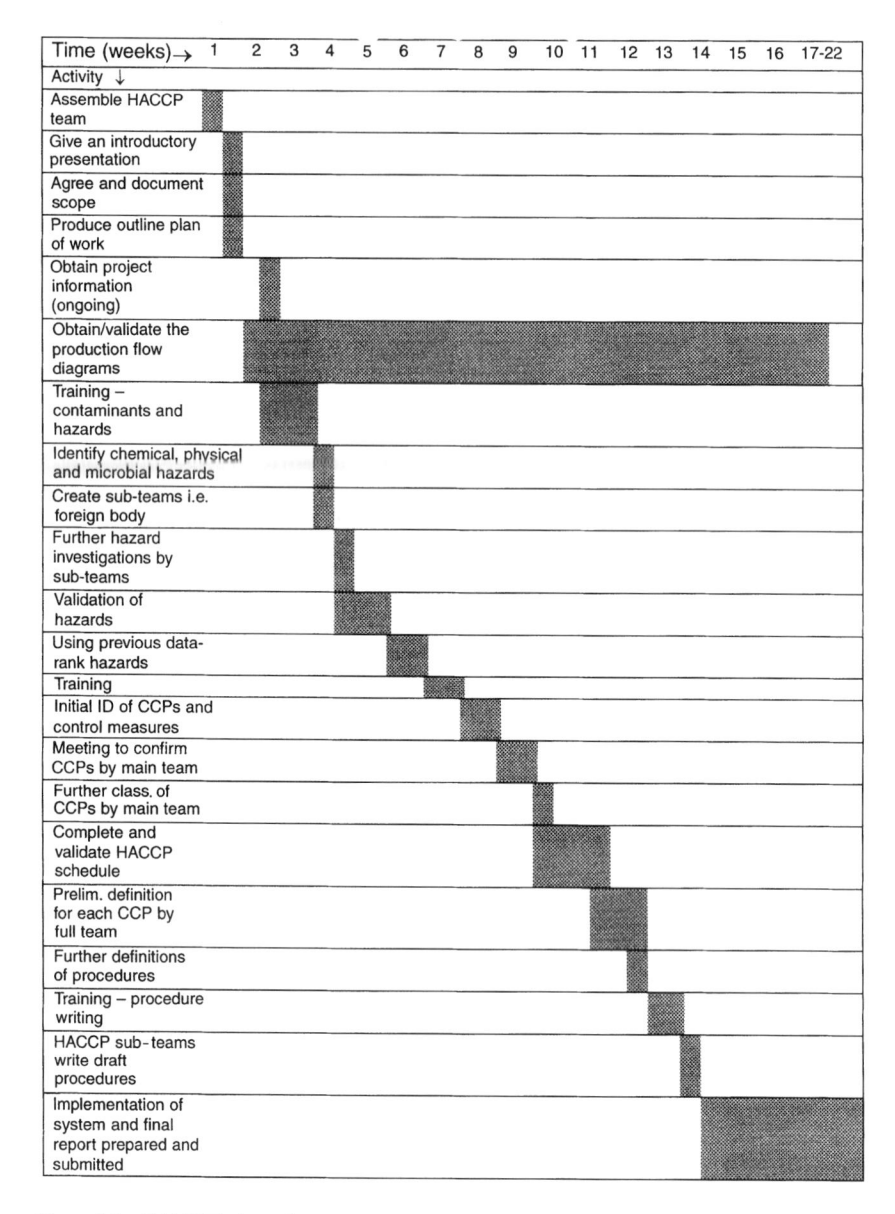

Time (weeks) →	1	2	3	4	5	6	7	8	9	10	11	12	13	14	15	16	17-22
Activity ↓																	
Assemble HACCP team	■																
Give an introductory presentation	■																
Agree and document scope	■																
Produce outline plan of work	■																
Obtain project information (ongoing)			■														
Obtain/validate the production flow diagrams			■	■	■	■	■	■	■	■	■	■	■	■	■	■	
Training – contaminants and hazards			■														
Identify chemical, physical and microbial hazards				■													
Create sub-teams i.e. foreign body				■													
Further hazard investigations by sub-teams					■												
Validation of hazards						■											
Using previous data - rank hazards							■										
Training								■									
Initial ID of CCPs and control measures								■									
Meeting to confirm CCPs by main team									■								
Further class. of CCPs by main team										■							
Complete and validate HACCP schedule										■	■						
Prelim. definition for each CCP by full team												■	■				
Further definitions of procedures													■				
Training – procedure writing														■			
HACCP sub-teams write draft procedures															■		
Implementation of system and final report prepared and submitted																■	■

Figure 8.1 GANNT chart for MD fisheries. CCP, critical control point; HACCP, hazard analysis critical control point.

The first section of the schedule (Figure 8.4) documents the work of the team in identifying the significant hazards, their sources and recommended controls – the highlighted section emphasises that either by judgement or the use of the recommended decision tree this process step has been agreed to be a CCP. The

HACCP AND QUALITY ASSURANCE OF SEAFOOD

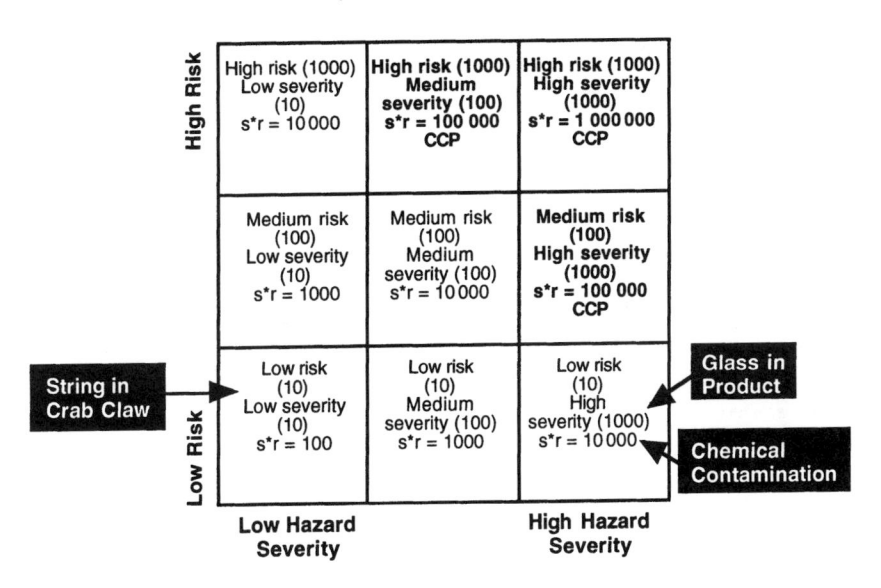

	Low Hazard Severity		High Hazard Severity
High Risk	High risk (1000) Low severity (10) s*r = 10 000	High risk (1000) Medium severity (100) s*r = 100 000 CCP	High risk (1000) High severity (1000) s*r = 1 000 000 CCP
	Medium risk (100) Low severity (10) s*r = 1000	Medium risk (100) Medium severity (100) s*r = 10 000	Medium risk (100) High severity (1000) s*r = 100 000 CCP
Low Risk	Low risk (10) Low severity (10) s*r = 100	Low risk (10) Medium severity (100) s*r = 1000	Low risk (10) High severity (1000) s*r = 10 000

String in Crab Claw → (Low risk, Low severity cell)

Glass in Product, Chemical Contamination → (Low risk, High severity cell)

Figure 8.2 Risk assessment matrix (CCP, critical control point).

next section of the HACCP schedule represented in Figure 8.5 records the HACCP team decisions in relation to the management of the system. Finally, examples of the necessary records which must be kept to ensure that the HACCP plan is working at the identified CCPs are shown in Figure 8.6.

Spencer-Garret and Hudak-Roos (1991) discussed the development of the mandatory seafood inspection system based upon HACCP system for the seafood industry commissioned by the US congress. The work was undertaken by the National Marine Fisheries Service and involved the investigation of pitfalls that occurred when implementing HACCP in food control systems. The adopted strategy minimised the effect of the perceived pitfalls by involving industry in commodity-based workshops, testing of the models in plants, and the summarised results reviewed by the industry steering committee. This was then reviewed by the core team and the regulatory HACCP model produced for use in seafood. The authors explained the key role of monitoring, corrective action and record-keeping elements of the HACCP system further (Hudak-Roos and Spencer-Garret, 1992).

The American approach has 'extended the use of HACCP strategy to combat fraud and has a specific sanitation CCP section which effectively controls many of the quality hazards' (Dillon and Horner, 1995). HACCP was subsequently adopted by the United States Food and Drugs Administration (FDA) as means of forcing companies to undertake a programme of internal inspection.

The Danish workers, Jacobson and Lillie (1992) compared HACCP and ISO 9000 for the fish sector and concluded that a combination of the two approaches

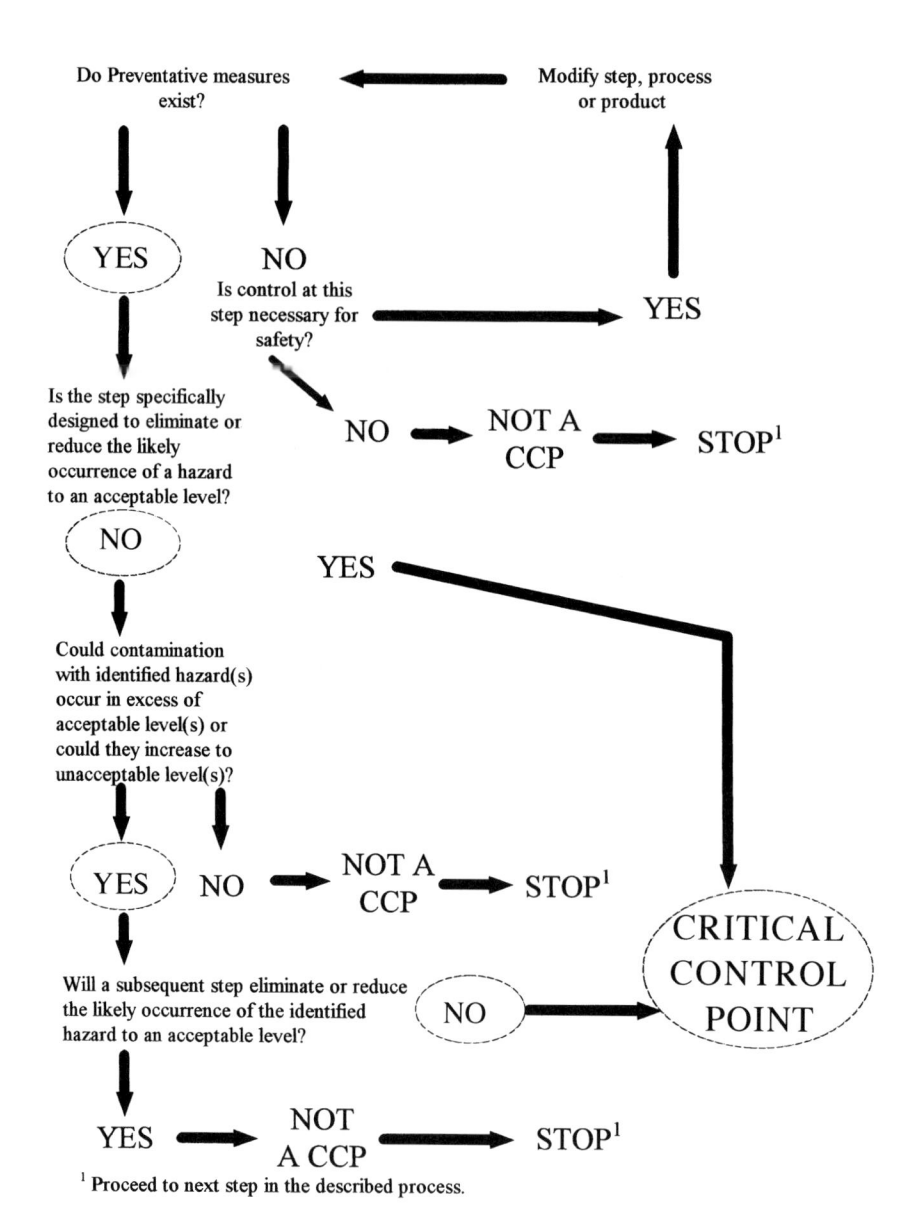

Figure 8.3 An example of the application of the Codex decision tree. The step of the process is taken as Step 5 – Picking claws with the hazard being *Staphylococcus aureus* and the source of the hazard being the staff. This example is shown in Figure 8.4. Alternatively, critical control points (CCPs) may be decided by experience and judgement.

[1] Proceed to next step in the described process.

Step No.	1 Potential hazard	2 Hazard source	3 Preventative measure(s)	4 Critical control point(s)
Step 5 – Picking claws	Microbial contamination Enteric pathogen *Staphylococcus aureus*	Staff	Hygiene training Medical screening Hygenic handling procedures	Step 5 Picking CCP4

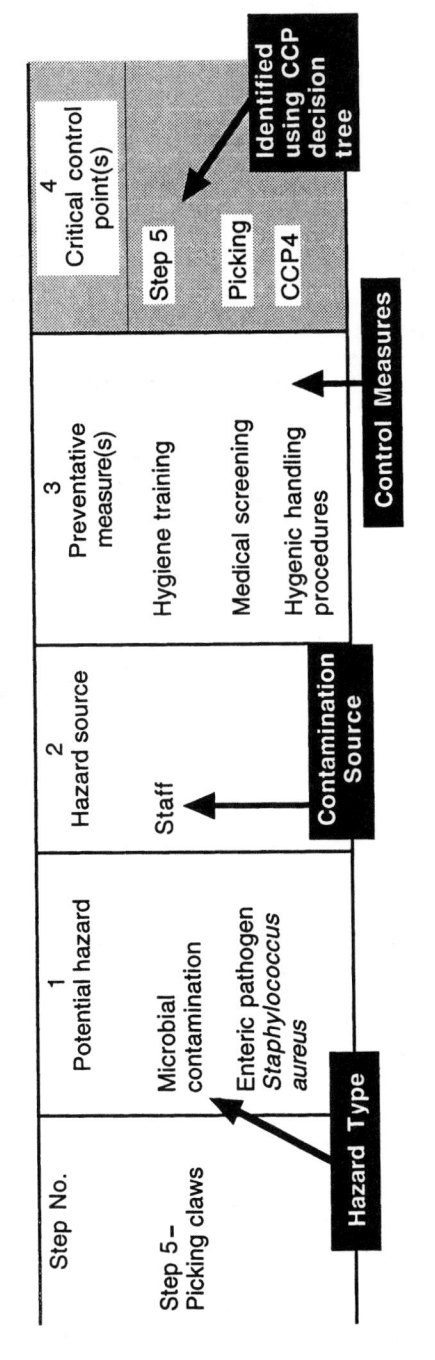

Hazard Type → (Potential hazard)
Contamination Source → (Hazard source)
Control Measures → (Preventative measures)
Identified using CCP decision tree → (Critical control points)

Figure 8.4 Section 1: HACCP analysis.

5 Target & Critical Limits	6 How Monitored	7 When Monitored	8 Who Monitors	9 Corrective Action
All staff trained	Training Assessment	Annually	Technical Manager	Re-train
Medical screening every 6 months	Medical Records Assessed	Quarterly	Technical Manager	Send to non-risk area
Comply with handling procedures	Visual Observation	Ongoing	Production Supervisor	Re-train and/or discipline

Zero tolerance. A time/temp control would have limits

Figure 8.5 Section 2: monitoring and control.

CCP 1	CCP 2	CCP 3	CCP 4
Incoming raw material microbial data physical, chemical. Vendor assurance data	Processing records e.g. temperature of cook, time of cook, etc.	Equipment hygiene, e.g. swab results, ATP data	Step 5 Picking claws, hygiene training, medical records, compliance records, end-product micro records

Figure 8.6 Example of critical control point (CCP) records.

Table 8.4 HACCP:QMP

HACCP	QMP
1 Conduct the hazard analysis and risk assessment	Hazard analysis of the major fish products and processes in Canada was performed by the QMP development group
2 Identify the critical control points (CCPs)	Generic CCPs were identified by the development group and set out in the QMP submission guide. Processors identify the CCPs specific to their product and processes using the QMP submission guide
3 Establish the critical limits	Equivalent
4 Establish the monitoring systems	Equivalent
5 Establish the corrective action systems	Equivalent
6 Establish verification procedures	Verification performed by DFO
7 Establish record-keeping system	Equivalent

HACCP, hazard analysis critical control point; DFO, Department of Fisheries and Oceans; CCP, critical control point; QMP, quality management programme.

was the most effective and envisaged certification through a registered quality system which encompassed all stages in the chain from capture to the table. McEachern (1992) reviewed the Canadian QMP programme with HACCP as shown in Table 8.4.

8.5 HACCP verification

Principle 6 of the HACCP plan is the verification stage which is increasingly conducted through a risk-based audit. Recent guidance provided to UK enforcement officers within their codes of practice has outlined a HACCP approach to the risk-based inspection of food processors (code of practice 8 and 9).

However, there are a number of ways in which an audit can be designed and will ultimately depend upon the requirements of the quality assurance system or legal standard in place. While principle 6 of HACCP specifies that the company shall implement a verification system, the Canadian QMP does not require this of its companies and places the onus of verification on the DFO inspection division. This, in turn, proceeds to place one system under the control of internal procedures and the other in the realms of external control. Prior to the implementation of QMP, the DFO inspected fish processing plants by traditional methods, the inspection of the processing environment and the final product. Figure 8.7 highlights sections of the traditional checklists used.

The Canadian QMP checklists are external audits conducted by a government body, and, as such, a brief look at the checklists will enable the reader to note very quickly that there is a strong pass or fail feel to them. This results

(a)

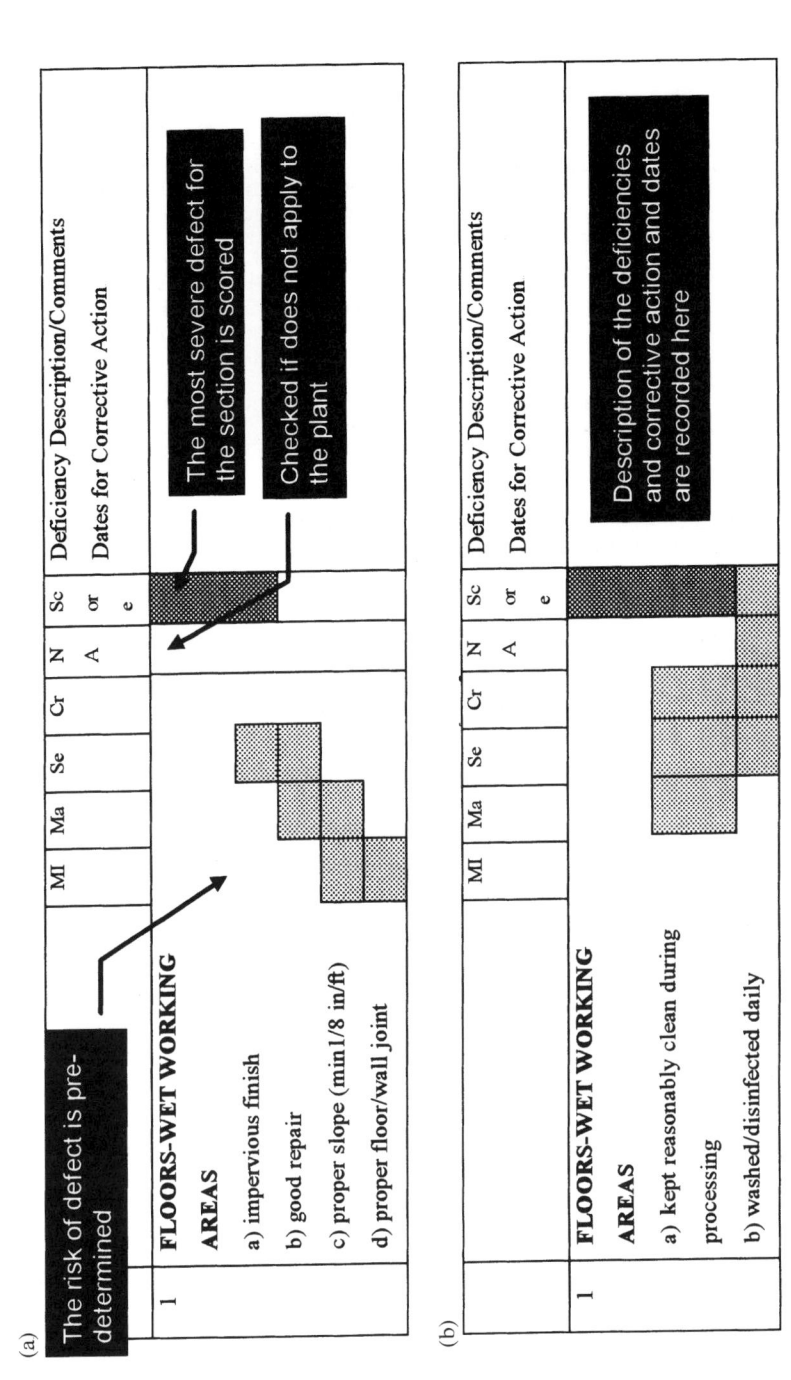

	MI	Ma	Se	Cr	NA	Score	Deficiency Description/Comments
							Dates for Corrective Action

The risk of defect is pre-determined

The most severe defect for the section is scored

Checked if does not apply to the plant

1	**FLOORS-WET WORKING AREAS**						
	a) impervious finish						
	b) good repair						
	c) proper slope (min 1/8 in/ft)						
	d) proper floor/wall joint						

(b)

	MI	Ma	Se	Cr	NA	Score	Deficiency Description/Comments
							Dates for Corrective Action

Description of the deficiencies and corrective action and dates are recorded here

1	**FLOORS-WET WORKING AREAS**						
	a) kept reasonably clean during processing						
	b) washed/disinfected daily						

Figure 8.7 Examples of (a) a schedule I requirement – construction and maintenance of floors and (b) a schedule II requirement – operation/sanitation of plant environment.

from the categories assigned to the different risks, based on severity. Within the Canadian plant audits the following are definitions of the deficiencies:

- A minor deficiency is one that is 'not in accordance with the requirements and not deemed to be major, serious or critical'.
- A major deficiency is one that would 'inhibit general sanitation and deterioration of the product and one that is not serious or critical'.
- In order for a failure-to-comply to be classified as a serious deficiency it must be such that it 'prevents proper plant sanitation, may result in tainted, decomposed or unwholesome product' but is not considered critical.
- Finally, a critical defect is one which 'results in unwholesome products, presents health and safety threats; not in accordance with good manufacturing practice (in respect to canneries)'.

Under the traditional inspection regime, DFO inspected over 80 specific items of the plant environment to determine if the plant was operating in compliance with construction and sanitation requirements set out in the Canadian Fish Inspection Regulations. Inspectors also sampled final product on a routine basis (1–2 weeks) to determine compliance with Canadian fish product standards. The introduction of QMP shifted the responsibility for the plant environment inspection, final product inspections and other control and assurance measures to the processing plants. Fish processing plants are now required to monitor their compliance with the regulatory requirements and perform their own self-checks such as the plant environment inspections.

The role of DFO inspectors has shifted under the QMP, where they have assumed an auditing role to verify that processing plants are operating in compliance with the regulations. Inspectors perform a QMP inspection that verifies the processing plant's written in-plant QMP to ensure the documented standards, monitoring procedures, record-keeping systems and guidelines for corrective action meet the minimum requirements as set by the Department of Fisheries and Oceans. Confirmation that the in-plant QMP is being followed is by observation and subsequent verification that the relevant processor's records are accurate. The QMP inspection rating and report form was therefore developed as a checklist to assist the inspector in performing the verification of controls at each CCP. In the course of performing the verification, inspectors review the records connected to the CCP, observe the monitoring procedures taking place in the plant if possible, review and follow-up on the corrective actions and perform a parallel inspection. Deficiencies in the programme are rated as Class I, II or III.

8.5.1 *Defect definitions*

8.5.1.1 *Class I.* A Class I deficiency means a situation where the registered establishment has deviated from the Department's QMP requirements in such a manner that the product fails to meet safety or minimum acceptable quality requirements, or is fraudulently presented or labelled or the establishment has five or more serious deficiencies or one or more critical deficiencies as per Schedule I and II of the FIR (Fish Inspection Regulations), or where there is non-compliance with a regulation that is intended to ensure the production of a safe product.

8.5.1.2 *Class II.* A class II deficiency means a situation where the registered establishment has deviated from the Department's QMP requirements in such a manner that the records are not reliable enough to demonstrate that Class I deficiencies have been or will be avoided or the product fails to meet requirements for grade, handling, composition or identity standards.

8.5.1.3 *Class III.* A Class III deficiency means a situation where the registered establishment has deviated from the QMP requirements but a Class I or II deficiency has not resulted. An example of extracts from a QMP checklist are given in Figure 8.8.

At the conclusion of the QMP inspection the deficiencies are noted on the checklists shown and the QMP rating calculated based on the 'Rating Table' illustrated in Table 8.5. An in-plant QMP can be rated as 'Excellent', 'Good', 'Satisfactory' or 'Fail'. The 'QMP Rating' is an indication of the level of compliance and determines the frequency that the DFO Inspectors perform verifications of the inspections.

The HACCP elements of the European Fish Directive were explained in EC 94/356 which has been represented as a simple annotated checklist in Table 8.6. This checklist has been adapted utilising the task/element approach described by Dillon and Griffith (1995) and is an example of a HACCP-based audit. As can be seen by this checklist there are two sections that the auditor may complete: the first is a simple compliance question requiring a 'Yes/No' answer and the second section is a comments section which allows the auditor to note any special observations or evidence recorded. This may prove especially useful if the company has only part way complied with the requirement or if the auditor notes something that may be a problem in the future. This audit checklist is not as generically pre-defined for selected CCPs as the Canadian checklists but must be developed by the auditor for each plant.

Properly planned and executed audits provide factual information for management decisions and enable the effective monitoring and development of systems. Audits are an essential requirement of quality management systems as described within the ISO series of quality management system standards and the verification of either HACCP or quality assurance seafood systems may be viewed as an essential element of a due diligence defence. Furthermore an increasing external scrutiny of seafood quality systems will occur worldwide.

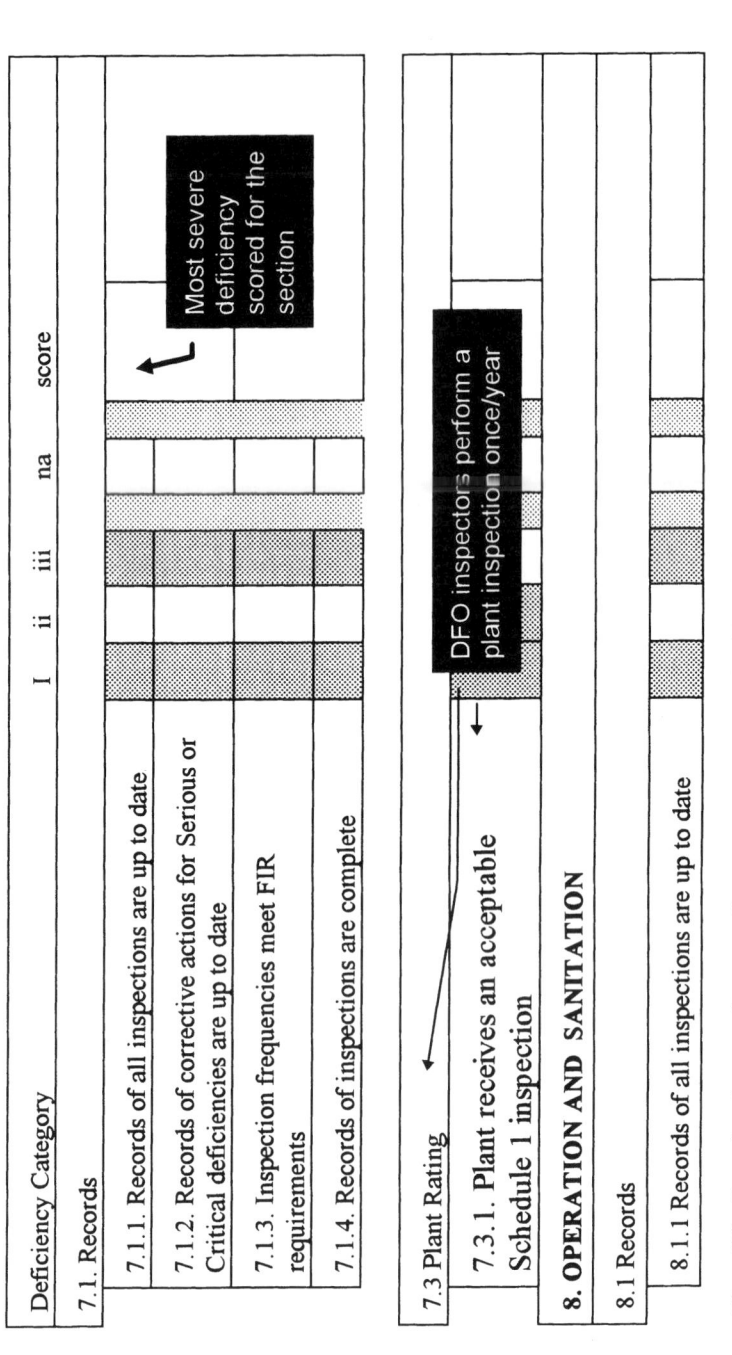

Deficiency Category	I	ii	iii	na	score
7.1. Records					
7.1.1. Records of all inspections are up to date					
7.1.2. Records of corrective actions for Serious or Critical deficiencies are up to date					Most severe deficiency scored for the section
7.1.3. Inspection frequencies meet FIR requirements					
7.1.4. Records of inspections are complete					

7.3 Plant Rating					
7.3.1. Plant receives an acceptable Schedule 1 inspection					DFO inspectors perform a plant inspection once/year

8. OPERATION AND SANITATION

8.1 Records					
8.1.1 Records of all inspections are up to date					

Figure 8.8 Examples of extracts from a quality measurement programme (QMP) checklist. DFO, Department of Fisheries and Oceans.

Table 8.5 Quality management programme (QMP) rating table

Rating	Number of deficiencies		
	Class 1	Class II	Class III
Excellent	0	0	0–3
Good*	0	1–2	4–5
Satisfactory	0	3–4	6 or more
Fail	1	5 or more	NA

*The total number of Class II and III deficiencies cannot exceed more than 5. NA, not applicable.

8.6 Future developments of seafood quality systems

Mitchell (1995) emphasised the increasing role of HACCP-based quality systems in ensuring consumer protection and facilitating international trade through its role within the Uruguay round of the General Agreement on Tariff and Trade. The Agreement on the Application of SPS (Sanitary and Phytosanitary Measures) has elevated the importance of Codex standards, guidelines and recommendations which now become the baseline for assuring consumer protection. The World Trade Organisation may fine or impose trade penalties on member countries who do not comply. Concern has been expressed of this increased requirement for HACCP-based systems by developing-country members during a Food and Agriculture Organisation (FAO) expert technical meeting reviewing the use of HACCP within food control. Lima dos Santos (1993) highlighted the commercial need for developed countries to assist developing countries in effectively implementing quality assurance systems as 50% of fish/fishery products on the international market are from this sector.

The major recommendations from the FAO expert technical meeting (1994) held in Vancouver included the need for the collation of HACCP models, establishment of agreed performance indicators to measure effectiveness, and to evaluate the quantitative evidence of benefits accrued by establishment of such systems.

Clarification and standardisation of the methods and approaches which underpin HACCP has been recommended by Dillon and Griffith (1995). Further investigation has been undertaken into the cost/benefit analysis of quality assurance systems based on HACCP (Spencer-Garret, 1989–1991, Ryder *et al.*, 1995). This work involved undertaking an investigation into the development of a model to cost the implementation of effective quality systems. A review of the existing approaches and methods of quality costing has been completed and data from a variety of sources was used to develop a prototype data base. The cost of implementing HACCP and quality programmes will continue to be measured and the predicted control and failure

Table 8.6 Hazard analysis critical control point (HACCP)-based audit checklist. 94/356/EC Guidelines. This describes specific HACCP requirements for seafood

Site details
UK Seafood Processor
Requirements (91/493 specified in 94/356 EEC)

Assessor name/date

Item No.	Requirements	Compliance	Comments/remarks
Article 6(1) 91/493	Explained and agreed	Is the person competent? Do they have the correct information?	
	Identification of Critical Control Points	Who by? Is the team suitably trained?	
	Establishment and implementation of monitoring and checking critical points	Is the correct equipment available? Do the team have the correct attributes and motivation?	
	Approved laboratories used	Who is responsible for checking labs?	
	Are formal written records kept?	Who is responsible for keeping the records? Where are the records kept?	
	Or is full documentation containing all information related to own-check system and results maintained?	Is the person who keeps the records well-motivated?	
	What approach was adopted for compliance with article 6(1)?	Is the approach correct? Is it based on correct information?	
Article 1	Explain 'own-checks'	Is the person competent?	
	Who are the personnel responsible for compliance?	Do they have the right training and motivation? Are they competent?	
Article 1.2	Have you employed any GMP guidelines?		
Article 1.3	Has a training programme been developed and used?	Who by? Is this person suitably qualified and experienced?	
Article 2.1	What are critical control points and how are they identified?	By whom? What information sources are used? Is the person responsible competent?	
Article 2.2	Has a generic model been used – if so, is it modified?	Whose model? Is it relevant?	
Article 3	How is monitoring and checking undertaken?	Is the person qualified/trained? Is the equipment suitable? Are records kept?	
Article 4.1	What sampling is undertaken?	What is the agreed frequency? By whom? Is the person competent?	
Article 4.2	Has the HACCP system been validated by sampling?	Microbial, chemical or physical test?	
	How are changes covered by system design?	Triggers for review	
	Are regular reviews of HACCP plan implemented?	How? Who by? Do the people have the right attributes?	

Examples of questions the auditor may ask to verify the HACCP system

Article 4.3	Is confirmation of system in line with Article III of the Annex available?	——— Is the information correct? Check information
Article 5	Are external laboratories approved in line with EN 45001 standards or equivalent requirements?	——— Examine accreditation certificates. Does the lab comply?
Article 5	Demonstrate how internal labs comply with Annex B of 88/320/EEC	——— Are the tests in the lab relevant? Are the personnel trained/qualified?
Article 6.1	How and where are the indelible records of implementation and verification of own checks kept? (Which are requested in article 6(1) 91/493 EEC)	——— Check Vendor Assurance records, record keeping system, swab records and compliance audits. Who responsible? Suitably trained/qualified?

costs will be assessed to assist in further development of the data base. The important economic aspects of developing systems has been thoroughly reviewed (Zugarramurdi et al., 1995).

The development and evaluation of HACCP-based systems is increasingly being assisted by software systems (e.g. UK Micromodel to enable assessment of the relevance of specific pathogens, FDA pathogen modelling package, UK HACCP development software from CFDRA (Chipping Campden Food and Drink Research Association) and Quality System Associates Ltd. (RAMAS), and the American software package 'doHACCP'). These software systems will increasingly be used to enable the development of standardised HACCP plans through, for example, the employment of agreed control and monitoring procedures held on approved data bases. The interrogation and approval of existing or modified HACCP plans via modem links either by purchasing companies or enforcement bodies may occur, and could become a necessary element of product liability insurance schemes.

The strategic role of training in effecting consistent change has been recognised worldwide and the United Nations held a meeting in Rome in early 1995 to develop a HACCP 'training of trainers' course. The United Kingdom Royal Institute of Public Health and Hygiene (RIPHH) produced a HACCP training standard (1994) in an attempt to ensure a uniform understanding and application of HACCP within the UK food sector. This system approves introductory courses against the training standard and provides an approved and validated HACCP exam for those participants who wish to certify their understanding and application of HACCP principles. The American and Canadian HACCP development strategies have also employed participative training elements jointly developed between industry and government to ensure consistent development and verification of HACCP plans.

This chapter has reviewed the approach to the systems developed in Europe and North America and highlights the need for further work to fill the gaps in existing scientific knowledge to promote the consistent development and assessment of systems. This will require coordinated international planning and partnership between industry, government and academia to resolve these problems. Resource restriction in small businesses, developing countries and reduced manpower within many inspection services necessitates the increased use and more effective development of such systems.

References

British Standards Institute (1989, 1991), *Quality Systems for the Food and Drink Industries. Guidelines for the Use of BS5750 Part 2.* British Standards Institute. BSI Standards, 389 Chiswick High Road, London.

Bryan, F. (1992), *Hazard Analysis Critical Control Point Evaluations.* WHO, Geneva.

Chipping Campden Food and Drink Research Association (CFDRA) (1992), *Technical Manual 38, HACCP – A Practical Guide.* Chipping Campden, Gloucestershire.

Corlett, D. (1991), Recommendations of the US national advisory committee on microbiological criteria for food. Report of the seafood working group. *Food Control* **3**(4), 202–211.

Corlett, D. and Pearson, M. (1992), *HACCP Principles and Applications*. Chapman & Hall, London.

Dillon, M. (1991), *Quality Systems and Standards – Challenge to Developing Countries*. IFST – National Conference, Keele University.

Dillon, M. (1993), *Due Diligence – The Implications of the Food Safety Act (1990) to the Development of Appropriate Quality Systems (Based on HACCP)*. International Conference – New Markets for Seafood Sept–Oct 1993. University of Hull. International Fisheries Institute.

Dillon, M. and Griffith, C. (1995), *How to HACCP: An Illustrated Guide*. Unit 43 Cleethorpes Enterprise Centre, Wilton Road Estate.

Dillon, M. and Horner, W.B.F. (1994), *Evaluation of a Practically Based Approach to the Implementation of HACCP Systems in the Manufacture of Seafood Products*. WEFTA Annual Conference, France.

Dillon, M. and Horner, W.B.F. (1995), *An Investigation of Quality Assurance Critical Control Point (QACCP) Systems*. International Seafood Conference, WEFTA 25th Annual Meeting, 13–16th November.

Food and Agricultural Organisation of the United Nations (FAO) (1994), Report of the FAO Expert Technical Meeting on *The Use of Hazard Analysis Critical Control Point (HACCP) Principles on Food Control*. 12–16 Dec. 1994, Vancouver.

Food and Agricultural Organisation of the United Nations (FAO) (1995), Summary of proceedings of the HACCP working group of experts convened to plan a training of trainers course in HACCP for developing countries.

Guidelines on the Statutory Defence of Due Diligence (Feb 1991), FDF, LACOTS, IEHO, NFU, NCC and Retail Consortium.

Howells, G., Bradgate, R. and Griffiths, M. (1990), *Food Safety Act 1990*. Blackstone Press Limited, London.

Hudak-Roos, M. and Spencer-Garret III, E. (1992), Monitoring, corrective actions and record keeping in HACCP. In *Quality Assurance in the Fish Industry*, H.H. Huss *et al.* (Eds), Elsevier, p. 521–32.

Irish Workshop. Fish Processing HACCP Workshop. The National Food Centre Training, in association with the Department of the Marine, Bord Iascaigh Mhara and the Irish Fish Processors and Exporters Association (1994).

Jacobson, M. and Lillie, A. (1992), Quality systems in the fish industry. In *Quality Assurance in the Fish Industry*, H.H. Huss *et al.* (Eds), Elsevier, p. 515–20.

Lima dos Santos, C. (1993), *Impact of Council Directive 91/493 on Seafood Trade from Third World Countries: The Case for Developing Countries*. International Fisheries Conference, at the University of Hall International Fisheries Institute, 29 Sept – 1 Oct 1993, Conference Proceedings.

McEachern, V. (1992), QMP in the Canadian Fish Processing Industry. *Infofish International* **4**, 92.

Mitchell, R. (1995), *HACCP – An International Overview*. Flair-Flow Europe Conference, Cardiff, June.

Mortimer, S. and Wallace, C. (1995), *HACCP – A Practical Approach*. Chapman & Hall. London.

Pallet, T. (1992), *Due Diligence*. International Food Hygiene.

RAMAS (Risk Assessment and Management Application Software), QSA Ltd, St Albans, Herts, UK.

Robertson, R., Prytz, K., Lovland, J. and Sorenson, N.K. (1992), Total quality management in the Norwegian fishing industry. In *Quality Assurance in the Fish Industry*, H.H. Huss, *et al.* (Eds), Elsevier, p. 561–70.

Royal Institute of Public Health and Hygiene (RIPHH) (1995) HACCP Training Standard, *HACCP Principles and Their Application in Food Safety*. 28 Portland Place, London, WIN 4DE.

Spencer-Garret, E. III and Hudak-Roos, M. (1991), *Developing An HACCP Based Inspection System for the Seaford Industry*.

Spencer-Garret, E. (1989–1991), *Economic Impact of HACCP Models. Breaded, Cooked and Raw Shrimp and Raw Fish. Economic Impact of HACCP models. Blue Crab, Breaded and Speciality Products, Molluscan Shellfish, Smoked and Cured Fish and West Coast Crab*. National Fisheries Education and Research Foundation Inc., Pascagoula, Mississippi.

Ryder, J., Dillon, M. and Spencer-Garret, E. (1995), *Development of Quality Cost Models*. WEFTA 25th Annual Meeting, 13–16th November. (This work was funded by the Post Harvest Fisheries Research Programme of the United Kingdom Overseas Development Administration as part of Grant R50027.)

Zugarramurdi, A., Parin, M.A. and Lupin, H.M. (1995), *Economic Engineering Applied to the Fishery Industry*. FAO Fisheries Technical Paper 351. FAO, Rome.

9 Temperature modelling and relationships in fish transportation
C. ALASALVAR and P. C. QUANTICK

9.1 Introduction

Throughout the world there is a high demand for fresh and processed fish which has led to efforts by many fishing industries and processors to improve the transportation and supply of a high-quality product to consumers.

Controlling the transportation temperature of fresh fish is important from the time it is caught until it reaches consumers because this time interval determines the extent to which enzymatic or bacterial degradation takes place. Fish spoilage is a direct function of the temperature under which it is preserved (Ronsivalli and Baker, 1981). Controlled temperatures, coupled with a variety of pre-treatments, can extend the shelf-life of fish and the role of good temperature control during transportation of some fish products in protecting the public is likely to increase in importance. In other words, temperature control is the single most important factor in the safety and quality of fresh fish and their products.

Although there are many coolants such as ice, dry ice (solid carbon dioxide) and cooling gels ('ice-berg', 'ice-keeper', 'sorbagel'), etc., the best method of chilling during transportation of fresh fish is by direct contact refrigeration using ice (the functionality of these other coolants is discussed later). Ice is commonly used in developed and some developing countries to distribute fresh fish in insulated and non-insulated containers due to its large cooling capacity ($334.5 \, J \, g^{-1}$) for a given weight or volume; it also keeps fish cold, moist and glossy (FAO, 1992).

Insulated fish containers and transporting fresh fish for handling, such as expanded polystyrene boxes, are being used more frequently in the UK and other developed countries. In developing countries, locally-made containers (with or without insulation) are often used due to the high costs of good insulation. However, the losses associated with poor temperature control are often high. The effectiveness of an insulating material is measured by its thermal conductivity. The lower the numerical value of thermal conductivity the better the insulator. Good insulating materials can significantly reduce heat leakage and therefore melting of ice. This is extremely important in developing countries where ice is 10 times more expensive than developed countries (Myers, 1981). The amount of ice needed to keep fish fresh is therefore economically important in tropical and sub-tropical countries, since in warmer climates ice melting rates are high.

Modelling of time–temperature during transportation is of great importance generally in deducing how much ice may be needed to keep fish at or around 0–8°C and how long fish will be maintained at this temperature before spoilage occurs, since temperatures >8°C may allow rapid microbial growth and lead to health problems for consumers (EC, 1990). Using computer programs, such as the Mailprof (Alasalvar and Nesvadba, 1995), it will therefore be easier to calculate and maintain the correct ice requirement for any shape of fish and their products, and also calculate the expected time–temperature regime to ensure that products are kept below 8°C for any length of transportation.

This chapter is aimed at discussing the main problems of fresh fish transportation in developing and developed countries and modelling the behaviour of chilled fish using appropriate computer programs. Heat transfer and its application in fish transportation is also assessed as an example of temperature modelling/predictive modelling for fish and their products. The advantages of using Food MicroModel are also discussed.

9.2 Transportation of fish

Fish can be transported by road, air and sea. Railroads do not offer a common means for seafood transportation in most developed countries, so road, air and sea transportation of fish are reviewed below.

9.2.1 *Road transportation*

Refrigerated (temperature-controlled) transport is expensive and not widely used in most developing countries. Therefore, fish are commonly transported in unrefrigerated vehicles over large distances inside non-insulated containers (*e.g.* wooden or other locally-made materials). Sometimes upon arrival at the retailer or consumer, the fish is in an unacceptable state or nearly so. However, the use of insulated or semi-insulated containers is spreading in many developing countries for use in fish transportation (Makene and Mgawe, 1991).

In developed countries, fresh fish are often transported by temperature-controlled vehicles inside insulated containers with ice or cooling gels. There are many types of temperature-controlled vehicles used in the transportation of fresh fish and their products, ranging from large 12 m heavy goods vehicles with independent cooling units, to light goods vehicles relying on insulated containers to maintain the temperature of chilled fish. The purpose of these vehicles is to maintain the chill temperature, and not to provide cooling. In such vehicles, journey times may vary from 2 h to several days. Temperature-controlled vehicles reduce the use of ice during transportation.

Heavy goods vehicles have an independent refrigeration unit powered by diesel, often with an auxiliary electric motor, used to circulate cold air around the vehicle chamber from an evaporator unit at the front of the vehicle. The cold air is distributed in different ways within different vehicles, but most have cold air leaving from the top of the cooling unit near the roof, and returning via the base to the front of the vehicle and the return air intake. Some vehicles are cooled by direct evaporation of liquid nitrogen from a reservoir on the vehicle. Temperature read-out and single-channel chart records may be used on refrigerated vehicles to monitor the cold chain. If the load is not spaced correctly, circulation can be restricted and 'hot spots' can occur (Heap, 1992). In both heavy and light goods vehicles, fish boxes must be stowed correctly in order to allow cold air circulation around the box. Many light goods vehicles delivering chilled fish are fitted with refrigerated units driven from the vehicle engine or transmission which means that cooling is not possible whilst the vehicle is stationary (Woolfe, 1992).

9.2.2 *Air transportation*

With the increased acceptance of fresh fish and seafoods spreading throughout the world, air transportation is becoming more important than ever to the seafood industry. But, as the industry has moved towards more air transportation of fresh fish, leakage has become a serious problem. The drip from fish is smelly and damaging to aluminium and other high technical metals used in airline bodies (Regenstein and Regenstein, 1991; FAO, 1992). To alleviate leaking problems, it is necessary to improve container insulation and packaging integrity.

Many flight containers have been developed in order to prevent leakage and to keep the product chilled during air transportation. M-1, LD-3, LD-7, EH, AFC, E, Q, gas pack, tray pack, refrigerated airline and A, B and C containers, etc. are mainly used in the USA (Cox, 1982; Mignault, 1982; Rogers, 1982). They are all leak-proof and made from high density polyethylene according to the individual airline specification and most either belong to airlines or rental units, or can be leased to processors. More recently, moulded foam and expanded polystyrene boxes have been used in air transportation which, although waterproof and light, have poor impact resistance. Therefore, when such boxes are used they should be placed within strong external waterproof cardboard cartons to protect against damage (FAO, 1992).

Dry ice and cooling gels are frequently used to provide a source of cooling during air transportation. The product should be pre-chilled before packaging since it reduces the quantity of coolant required and, therefore, weight and space requirements. Most of the containers/boxes mentioned above, except moulded foam and expanded polystyrene, have been designed for using dry ice due to leakage problems. A separate section in the container holds dry ice but only a limited amount of dry ice can be used in the hold of an aeroplane because its escape may kill live animals being transported (Regenstein and Regenstein, 1991). Therefore, each container should be labelled showing the initial weight of dry ice used. If dry ice were placed on the top of fresh fish, it would cause 'freezer-burn' so a separate compartment in the container is necessary that holds the dry ice and keeps it away from the product. The compartment has a vent system which allows the carbondioxide gas to enter the product container.

Cooling gels can also be used in air transportation with moulded foam and expanded polystyrene boxes and gels have the advantage that they do not produce gaseous or liquid effluent during transportation. The combination of water and a gelatine-type material creates a water-ice coolant package that does not drip, even when thawed. The coolants are often dyed blue so that any leakage is obvious. Cooling gels are discussed in more detail in section 9.3.3. Ice is only rarely used in air transportation and more care has to be taken with ice and cooling gels to ensure that there is no partial freezing of water and no leakage.

In Europe, polystyrene boxes have been used for some years. Some users put a polyethylene bag inside, whilst others just pack fillets directly into the polystyrene box. Fish transportation is mainly carried out by road and seaways in Europe, whereas in the USA and Canada transportation is mainly by air due to the longer distances and is more advanced than other countries. In most developing countries, because of the lack of facilities and high fares, air transportation of chilled fish has not yet been widely adopted.

9.2.3 *Sea transportation*

Intermodal freight containers are normally used in the transportation of fresh fish at sea. These units are electrically driven from three-phase supplies from either mains or a diesel generator. In the UK, almost all fresh fish are transported to Europe by sea. Palletised polystyrene fish boxes (3:1 ratio of fish:ice) are stowed inside the container which is carried by tractor to the ship, where a chilled room is not necessary because the container has its own cooling units. When the product arrives at the harbour, the container is transported by road to its destination.

9.3 Containers and cooling gels

The problems of transporting fresh fish are numerous. The lack of appropriate containers and ice are major constraints under tropical and sub-tropical conditions. Temperature is the most crucial factor in controlling the speed of spoilage, as a few hours of ambient storage are sufficient to cause significant spoilage in fish. It is difficult to expect good-quality products from the retail market if the use of non-insulated containers is not discouraged. Therefore, the use of well-insulated containers with ice in fresh fish transportation is of paramount importance in supplying a quality product to consumers. Ice keeps fish fresh and insulated containers will help to maintain ice, making the whole operation economically feasible. Taking all aspects into account, the ideal transportation container should:

- prevent physical damage to the product;
- prevent desiccation of the product;
- prevent undesirable oxidative reaction;
- prevent outside contamination from reaching the product;
- have low thermal conductivity (good insulation);
- have low ice melting rate;
- retain the desired atmosphere around the product;
- be waterproof and easily cleanable;
- have a UV protector;
- not serve as a source of contamination by itself;
- be economical.

Quality and safety are becoming important issues and today most insulated containers should be made of polyethylene or polystyrene which are sanitary and help to extend product shelf-life. Wooden and non-insulated containers should not be used in fish transportation unless absolutely neces-sary. Whatever insulated container is used, the product must be stowed with a sufficient amount of ice. Containers used in fresh fish transportation in developing and developed countries are reviewed below.

9.3.1 In developing countries

In many developing countries, particularly in Africa, awareness is low about the advantages of using ice with an insulated container for prolonged storage life. Fresh fish transportation is often carried out in an inappropriate way leading to spoilage and loss of quality. There are many reasons for this, two of which are the high cost of ice and insulating materials.

In developing countries, fisherman often buy ice at a price 10 times greater than in developed countries (Myers, 1981). A review carried out in 1986 by the FAO/DANIDA Training Project on Fish Technology and Quality Control on fish and ice prices in 14 African countries demonstrated that in all cases and for all fish species, 1 kg of ice increased the price by at least twice the increase recorded for developed countries. The cheaper the fish species, the worse the situation. For instance, in the case of small pelagic fish, the percentage increase in the fish cost per kilogram of ice added, was of 66% for sardinella in Mauritania and for anchovy in Togo. Ice consumption, due to the high ambient temperature ($>30°C$) is considerable under tropical and sub-tropical conditions; 5–6 kg of ice per kilogram of fish from catching to sale is required to keep the fish chilled during transportation in non-insulated containers. A 1:5 or 1:6 (fish:ice ratio) is uneconomic and is a reason for the lack of fish icing in many developing countries (FAO, 1994).

Because of the high ice requirement under tropical and sub-tropical conditions, insulated fish containers for handling and transportation of fish are becoming increasingly important. However, imported insulated containers are expensive and not affordable by artisanal fishermen (Makene et al., 1988). Therefore, efforts have been made to introduce industrial insulated fish containers and to develop artisanal insulated containers. The utilisation and development of locally-made insulated fish containers is spreading in many developing countries. Such containers can significantly reduce the amount of ice needed, so reducing the market price of fish. A 1:1 fish:ice mixture can be maintained at 30°C for about 5.5 days; under the same conditions using a non-insulated plastic box it will last less than 3.5 h (FAO, 1994). Therefore, keeping fish fresh is practically impossible under most tropical conditions if non-insulated containers are used (Lupin, 1985).

'Bag type' insulated fish containers (Figure 9.1) are widely used in African countries if the quantity of fish is fairly small. The walls of the bag are constructed with woven palm which is cheap, locally available and can be replaced when necessary. Insulation can also be made from sawdust, coconut fibres, dried grass or rejected cotton fibres; however, the use of such materials presents some technical problems when wet as they tend to rot and lose their insulating capacity. To overcome such problems, the concept of insulated pillows was developed in various FAO/DANIDA workshops. The insulating

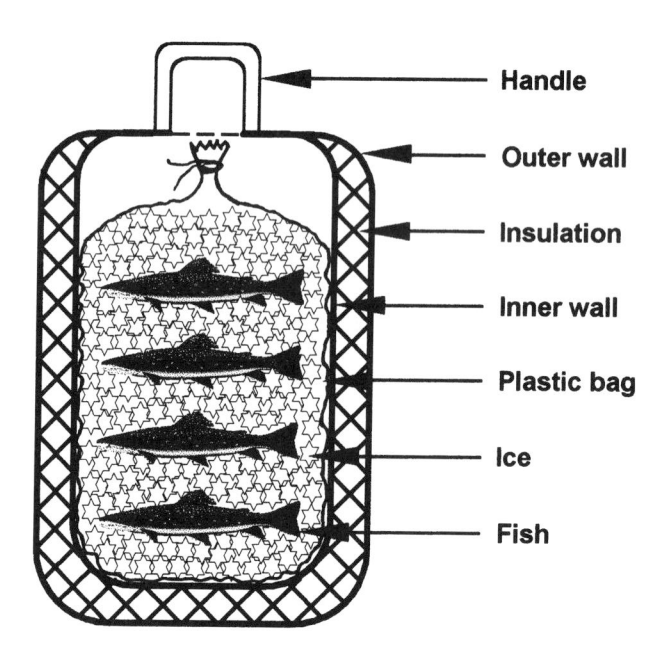

Figure 9.1 Insulated 'bag type' fish container.

material (e.g. coconut fibres) is placed inside plastic tubes of a type normally used to produce ordinary plastic bags (FAO, 1994).

Many types of locally available insulated fish containers such as 'Kikapu', 'Tenga', etc. are also used in African countries (Makene and Mgawe, 1991) and can extend the shelf-life of fish and reduce the amount of ice necessary during transportation. In other developing countries apart from African, locally made and imported insulated fish containers with ice are becoming widely used in fresh fish transportation although non-insulated fish containers are also still in use.

9.3.2 In developed countries

By contrast, fresh fish cannot be transported at ambient temperature without ice or cooling gels in many developed countries. According to EC (1990) food safety legislation (see section 9.4.2), fresh fish and their products should not exceed 8°C during transportation and storage. This legislation has led to manufacturers producing insulated containers which, when used with proper icing, optimise fish quality. Several types of fibre-board, plastic-coated cartons, wax-coated cardboard, plastic, metal, high- and low-density polyethylene, wooden and expanded polystyrene boxes are widely used (Regenstein and Regenstein, 1991; FAO, 1992; Wills, 1982; McDonald and Graham, 1985). In the UK, expanded polystyrene boxes are widely used due to their excellent insulating properties – they are also waterproof and reusable. Almost all processors use these boxes with an appropriate amount of ice during transportation and delivery.

9.3.3 Use of cooling gels in fish transportation

The US FDA (1992) acknowledges that a cooling gel (a water-soluble synthetic co-polymer with cold retention ability) should be non-toxic and non-hazardous and could be used as a coolant for perishable products such as fresh fish and mail-order fish, etc. during transportation. The cooling gel is placed in a bag specially designed to withstand the stress of freezing. The bag is usually heat-sealed or mechanically clenched and designed to withstand rough handling, vibration in transit against other packaging materials, and protruding fish fins.

Figure 9.2 shows a range of commercially available cooling gel packs. They are either filled with water ('ice-keeper') or soaked in water ('ice-berg' and 'sorbagel') so that the powder can form gel. Further sealing is not necessary since gel consistency does not allow water leakage.

The gel packs can also be prepared by mixing sufficient dry cooling gel (in powder form) with water to fill a polyethylene bag (60 g makes typically about 1 l gel). Once sealed, it can be stored at room temperature until used. Cooling gel packs are frozen down to $-30°C$ before dispatching the product. They are

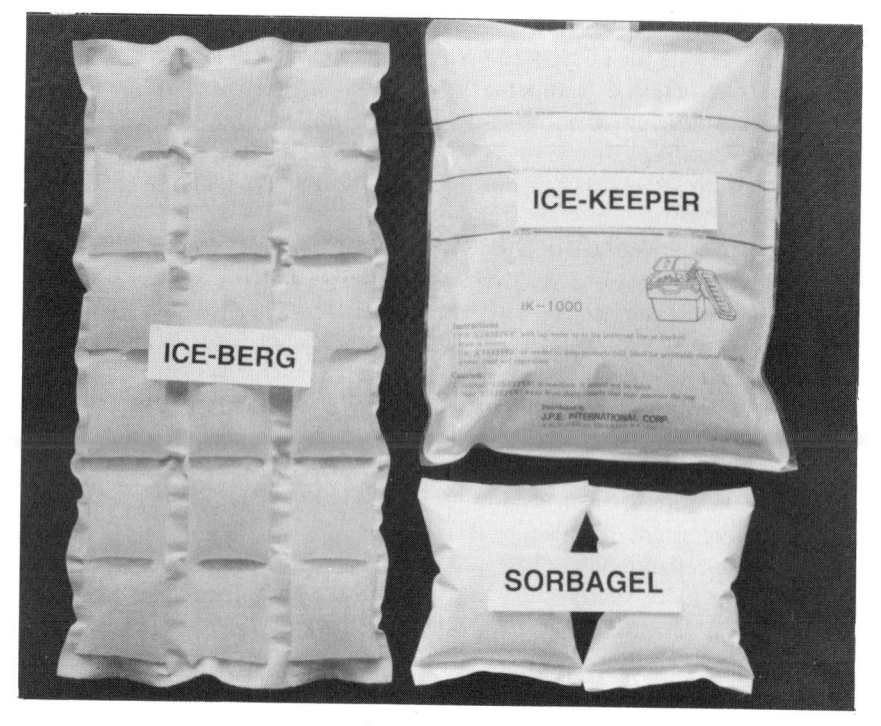

Figure 9.2 Commercially available cooling gel packs.

usually placed on the top and under the product. Making cooling gel packs is laborious and time-consuming and commercially available cooling packs are therefore more advantageous. The amount of cooling gel needed to keep the fish chilled during a certain time of transportation can be calculated theoretically by the Mailprof computer program as discussed later.

Cooling gel packs in small quantities cannot be used to reduce the temperature inside boxes of seafood but are used to maintain their temperature and protect them from the invasion of heat from outside of the packages (Harris, 1982).

Some companies (Shetland Smokehouse and Ace Dry Ice & Gel Packs, UK) indicate that 'ice-berg' has a cold retention higher than any starch-based gel or water. Their claim that the 'ice-berg' cold pack rating was $409.5 \, J \, g^{-1}$, was investigated by Alasalvar and Nesvadba (1995). They found that cooling gels ('ice-berg', 'ice-keeper' and 'sorbagel') had lower latent-heat capacities than ice, which could be explained by the gels containing hydrated organic co-polymers which could have decreased latent heat. In other words, cooling gels do not contain more energy or heat-absorbing capability than ice. The main advantage of using either of these cooling gels compared to ice is that even where there are minute punctures or holes in the pack, this does not cause any water leakage due to the gel's consistency. They also reduce the convection currents from external temperature differences. The solution is usually designed to withstand repeated freezing and thawing and can be used several times unless damage to the surrounding packs occurs, but whether or not it is worth transporting the gel pack back needs to be evaluated because of its cost.

If fish and fish products (including smoked fish) are transported either by mail or air, frozen cooling gel packs inside the insulated boxes (polystyrene box) are more advantageous than ice because no leakage occurs. In contrast, if the fish are transported by lorries (where leaking is not a problem), using ice is better since melt-water washes the fish and keeps them clean and moist until they reach the consumer. Removing contaminated water also prevents rapid spoilage. One much needed area of research is whether or not frozen gels or ice ($-30°C$) damage or burn the surface of fresh fish. This is not a particular problem in packed salmon but might be in fresh fish and thus is very important for consumer acceptability.

9.4 Safety, quality and spoilage of fish during transportation

In this section, the effect of temperature, legislation in fish transportation, the hazard analysis critical control point (HACCP) system and shelf-life of fish are reviewed.

9.4.1 *Effect of temperature on the growth of micro-organisms during transportation*

Fish is a very perishable foodstuff and spoilage is usually rapid since fish allow good microbial growth being highly nutritious, having a high moisture content and a relatively neutral pH value. Spoilage changes occur as a result of endogenous chemical and enzymatic activities but most commonly bacterial activity (Matches, 1982). These changes are influenced by many factors, the most important being temperature.

Depending on their response to temperature, micro-organisms can be divided into four main groups: psychrophile, psychrotroph, mesophile and thermophile (Table 9.1). Most commonly, the cardinal temperatures for growth (minimum, optimum and maximum growth temperatures) are used. With chilled fish, the factor of most concern is the minimum growth temperature (MGT), which represents the lowest temperature at which growth of a particular micro-organism can occur. If the MGT of a micro-organism is greater than 10°C, then this micro-organism will not grow at chill temperatures. Therefore, the groups of most concern in chilled fish transportation are the psychrophiles and psychrotrophs.

Table 9.1 Classification of micro-organisms according to growth temperature. From Morita (1973), Walker and Stringer (1990) and Jay (1992)

Temperature (°C)	Classification			
	Psychrophile	Psychrotroph	Mesophile	Thermophile
Minimum	<0–5	<0–5	(5–)[a] 10	(30–)[a] 40
Optimum	12–18	20–30 (35)[a]	30–40	55–65
Maximum	20	35 (40–42)[a]	45	(70–)[a] > 80

[a] Figures in parentheses are occasionally recorded for micro-organisms assigned to a particular classification.

As mentioned above, the ability of individual micro-organisms to grow and their rates of growth are affected by temperature. The effect of reducing temperature is generally to decrease the rate of enzyme activity. This applies not only to the chemical and biochemical changes in fish but also to the activities of micro-organisms. Psychrophiles and psychrotrophs are better adapted to growth at chill temperatures. Therefore, chilling alone cannot be relied upon to prevent all microbial growth, but can prevent the growth of some types and retard the rate of growth in others.

Temperature is, of course, one of the prime factors in controlling microbial growth of stored fisheries products. The rate of bacterial growth (generation time) at different temperatures is shown in Table 9.2. The time from catch to spoilage therefore decreases as the holding temperature increases.

The effect of temperature on microbial growth is shown in Figure 9.3. As the storage temperature decreases, the lag phase before growth (time before an increase in numbers is apparent) extends and the rate of growth decreases. Also, as the minimum temperature for growth is approached, the maximum population size often decreases.

There are two categories of micro-organisms which need to be considered throughout the storage and transportation of chilled fish: pathogenic (health-risk) and spoilage micro-organisms. Pathogenic micro-organisms are those

Table 9.2 Rate of bacterial growth (generation time) at different temperatures. From Matches (1982)

Temperature (°C)	Generation time (h)
33	0.5
22	1
12	2
10	3
5	6
2	10
0	20
−3	60

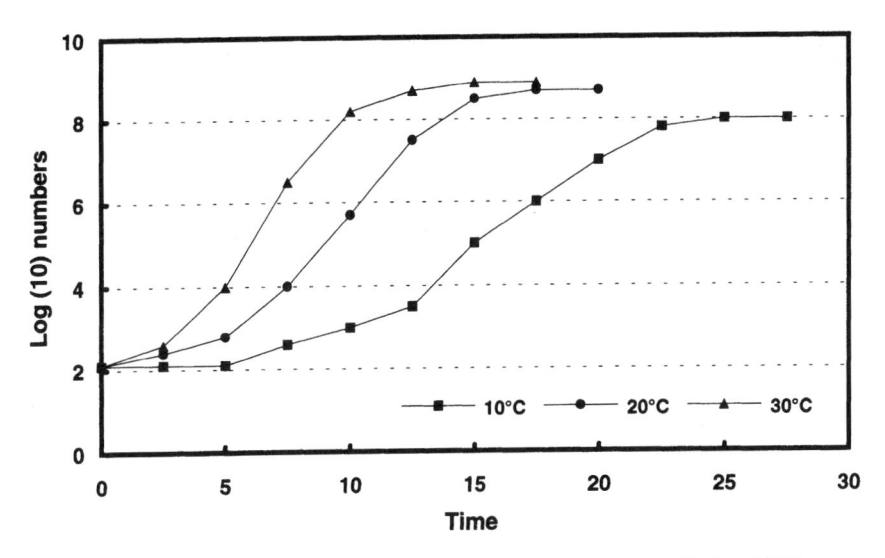

Figure 9.3 Effect of temperature on the growth of bacteria. From Walker (1992).

which may cause illness in people eating the fish and its products, while spoilage micro-organisms are those which make the fish go 'bad' from a quality perspective.

Hygienic practices and proper temperature control from catching to when the fish finally reaches the consumer reduce the microbial spoilage of fish and safety problems. The lower the initial level of contamination, the greater the time until microbial spoilage is evident. Temperatures over 8°C during transportation for long hours may allow rapid microbial growth and lead to health problems for consumers. Therefore, fresh fish should be kept below 8°C throughout the transportation in order to reduce spoilage rate of fish significantly.

9.4.2 *Temperature control and legislation in fish transportation*

With chilled fish and fish products, good temperature control is essential, not only to maintain the microbiological safety and quality of fish, but also to minimise changes in the biochemical and physical properties of fish. During the life of chilled fish, considerable opportunities exist for temperature abuse to occur. The greater this is, the greater the potential for microbial growth to occur. This may result in a product becoming unsafe and/or a loss in product quality. Temperature control is therefore one of the key issues with regard to fresh chilled fish.

There is an agreement between more than 20 countries to facilitate international traffic in certain perishable foodstuffs, including fresh fish (ATP, 1987). According to this agreement, fresh fish must always be transported in ice and the fish temperature must not be above 2°C. This agreement lays down common standards for the temperature-controlled equipment in which fresh fish is carried and it was intended to apply chiefly to all means of surface transport within Europe. It does not apply to air transport or to journeys by sea exceeding 150 km. The ATP (Agrément de Transport de Périssables) agreement has been published by the UK Department of Transport (Dept of Transport, 1988).

Although the ATP agreement was accepted by many countries, it is difficult to keep the fish below this temperature for a long journey, therefore, UK (1990) food law has specified chill temperatures for certain foods including fresh fish and their products (including mail-order smoked salmon). With this legislation, the temperature of the fresh fish and their products transported at ambient temperatures should not exceed 8°C, and this applies to all kinds of transportation. This legislation results from increasing consumer demands for a wider variety of chilled fish, the rising incidence of food-borne illnesses in the UK and many other countries, and increasing awareness of the role temperature can play in preventing food-borne illnesses (Dept. of Health, UK, 1990; EC, 1990; UK, 1990). Other countries have different maximum safety limits for fresh fish and their products, e.g. the USA (7.2°C), France (0–3°C), the Netherlands (7°C) and Sweden (8°C) (Turner, 1992). However, in most developing and developed countries, this legislation does not yet exist.

The European Union (EU) food labelling directive already impinges on chilled food including chilled fish in that foods which, microbiologically, are highly perishable must be labelled with a 'use by' date and the date must be followed by a description of the storage conditions which must be observed (EC, 1989).

9.4.3 *Application of HACCP in seafood*

Microbiological analysis of seafoods and their production plays a key role in assessing the safety, quality, shelf-life and compliance with regulatory specifications. Traditional microbiological testing methods are often slow, laborious and provide only retrospective data on finished product. There is therefore a need for quick, reliable and cost-effective methods to control organisms at the point of production and preparation. One recent approach to this is the HACCP system (see Chapter 8 for a more detailed description).

HACCP is a systematic approach to food safety management. It is a cost-effective system and targets resources to critical areas of processing, so reducing the risk of manufacturing and selling unsafe products. It emphasises control of microbiological hazards, as well as chemical and physical hazards (Bryan, 1992; Hall, 1994; Mortimore and Wallace, 1994). HACCP systematically assesses all aspects of food safety from raw material sourcing, through processing and distribution, to final use by the consumer. Use of a HACCP system also facilitates the move towards a preventive quality assurance approach within a company, reducing the traditional reliance and end-product inspection and testing (Mortimore and Wallace, 1994). HACCP can be used by astute processors to gain a competitive advantage.

Seafood safety regulation is on the verge of a worldwide 'paradigm shift' to HACCP. There is wide agreement (Huss, 1992; Lima dos Santos *et al.*, 1994) that HACCP is an effective means of achieving the stringent hygiene and safety controls required by Council Directive 91/493/EEC for the manufacture and marketing of seafood products. This directive and its successors require that risk analyses are performed and a critical control point approach to assuring fishery product safety and wholesomeness is adopted. HACCP, with its seven principles, forms a framework for the evaluation of actual hazards in seafood. With HACCP, processors use their own experience and scientific/technological principles to establish the hazard prevention system that works for a particular processing situation (Evans, 1995).

The development of a HACCP system for a seafood establishment begins with the construction of a flow-chart for the entire process. This chart should begin with the acquisition of raw materials and include all steps through gutting and subsequent transportation. A flow-chart for catching and transportation of fish is shown in Figure 9.4. Selection of fishing grounds is listed

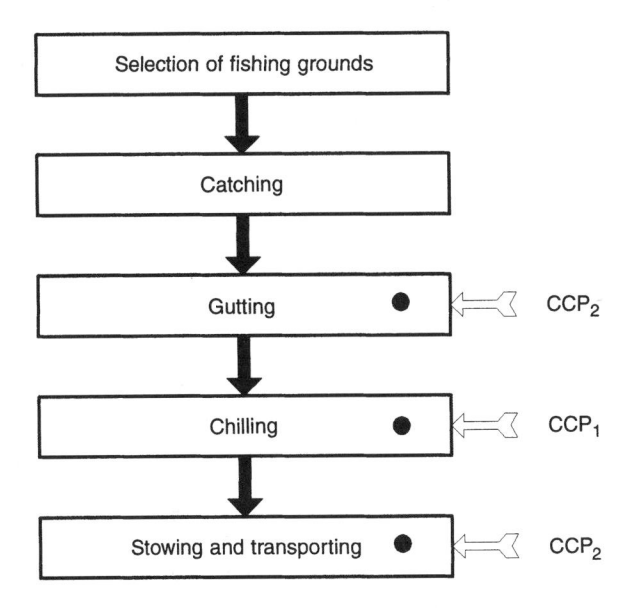

Figure 9.4 Flow diagram for the catching and transportation of fresh fish. CCP_1, control of hazard by elimination or prevention; CCP_2, minimisation of hazard but offers no guarantee of control.

because environmental factors, particularly temperature and in some cases water pollution, have a major influence on the composition of the fish microflora (ICMSF, 1988).

A properly documented HACCP system in fish transportation provides consumers with safe and more wholesome products and leads to a decrease in food-borne illness. Temperature control emerges as an important element in the maintenance of fish safety and quality, and as part of a HACCP system monitoring temperature effectively is of critical importance. Temperature modelling also has an important role to play for fish processors and fish product manufacturers in the prediction of shelf and/or distribution life.

In the US, it has been proposed that the seafood industry applies HACCP plans to their process as a mandatory requirement. An outcome of the National Marine Fisheries Service's (NMFS) Model Seafood Surveillance Project (1987), initiated in response to the US Congress Directive, was to design a programme of certification and surveillance for seafood consistent with HACCP systems. The NMFS and USFDA have concluded that HACCP is a viable approach for ensuring the safety of seafood, supported by the vast majority of the seafood industry. Most seafood processors in the US are implementing HACCP according to the proposed regulations of US FDA (1994).

In addition to the US initiatives, the United Nation's Codex Alimentarius Commission, as well as the EU, are adopting HACCP as the international standard for producing safe food. Under proposed rulemaking, beginning in 1995, all seafood exported to the EU has to be produced under standards certified by the exporting country and accepted by the EU as complying to their HACCP standards (Taylor, 1993). Other international food regulators have also adopted HACCP as a food safety standard (e.g. the FAO Codex Alimentarius Committee on fish and fisheries products, fresh and frozen fish and the Asian Pacific Economic Co-operation organisation – APEC) (Evans, 1995).

9.4.4 Factors affecting the shelf-life of fish

Fish deterioration is often thought of as simply the development of bad odour and bad flavour due to microbial spoilage, autolytic enzyme activity, and adverse chemical changes (mainly trimethylamine oxide (TMAO) breakdown, lipid oxidation, lipid hydrolysis and protein denaturation). Other causes of quality loss such as physical damage, dehydration and contamination can also be observed. Temperature control is the most important factor affecting the speed at which fish 'go off'.

Fish of any species generally undergo microbial spoilage in much the same way, but there are wide differences in the way that fish of different families, and even different species in a family, deteriorate (Connell, 1990). The method of capture, location of fishing grounds, seasons of the year, fat content and fish size can affect the keeping quality of fish.

Fatty fish such as mackerel and herring, with a fat content of 15% or more, spoil more rapidly than lean fish (Leu *et al.*, 1981). Bennour *et al.* (1991) indicated that shelf-life of iced mackerel was unacceptable for consumption after 6, 8 and 9 days for ice:fish ratios of 1:4, 1:3 and 1:2, respectively. Similar keeping times have been reported for the same species (Fernandez-Salguero and Mackie, 1987; Smith *et al.*, 1980). Lean fish in ice keep for about 12–18 days after which time they is likely to become inedible, or nearly so (Connell, 1990). Research workers (Boyd and Wilson, 1977; Lupin *et al.*, 1980; Bilinski *et al.*, 1983) have reported that gutting of fish extends shelf-life. In contrast, other workers have reported that gutting does not significantly extend the shelf-life (Avdalov and Ripoll, 1981; Maia *et al.*, 1983). The combined effect of heading and gutting on shelf-life of fish species has not been widely reported.

In general, flat fish keep longer than round fish; red-fleshed fish keep longer than white-fleshed fish; low-fat fish keep longer than high-fat fish; and teleost fish keep longer than elasmobranch fish (FAO, 1992). Fish from tropical water can be kept longer in ice than fish from cold water because psychrotrophic spoilage micro-organisms are virtually absent in warm waters. In other words, fish from cold-water regions have a larger proportion of psychrotrophs among their natural microflora, which can shorten the chill shelf-life appreciably (Børresen and Strøm, 1983).

9.5 Types of predictive modelling in fish transportation

Predictive modelling can be divided into three main groups:

- time–temperature function integrators (TTFI);
- mathematical approach/heat transfer;
- Mailprof computer program.

The effect of temperature on the rate of spoilage of fish is well recognised and great interest is being taken in the use of temperature function integrators for spoilage studies. TTFI provide data which can be used to predict the spoilage of fish for any length of time. Mathematical approach/heat transfer allows prediction of the temperature of fish and the amount of ice needed to keep the fish chilled for any length of time but it requires good accurate data. The Mailprof computer program allows prediction of the time–temperature profiles of fish to ensure that the product is maintained at chill temperatures for any length of time. Predictive modelling is reviewed later.

9.5.1 *Time–temperature function integrators and rate of spoilage*

TTFI have the potential for monitoring the time–temperature history of perishable foods including fresh fish and fish products. If the rate of spoilage at any given temperature is known in addition to the time temperature history, the sum effect on spoilage for such a history can be computed. By continuously monitoring temperature and continuously summing the product of the rate of spoilage and time over small time intervals, TTFI is achieved. Such integrals can provide a measure of how much spoilage has occurred during the period of observation and thus how much shelf-life has elapsed (Storey, 1985). It is possible to determine the fish quality at the end of the distribution chain by using TTFI probes, which display the equivalent time at 0°C corresponding to the time and temperature history of the chilled fish. TTFI probes are inserted into the fish on capture, through storage and during transportation so that data can be obtained. These stored data can readily be retrieved directly into a computer. Instruments are available which integrate continuously and display the integral as a equivalent time at an appropriate temperature.

Spencer and Baines (1964), Charm et al. (1972), Olley and Ratkowsky (1973), Ronsivalli and Charm (1975), Olley (1978) and Olley and Quarmby (1981) studied the relationship between temperature and rate of spoilage in fish. Ratkowsky et al. (1982) proposed a new model to describe the relationship between temperature and bacterial growth. This model is a linear relationship between the square root of growth rate of bacterial cultures and function of temperature over a significant range of up to about 15°C. This relationship is expressed mathematically by the following equation.

$$\sqrt{R} = b(T - T_0) \qquad (9.1)$$

where R is the root of growth per unit time, b is the slope of the regression coefficient, T is the absolute temperature at which growth is measured (K), and T_0 is an hypothetical temperature which is characteristic of each organism (K).

T_0 is a notional temperature found to be close to 263°K for psychrotrophs important in chilled spoilage. At 0°C, $R = b^2$ (100) and at 10°C, $R = b^2$ (400), for a given organism. Thus the relative rate of growth at 10°C is four times that at 0°C (Owen and Nesbitt, 1984).

Equation 9.1 can be applied to a large number of bacteria, moulds and yeasts, and also to the breakdown of nucleotides. The minimum temperature at which chilled fish is normally stored is 0°C, melting ice being an ideal refrigeration medium. It is convenient therefore to simplify equation 9.1 and redefine growth rate, r, as the rate relative to that at 0°C. Equation 9.1 thus becomes:

$$r = (0.1\,T + 1)^2 \qquad (9.2)$$

where r is the rate of spoilage relative to the rate of 0°C, and T is temperature of storage (°C).

Equation 9.2 is equally applicable to bacterial spoilage as it is to bacterial growth. Using equation 9.2, the spoilage rate at any temperature relative to the spoilage rate at 0°C can be calculated. It is also possible to predict the likely storage life of fish which have been maintained for some time at temperatures higher than the ideal 0°C (FAO, 1992). Ratkowsky *et al.* (1983) proposed a further empirical relationship for predicting the effect of temperature on growth rate over the range of temperatures at which bacteria grow.

Gibson (1985) has suggested a linear model to predict shelf-life. It is possible to estimate the remaining shelf-life at any temperature using equation 9.3.

$$\text{Remaining shelf-life (in days)} = \frac{DT - DT_0}{dDT/dt_0} \quad (9.3)$$

where DT is the detection time (h), DT_0 is the value of DT at cut-off quality level (at the rejection time), and dDT/dt_0 is the daily rate change in DT for fish stored at 0°C.

Jørgensen *et al.* (1988) working with whole cod, found that using DT in the predictive equation 9.3, the remaining shelf-life of whole cod could be predicted relatively precisely.

Monitoring of time–temperature during transportation can provide a useful indication of spoilage, provided that live data at some specific temperature are available, preferably, but not necessarily, at 0°C. The shelf-life of each fish can then be calculated at different temperature regimes. Shelf-life depends on the initial bacterial load of the fish since TTFI is a function of the rate of growth of micro-organisms. Temperature fluctuations during transportation will also affect the shelf-life prediction.

9.5.2 Heat transfer/mathematical approach

Before discussing the application of heat transfer in fish transportation, it is important to define what heat is. Heat is a form of energy which can be measured, cold cannot. It is the addition or removal of heat which results in a temperature change or change of phase. Heat transfer occurs in the direction of decreasing temperature. In other words, fish cannot be cooled without using something colder to act as a recipient for the heat to be removed from the fish. Heat is transferred in three basic ways: by conduction, convection and radiation.

Conductive heat transfer is achieved by direct contact so fish being cooled by direct contact with ice will experience heat transfer by conduction. Convective heat transfer is forced or natural movement of a fluid (liquid or gas). Fish in a chill room can be cooled by convective heat transfer due to natural circulation or fan circulation of the air. Radiative heat transfer from a heat source to a body is achieved without heating the intermediate space and without the need of an intermediate material. Fish will be exposed to radiated heat from the sun if they are left uncovered outdoors. Fish exposed to a light source indoors will also experience a radiant heat transfer (FAO, 1992). Conductive heat transfer, which is the most important in cooling of the fish inside containers, is discussed later.

9.5.2.1 *Heat transfer equations and units used to predict ice requirement.*

In mainly tropical and sub-tropical countries where the ambient temperature is >30–40°C during Summer, it becomes very difficult to estimate the correct level of ice consumption during fresh fish transportation. Using a large quantity of ice is not economical and re-icing is usually inappropriate. Sometimes, upon arrival at the retailer or consumer, the fish is in an unacceptable state or nearly so because of inadequate icing. In order to overcome such problems, many researchers have worked on the predicted estimation of correct level of ice consumption in fresh fish transportation (Boeri *et al.*, 1985; Chattopadhyay *et al.*, 1975).

Ice requirement for a given time inside a container depends upon many varying factors. It requires knowledge of possible ways of heat transfer and thermal properties of fish and container under various environmental conditions. Figure 9.5 shows factors affecting melting ice when stored with fish

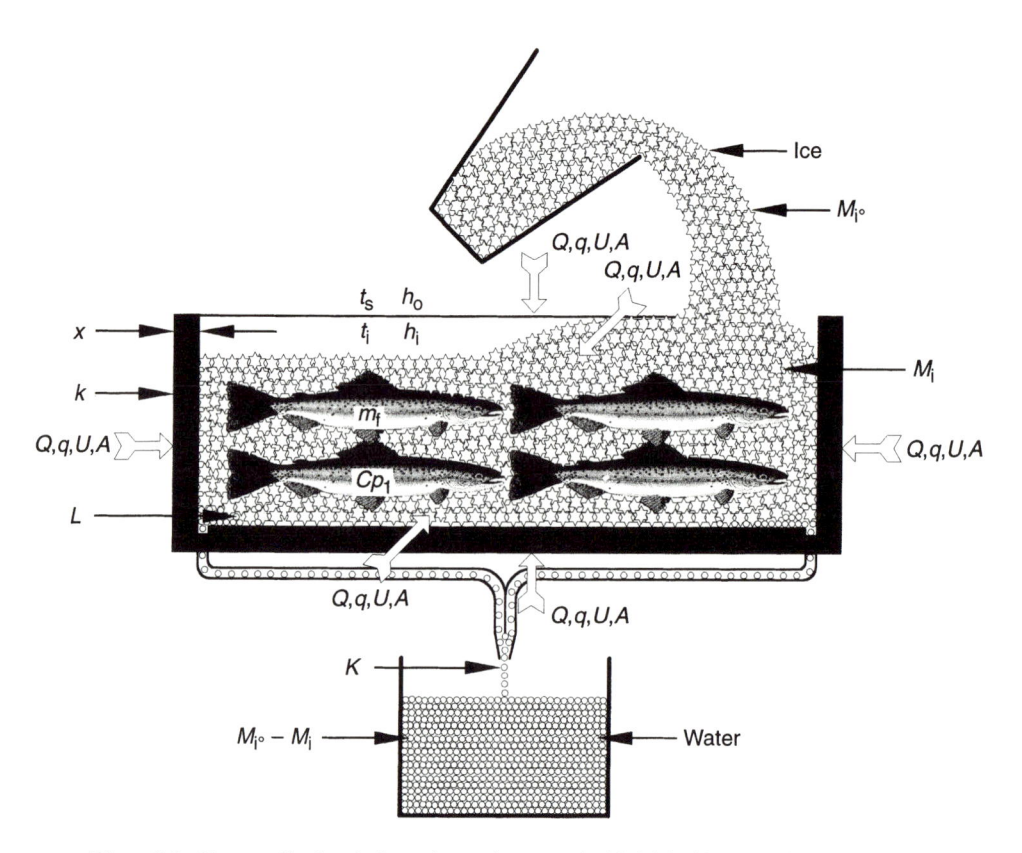

Figure 9.5 Factors affecting the ice meltage when stored with fish inside a container (all symbols are defined in the text).

inside a container. It is therefore important to know related heat transfer equations and units used to predict the correct ice requirement in fish transportation. These equations can also be used for cooling gels. All this information helps to better determine the amount of ice or cooling gels needed to bring the fish to 0°C and maintain it at that temperature during transportation.

Measurement of rate of heat transfer. The rate of heat transfer indicates how much heat is entering the walls of the container/box to melt ice per hour and is obtained using equation 9.4.

$$q = UA(t_s - t_i) \quad (9.4)$$

where q is the rate of heat transfer (W) (1 kcal h^{-1} = 1.163 W), U is the overall heat transfer coefficient (W m^{-2} °K^{-1}), A is the area of exposed surface (m^2), t_s is the surface temperature of the container (°C), and t_i is the melting temperature of ice (°C).

Measurement of overall heat transfer coefficient. Accurate calculations of ice melting rate, quantity of heat, rate of heat transfer and thermal conductivity of the fish container can be assessed if the overall heat transfer coefficient, U, of the fish container is known. The overall heat transfer coefficient of the fish container was calculated by Boeri et al. (1985), using equation 9.5. Assuming that t_s is constant and $t_i = 0°C$:

$$U = \frac{L(M_{i^0} - M_i)}{A(t_s - t_i)t} \quad (9.5)$$

where L is the latent heat of fusion of ice (kJ kg^{-1}) (1 kcal = 4.1868 kJ), M_{i^0} is the initial mass of ice (kg), M_i is the remaining mass of ice (kg), $M_{i^0} - M_i$ is the melted mass of ice (kg), and t is the time elapsed since icing (h).

Measurement of quantity of heat. Measuring the quantity of heat is important to assess how much heat is entering the walls of container to melt the ice. Ice needs a large amount of heat to melt it, because it has large reserves of 'cold' and this is why ice is so widely used in fish transportation. The quantity of heat, Q, is calculated from equation 9.6. This formula can only be used for steady-state heat transfer.

$$Q = \frac{k}{x} A(t_s - t_i)t \quad (9.6)$$

where Q is the quantity of heat (J), k is the thermal conductivity of the container (W m^{-1} °K^{-1}), and x is the overall thickness of the container (m).

Measurement of latent heat of fusion of ice or cooling gels. Latent heat is the amount of heat absorbed or evolved by a unit mass of material during a phase change. If the latent heat of coolant is known, it is possible to estimate the mass of coolant needed to keep the fish at or around 0°C during transportation. Although the best method of measuring the latent heat of fusion of a coolant is differential scanning calorimetry, it is also possible to calculate it by thawing a coolant against time. The latent heat of coolant, L, can be calculated using equation 9.7:

$$L = \frac{Q}{M_{i^0} - M_i} \qquad (9.7)$$

Measurement of ice melting rate. The effectiveness of the fish container can be assessed by measuring the ice melting rate which indicates how long a given quantity of ice will last inside the container. The ice melting rate, K, is calculated from equation 9.8:

$$K = \frac{M_{i^0} - M_i}{t} \qquad (9.8)$$

where K is the melting rate of ice ($kg\,h^{-1}$).

Boeri et al. (1985) proposed a formula to verify the relationship given by equation 9.9. Assuming that $t_i = 0°C$, this formula can be used for both experimental and predictive calculations.

$$K = \frac{AU}{L} t_s - \frac{AU}{L} t_i \qquad (9.9)$$

Measurement of thermal conductivity of the fish container. The ice melting rate can be significantly reduced with a good insulating material. It is therefore important to measure the thermal conductivity of the fish container used in fish transportation. Although the best method of measuring thermal conductivity of rigid material is the standard guarded hot-plate method (BSI 874, 1986), it is also possible to calculate it by thawing ice inside the container during storage. The thermal conductivity of the container, k, is calculated from equation 9.10. This formula can only be used for steady-state heat transfer.

$$k = \frac{q}{A(t_s - t_i)} x \qquad (9.10)$$

Area measurement of the fish container. The ice melting rate also depends on the surface area of the container. The larger the surface area the greater the ice melting rate. It is of great importance to measure the area of the container in order to calculate the right ice consumption during transportation. Four sides, bottom and lid of the container (outside and inside) are measured, using equation 9.11:

$$A = \frac{A_i + A_o}{2} \qquad (9.11)$$

where A_i is the inside area of the container (m^2), and A_o is the outside area of the container (m^2).

Specific heat estimation of fish. Knowledge of the thermophysical properties of fish including specific heat is essential in predicting heat transfer. Specific heat defines how much heat is required to change the temperature of the fish. Where there is no change in mass the specific heat is defined as the amount of heat necessary to raise the temperature of 1 kg of the product by 1°C or °K. If the water content of fish is high (80%), the amount of ice necessary for cooling fish to 0°C will also be high since specific heat is mainly proportional to water content. In contrast, if the water content of fish is low (60%), less ice will be needed to cool the fish to 0°C. Therefore, it is important to estimate the specific heat of fish for proper calculation of ice requirement. Specific heats above (Cp_1) and below (Cp_2) freezing are calculated using Siebel's formula (1982):

$$Cp_1 = 0.837 + 3.349 \, X_w \tag{9.12}$$

$$Cp_2 = 0.837 + 1.256 \, X_w \tag{9.13}$$

where Cp_1 is the specific heat of fish above freezing point (kJ kg^{-1} °K^{-1}), Cp_2 is the specific heat of fish below freezing point (kJ kg^{-1} °K^{-1}), and X_w is the mass fraction of water (%).

Predicted ice requirement for cooling fish to 0°C. To determine the correct ice requirement in fish transportation, it is necessary to calculate the mass of ice needed to cool the fish to 0°C, which is calculated by FAO (1992), using equation 9.14:

$$m_i = \frac{Cp_1 t_1}{L} m_f \tag{9.14}$$

where m_i is the mass of ice necessary to cool fish to 0°C (kg), t_1 is the temperature of the fish just before being iced (°C), and m_f is the mass of fish (kg).

Predicted overall heat transfer coefficient. Where it is difficult to carry out experimental work, the predicted overall heat transfer coefficient of the fish container can be used to calculate the approximate ice requirement during transportation of fish. The predicted overall heat transfer coefficient, U_t, of the container is calculated by ASHRAE (1981), using equation 9.15:

$$U_t = \frac{1}{\frac{1}{h_i} + \frac{x}{k} + \frac{1}{h_o}} \tag{9.15}$$

where U_t is the theoretical overall heat transfer coefficient (W m^{-2} °K^{-1}), h_i is the internal heat transfer coefficient (W m^{-2} °K^{-1}), and h_o is the external heat transfer coefficient (W m^{-2} °K^{-1}).

9.5.2.2 Application of heat transfer in fish transportation.
Application of heat transfer in fish transportation is of paramount importance in determining the best ratio of fish and ice at different ambient temperatures. Any measures that can be taken to reduce the amount of ice needed to get the fish to the consumer in good condition would clearly be worthwhile.

The amount of ice needed to chill the fish and keep it chilled during transportation is affected by a number of factors, one of the major ones being the amount of heat leaking into the container from the environment. As mentioned earlier, heat is conveyed from the environment to a cold place in one or more of three ways: conduction, convection and radiation; the former being the most important in cooling of fish during transportation. With conduction, ice can melt in two ways inside the container: first, when ice is placed in a container, the heat is transferred from the warm fish to the ice; second, if all ice has been put on the top of the fish, some will melt in cooling the fish, and some by heat coming in through the walls of the container. Heat from the warm air outside passes through the walls of the container to warm the fish or melt the ice (FAO, 1992).

The amount of heat conducted through the walls of the container depends on four factors: the surface area of the container; its thickness; the difference in temperature between the outside and inside; and the material of which it is made. The surface area of the container is proportional to the entry of heat from the environment. Doubling the area doubles the quantity of heat flow, whereas doubling the thickness of the container will halve the flow of heat. The amount of ice that melts in a given time during transport depends on the rate of heat flow from the outside temperature. If the outside temperature is lower than 0°C, the rate of heat flow from the outside through the container will be negligible and ice will only receive heat from the fish. Finally, heat flow is also proportional to the thermal conductivity of the material, k, which is one of the key properties of the insulating packaging.

The sections below will assist in answering the following questions: How much ice is necessary to cool the fish to 0°C? How long will the ice remain inside the container and how much heat is entering the container from its surroundings to melt the ice? Experimental and predicted measurements are also assessed.

9.5.2.3 Prediction of ice requirement for cooling fish to 0°C.
The ice packed in a container of fish is intended to do two things: first, to cool the fish to 0°C; and second, to maintain it at that temperature by absorbing the heat entering the container from its surroundings. To determine the ice requirement, it is necessary to find out how much ice has to be used to cool fish to 0°C for a given time. The ice required for cooling fish from the initial temperature is fixed and cannot be reduced, but the use of a well-insulated container can considerably reduce the ice requirement during subsequent storage and transport.

As soon as ice is placed on the warm fish, heat from the fish flows out into the ice and melts it. Heat keeps on flowing until there is no difference in temperature between the two. Any further melting of ice that occurs is due to heat from the surroundings – the air, the fish container, and on so. Once the fish is cooled to 0°C, the only sources of the heat are the surroundings (Regenstein and Regenstein, 1991). It is therefore important to place more ice on the top, bottom, and sides of the fish in their containers rather than placing it only on the top of the fish.

Predictive calculation of the ice requirement for cooling fish down to 0°C can provide valuable information at the planning of fish storage and transportation and also promote a better understanding of the relative effect of the various elements which influence the rate of ice meltage. In addition, by considering all possibilities and predicted ice requirements, a more rational judgement can be made when selecting procedures to be used. Alasalvar (1994) compared the experimental and predicted mass of ice necessary to cool down fish to 0°C inside a polystyrene box at different ambient temperatures (Table 9.3). The experimental mass of ice needed to cool down the fish to 0°C was calculated from melting of ice using the time–temperature profile of fish. Equation 9.14 was used for predictive calculations. The experimental results generally agreed with the predictive calculations for fresh salmon with ice (3:1 ratio), although experimental results showed lower values. This might be due to higher cooling rate of fish because of the thickness of layers.

The time taken for a fish or fillet to cool depends on its distance from the ice layer, so that fish at the bottom of the container cool very slowly indeed even when there is ice remaining on top of the fish at the end of the journey. There may still be fish or fillets at 5°C or above in a such container (FAO, 1992).

It is possible to roughly estimate the amount of ice required to convey a given quantity of fish safely on any length of distribution. Obviously, to

Table 9.3 Experimental results and predicted calculations of ice necessary to cool down fish to 0°C inside a polystyrene box at different ambient temperatures. Values are means of triplicate determinations. From Alasalvar (1994)

	Ambient temperature (°C)	Initial fish temperature (°C)	Ice necessary to cool fish to 0°C (kg)	Time elapsed since icing (h)	Percentage of ice melted since icing (%)
Experimental	1.1 ± 1	2	0.425	26	5.26
	13.8 ± 1.3	5	1.17	6	14.48
	21.1 ± 0.5	5	1.04	9	12.87
Predicted	1.1 ± 1	2	0.49	36	6.06
	13.8 ± 1.3	5	1.21	7	14.98
	21.1 ± 0.5	5	1.21	9.5	14.98

distribute fish at 0°C, more ice is used than required simply for cooling it down. It is also advantageous to cool the fish before they are packed in order to reduce the amount of ice required on the journey as well as to prevent spoilage before dispatch.

9.5.2.4 *Ice melting rate.* The effectiveness of insulated fish containers can be assessed by measuring the ice melting rate. The ice melting rate tells how long a given quantity of ice will last inside the container and therefore how long the fish will stay at or around 0°C. The ice melting rate can be reduced by using a well-insulated container that helps to stabilise storage conditions and it is easier to predict and maintain the correct ice requirement. In spite of the obvious difficulties and likely inaccuracies, a calculation of the ice melting rate can be useful in planning storage to enable comparison to be made between different options, and to allow preliminary estimates of quantities, costs and equipment to be made.

Experimental and predicted relationships between ice melting rate (K) and surrounding surface ambient temperatures (t_s) in a polystyrene box containing gutted salmon (3:1 ratio) are shown in Table 9.4. The experimental ice melting tests showed that the actual ice melting is close to that predicted theoretically (15% error). The difference between the experimental and predicted ice melting rate can be attributed to the difficulty in keeping constant ambient temperatures under industrial conditions, and also the difficulty in approximating the relationship between K and t_s by a straight line which, however, can be accepted for the purpose of estimating the amount of ice melting inside a polystyrene box. It is possible to estimate the consumption of ice if the surrounding temperature of the box is known. Prediction of ice melting rate gives an accurate indication of the ice requirement, since reliable data on both materials and conditions are often not readily available.

9.5.2.5 *Rate of heat transfer.* When the rate of heat transfer of the container is known, it is possible to calculate the mass of ice needed for any length of transportation. The mass of ice that melts inside a container depends upon the rate of heat transfer from the surrounding temperature. Rate of heat transfer, which is also proportional to the surface temperature, is higher for ambient stored fish than for chilled stored fish. As shown in Table 9.4A, a rate of 10.026 W would mean that nearly 0.1 kg of ice would melt every hour as a result of heat entering the walls of the polystyrene box, since each kilogram of ice requires 334.5 J g^{-1} to melt it. The rate of heat transfer is smaller when the surface temperature is lower (0.533 W at 0.55 ± 0.5°C). There is also reasonable agreement between the experimental and predicted calculation of rate of heat transfer (Table 9.4A and B).

9.5.2.6 *Overall heat transfer coefficient.* More accurate calculation of the ice requirement can be made if meltage tests are used to determine the overall heat

Table 9.4 Experimental (A) and predicted (B) measurements of ice melting rate and heat transfer of an expanded polystyrene box at different ambient temperatures. Values are means of triplicate determinations. Standard deviations are given in parentheses. Predicted overall heat transfer coefficient, U_t, was used throughout the predicted calculations. From Alasalvar (1994)

(A)			Surface temperature (°C)		
	Symbol	Unit	0.55 ± 0.5	10.5 ± 1.4	17 ± 0.9
Ice melting rate	K	$kg\,h^{-1}$	0.006 (0.000)	0.108 (0.002)	0.172 (0.004)
Rate of heat transfer	q	W	0.533 (0.002)	10.026 (0.133)	16.126 (0.185)
Overall heat transfer coefficient	U	$W\,m^{-2}\,°K^{-1}$	1.034 (0.005)	1.019 (0.014)	1.013 (0.012)
Latent heat	L	$J\,g^{-1}$	345.88 (1.57)	350.59 (4.68)	355.71 (7.57)
Thermal conductivity	k	$W\,m^{-1}\,°K^{-1}$	0.031 (0.000)	0.031 (0.001)	0.030 (0.001)

(B)	Symbol	Unit	0.55 ± 0.5	10.5 ± 1.4	17 ± 0.9
Ice melting rate	K	$kg\,h^{-1}$	0.005	0.091	0.148
Rate of heat transfer	q	W	0.448	8.547	13.839
Overall heat transfer coefficient	U_t	$W\,m^{-2}\,°K^{-1}$	0.869	0.869	0.869

transfer coefficient of the container. Alasalvar (1994) compared the experimental and predicted measurements of overall heat transfer coefficient of the polystyrene box. He found that although there are statistically highly significant differences ($P < 0.01$) between the experimental and predicted measurements, the differences could be due to the internal and external heat transfer coefficients of the polystyrene box (Table 9.4A and B). This low value is due to the low thermal conductivity of expanded polystyrene (Table 9.5). McDonald and Nesvadba (1988) have found the overall heat transfer coefficient of high density polyethylene in double skin (18.5 mm) and dry wood (12.5 mm width of gap) fish boxes with thermal conductivities of 0.175 and 0.197 W m^{-1} °K^{-1} to be 3.07 and 3.51 W m^{-2} °K^{-1}, respectively. The experimental value of U is also sensitive to small variations in the value assigned to the area and surface temperature.

Lupin (1985) mentioned that as the external area is bigger than the internal and wall thickness is not uniform, the corners being thicker, this leads to a different local heat transfer coefficient. The overall heat transfer coefficient and the rate of heat are affected by the position of the container. If the container is on the floor, heat lost through the bottom will vary considerably from heat lost through the top and sides. Heat can also be lost through the lid, if it is not properly closed. Boeri et al. (1985) observed that the location of the container in a stack is important. The top and bottom containers in a stack are the most affected by heat transfer, due to the greater surface area; also, the air speed at the top is higher than around the intermediate containers. The ice melting rates in the top containers was always higher than in the middle or bottom containers. The difference with intermediate containers is not only due to the smaller area exposed, but also to the different air temperature gradients existing for the top and bottom containers and the different heat transfer coefficient for horizontal and vertical surfaces.

If a true value of overall heat transfer coefficient is sought, the dimensions of the container and surface temperature should be measured carefully. Although it is easy to estimate the overall heat transfer coefficients, experimental trials are recommended when materials and conditions are not well known.

9.5.2.7 *Latent heat of fusion of ice.* Measuring latent heat is important to estimate the mass of coolant needed to cool down fish to 0°C and maintain it at that temperature during transportation. Because of the large latent heat capacity of ice, it is widely used in fish transportation as a means of chilling. Cooling gels can also be used since they too have a large latent heat capacity (Alasalvar and Nesvadba, 1995). Table 9.4A shows the latent heat measurement of ice at different melting rates which are close to the latent heat of fusion of ice (334.5 J g^{-1}). Getting a value close to 334.5 J g^{-1}, by measuring latent heat of fusion of ice, will justify the true calculations of rate of heat transfer, ice melting rate, overall heat transfer coefficient and thermal conductivity of the container.

Table 9.5 Thermophysical properties of commercially available expanded polystyrene boxes. Values are means of triplicate determinations. From Alasalvar and Nesvadba (1994)

Sample	Mass of sample (g)	Thickness (mm)	Density (kg m^{-3})	Thermal conductance (W m^{-2} °K^{-1})	Thermal conductivity (W m^{-1} °K^{-1})
PSS	88.36 ± 1.30	40.0 ± 0.2	23.7 ± 0.35	0.7548 ± 0.0067	0.0302 ± 0.0003
TS	96.51 ± 1.07	47.0 ± 0.3	22.07 ± 0.25	0.6840 ± 0.0122	0.0321 ± 0.0006
PS	92.76 ± 0.95	45.0 ± 0.2	22.16 ± 0.23	0.7176 ± 0.0044	0.0323 ± 0.0002

PSS, Plasboard smoked salmon; TS, Thulcraft salmon; PS, Plasboard salmon.

9.5.2.8 *Thermal conductivity of fish container/box.* The effectiveness of an insulating material is measured by its thermal conductivity. Good insulating materials can significantly reduce heat leakage and therefore melting of ice. For this reason, insulated fish containers (especially polystyrene boxes) for handling and transportation of fresh fish are becoming more important in the fish industry.

Thermophysical properties of commercially available expanded polystyrene boxes used in fish transportation in the UK are shown in Table 9.5. There is a reasonable agreement between the experimentally determined thermal conductivity of polystyrene boxes and standard guarded hot-plate results (Tables 9.4A and 9.5). If the thermal conductivity of the packaging materials is known together with the latent heat capacity of the ice then it is possible to estimate the ice melting rate and the time for which the temperature remains below 8°C during transportation.

9.5.3 *Computer modelling of time–temperature*

It is very difficult to estimate the amount of ice needed to keep the fish below 8°C during transportation under various conditions. Although it is possible to calculate the ice requirement for any length of transport using mathematical calculations, this might be difficult for people who work in the fishing industry. Therefore, there is a need for a method that can easily be used by processors or manufacturers. Computer modelling of time–temperature will be of benefit to calculate the correct ice requirement to keep the fish chilled for any length of transport under various conditions and will also help to quantify the safe limits of transportation and ensure the microbiological safety of the fish.

9.5.3.1 *Importance of time–temperature monitoring of fresh fish during transportation.* Monitoring time–temperature profiles of the fish provides valuable information that can be used in the modelling of time–temperature, application of HACCP, calculation of rate of spoilage and monitoring of safety of product. It is therefore important to determine the time for which the product temperature could be maintained below 8°C during transportation.

276   FISH PROCESSING TECHNOLOGY

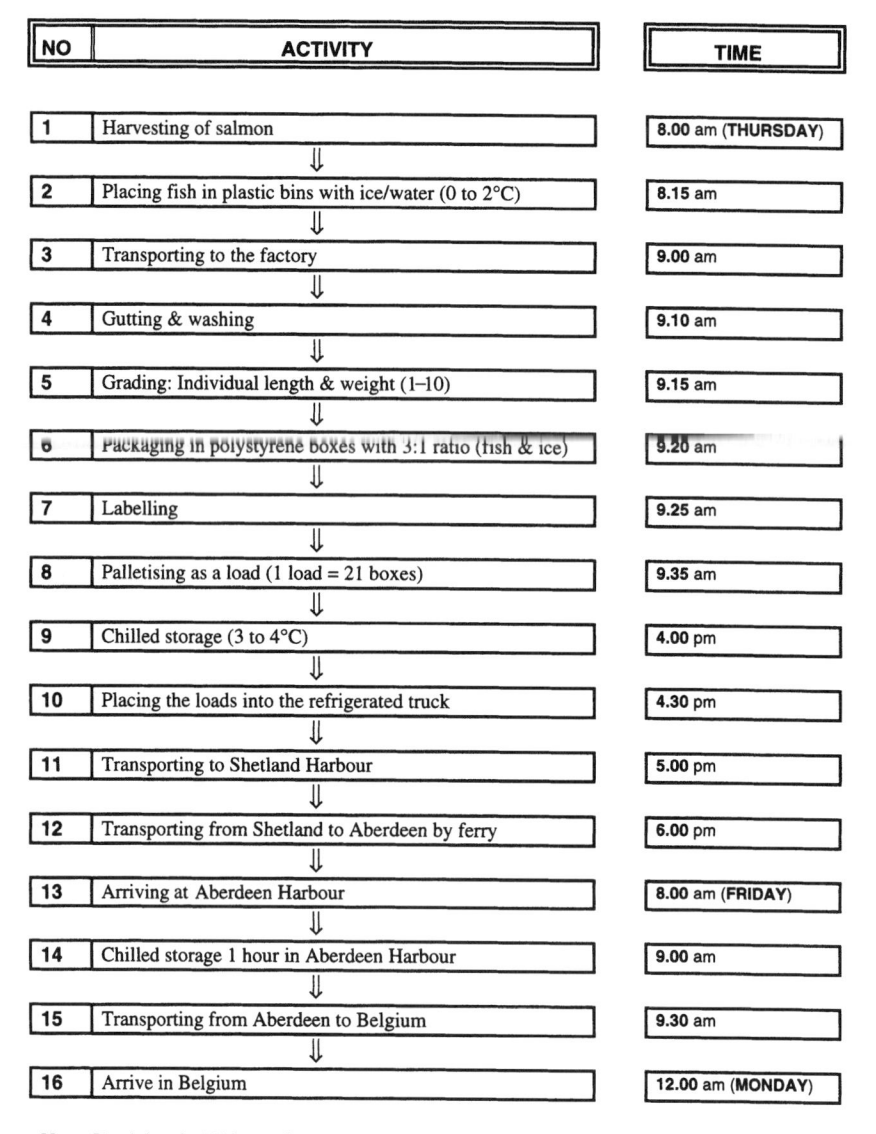

NO	ACTIVITY	TIME
1	Harvesting of salmon	8.00 am (THURSDAY)
2	Placing fish in plastic bins with ice/water (0 to 2°C)	8.15 am
3	Transporting to the factory	9.00 am
4	Gutting & washing	9.10 am
5	Grading: Individual length & weight (1–10)	9.15 am
6	Packaging in polystyrene boxes with 3:1 ratio (fish & ice)	9.20 am
7	Labelling	9.25 am
8	Palletising as a load (1 load = 21 boxes)	9.35 am
9	Chilled storage (3 to 4°C)	4.00 pm
10	Placing the loads into the refrigerated truck	4.30 pm
11	Transporting to Shetland Harbour	5.00 pm
12	Transporting from Shetland to Aberdeen by ferry	6.00 pm
13	Arriving at Aberdeen Harbour	8.00 am (FRIDAY)
14	Chilled storage 1 hour in Aberdeen Harbour	9.00 am
15	Transporting from Aberdeen to Belgium	9.30 am
16	Arrive in Belgium	12.00 am (MONDAY)

Note: Total time is 100 hours from harvesting to the destination.

Figure 9.6 Sequence of events during transportation of fresh salmon from Shetland to Belgium.

The sequence of events for the preparation of fresh salmon and its transportation is shown in Figure 9.6. Figure 9.7 shows the time–temperature changes of gutted salmon with ice (3:1 ratio) in chill transportation from Shetland (UK island) to Belgium (Alasalvar, 1994). The fish temperature, which was under

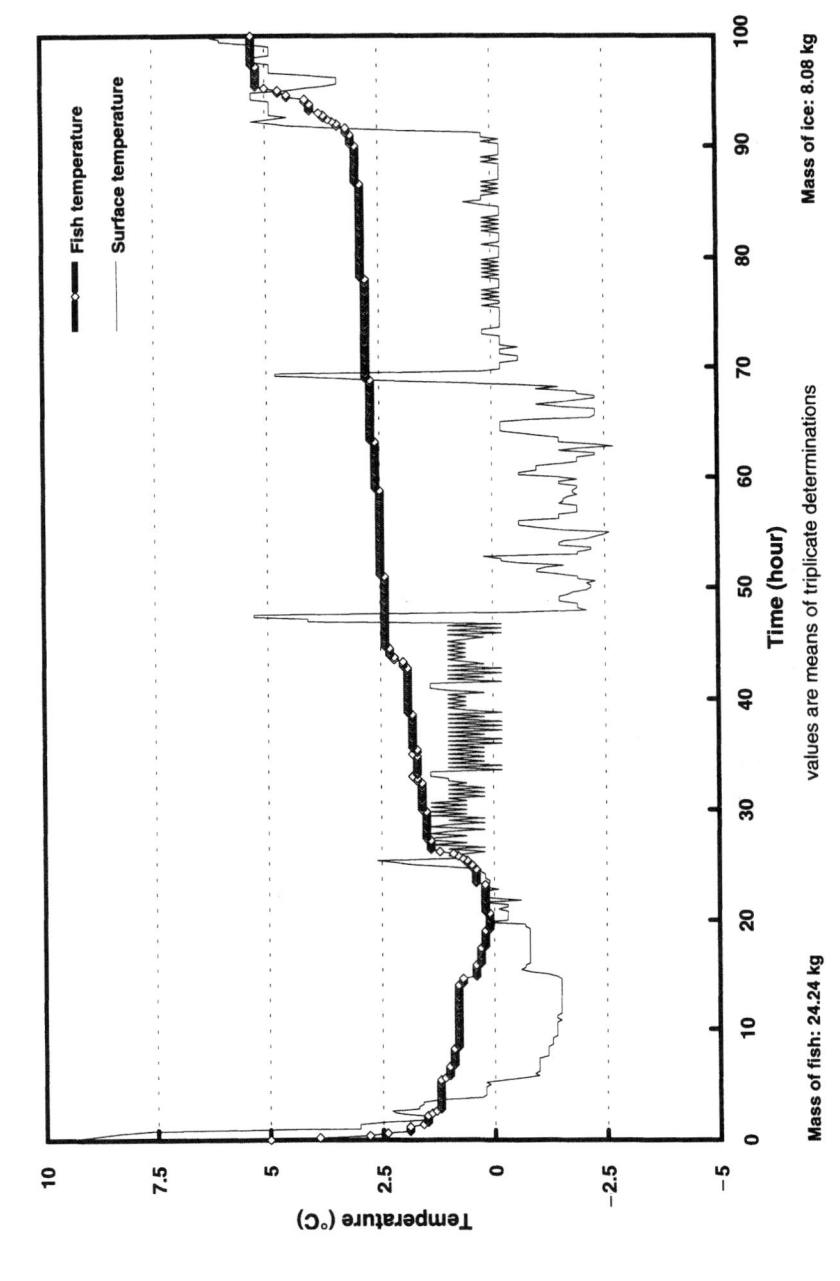

Figure 9.7 Temperature changes of gutted salmon with ice (3:1 ratio) inside a polystyrene box during transportation from Shetland to Belgium.

the safe limit (8°C) at all times, was between 0.6 and 1°C for 30% of the time in transportation, between 1 and 3°C for 60% of the time and between 3 and 5°C for 10% of the time. There are temperature fluctuations (up to 5°C inside the truck) after 27, 48, 70 and 93 h, respectively, and this could be explained by the lack of refrigeration during transfer to the ferry and lorry. A total of 5 h temperature fluctuations during transportation caused 0.2 kg more ice to melt since the rate of heat transfer of the polystyrene box is five times greater at 5°C than at 1°C. In other words, 0.05 kg of ice would melt every hour as a result of heat entering the walls of the box at 5°C, whereas only 0.01 kg of ice would melt at 1°C. There are also temperature fluctuations that are due to the varying temperature of the truck resulting from the different refrigeration systems of the ferry and lorry. The surface temperature of the polystyrene box varied from about 1 to -2.5°C during transportation.

9.5.3.2 *Mailprof computer modelling program.* The best method of chilling during transportation of fresh fish is by direct contact refrigeration using ice or cooling gels. For effective and economical utilisation of ice, it is necessary to know the time required for a quantity of ice needed to maintain the temperature of the fish below 8°C inside the container during transportation. It is possible to calculate this time using the Mailprof computer program. The impetus for the Mailprof computer program came from a study of foods including smoked salmon distributed by mail order (MAFF, 1991). The study found that the temperature of these foods during transit may allow rapid bacterial growth due to lack of time–temperature control. Alasalvar and Nesvadba (1995) modelled the time–temperature profiles of smoked salmon packaged with different cooling gels or ice inside a polystyrene box and shipped at ambient temperatures. This program has also been used (Alasalvar, 1994) to predict the time–temperature profiles of fresh salmon and to estimate the ice requirement during transportation. It runs on a personal computer, (PC), and requires input values of surface heat transfer coefficient of the container, initial temperature of fish and gels (or ice), surface temperature of the container (it is usually lower then ambient temperature), freezing point of fish, mass of fish and gels (or ice), protein, fat and moisture content of the fish (Table 9.6).

9.5.3.3 *Estimation of time–temperature profiles of fish during transportation using the Mailprof computer program.* The accuracy of the Mailprof computer program is assessed in comparison with experimental profiles. Mailprof-predicted time–temperature profiles of the fish were compared with the experimental profiles at ambient temperatures of 1 ± 1, 13.8 ± 1.3 and 21.1 ± 0.5°C by Alasalvar (1994). The temperature profile of the fish predicted by Mailprof agreed with the experimental profile (Figure 9.8) at ambient temperatures of 21.1 ± 0.5°C. Between the Mailprof-predicted and experimental time–temperature profiles of fish, there was 5–10% error in reaching 8°C.

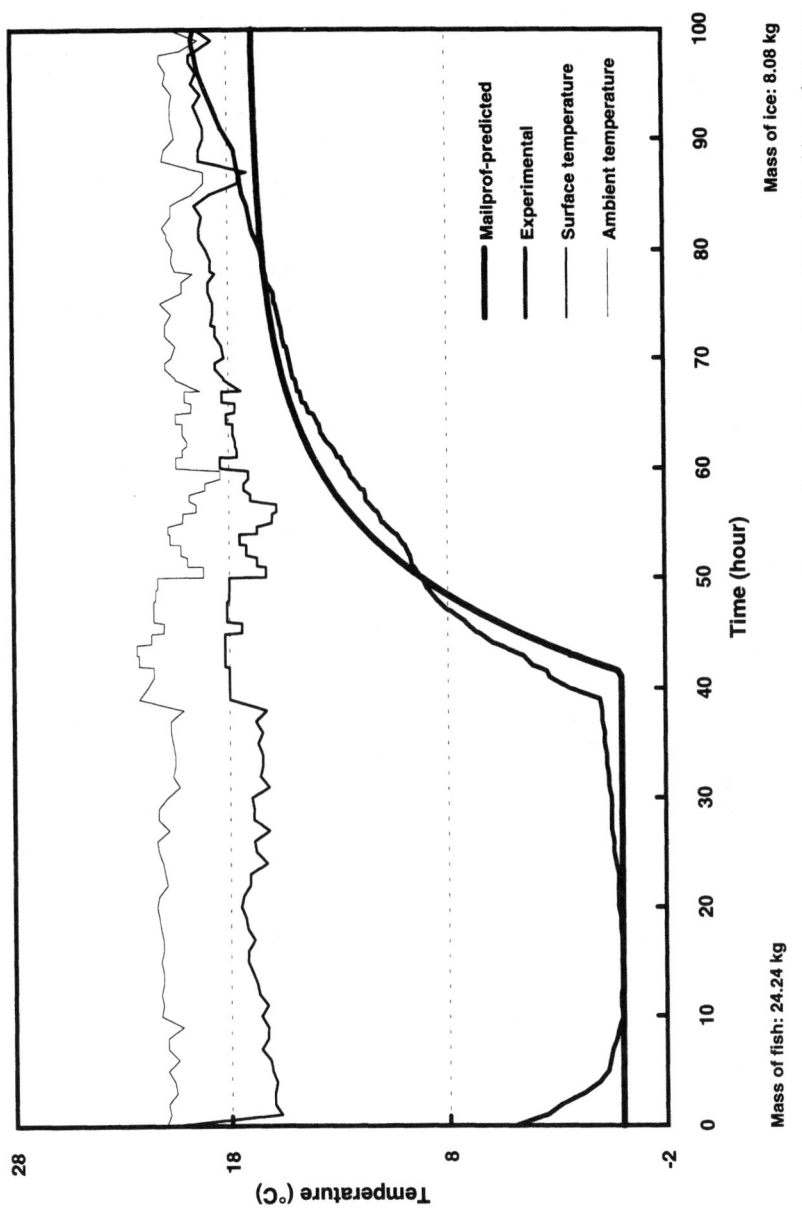

Figure 9.8 Predicted (Mailprof) and experimental time–temperature changes of gutted salmon with ice (3:1 ratio) inside a polystyrene box stored at $21.1 \pm 0.5°C$.

Figure 9.9 Predicted (Mailprof) and experimental time–temperature changes of minced smoked salmon with gel inside a polystyrene box stored at $25 \pm 1°C$.

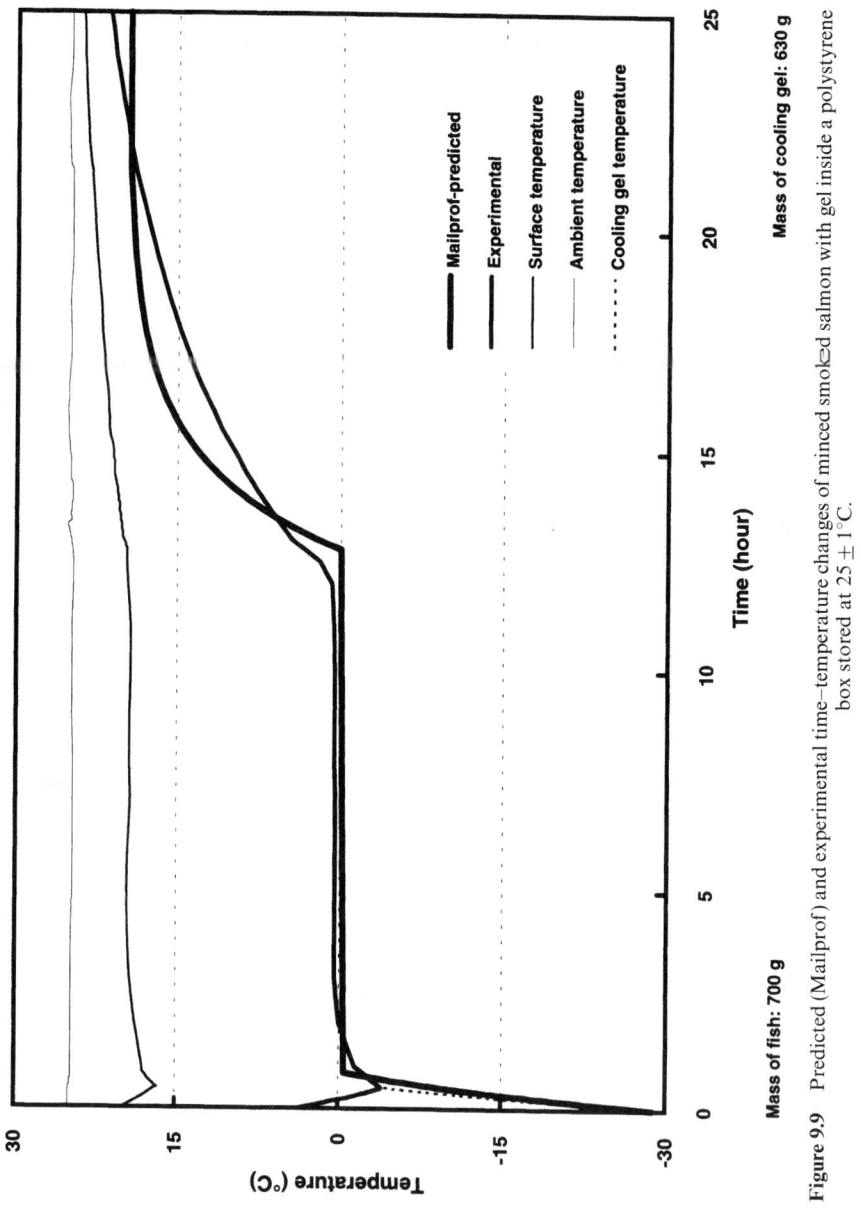

Experimental conditions as well as inputs used to predict the time–temperature profiles of fish affect this error. Better accuracy can be obtained when all the data and experimental conditions are measured carefully. As shown in Figure 9.8, when ice completely melted after 47 h, fish temperature was at the safe limit (8°C). At higher ambient temperatures, this increase will be more rapid. Especially in tropical and sub-tropical countries (temperature > 30–40°C), ice should always be available inside the container until it reaches the retailer or consumers.

As mentioned above, Alasalvar and Nesvadba (1995) also found that the temperature profile predicted by Mailprof agreed with the experimental profile (Figure 9.9) for smoked salmon with a cooling gel ('ice-berg') inside a polystyrene box, both curves reaching 8°C almost at the same time.

Inputs used for running the Mailprof computer program are listed in Table 9.6. Using this program, it is possible to estimate the time of melting of ice and subsequent increase in fish temperature in iced fish containers in order to ascertain the total time required for the fish to increase from 0 to 8°C during transportation. This helps to quantify the safe limits of operation and ensure the microbiological safety and quality of the product.

It is possible by computer modelling to predict the temperature of product in a time–temperature regime, provided the surface heat transfer coefficient of the packaging is known. Chill temperatures can be maintained for a predetermined time using optimum amounts of gels (or ice) in relation to the degree of insulation calculated using the Mailprof computer program (MAFF, 1991). The time during which the product is at low temperature depends on the mass

Table 9.6 Experimental data for Mailprof computer run (obtained from experimental results with the exception of the surface heat transfer coefficient, which was estimated by matching the predictions of the Mailprof computer program with experimental temperature records). From Alasalvar (1994) and Alasalvar and Nesvadba (1995)

Parameters	Unit	Imputs Figure 9.8	Imputs Figure 9.9
Water content of fish	%	75.14	59.60
Protein content of fish	%	21.38	22.69
Fat content of fish	%	2.28	16.21
Freezing point of fish	°C	−2.56	−3.5
Water content of ice/gel	%	100	100
Freezing point of ice/gel	°C	0	0
Thickness of the container	mm	30	20
Mass of fish	kg	24.24	0.7
Mass of ice/gel	kg	8.08	0.63
Thickness of fish layer	mm	75	10
Thickness of ice/gel layer	mm	60	10
Initial fish temperature	°C	5	5
Initial ice/gel temperature	°C	0	−35
Surface temperature of the container	°C	17	20
Surface heat transfer coefficient	$W\,m^{-2}\,°K^{-1}$	3.18	2.86

and initial temperature of the product, the packaging materials and the time and temperatures expected during transportation. The use of ice or cooling gels helps to maintain low temperatures during transportation. This program can be used for any fish:ice ratio provided that the parameters are known. The system using either cooling gels or ice during chill transportation in the UK could be adapted to other developed and developing countries.

9.6 Food MicroModel

Recently, several groups of workers have begun to move away from traditional challenge testing and to apply modern predictive modelling techniques to the growth and survival of micro-organisms in foods (Gould, 1989; Roberts, 1990; Cole, 1991). A UK coordinated programme of research on the growth, survival and thermal death of food-poisoning bacteria in conditions relevant to food was initiated and funded by the Ministry of Agriculture, Fisheries and Food (MAFF). This predictive modelling technique is called Food MicroModel. It is a software package that will predict the growth, survival and thermal death of the major food pathogens and food spoilage organisms in a wide range of foods. Using this model, users can simulate new food formulations or reformulations to screen for potential microbiological problems. The effects of storage and packaging conditions on product safety can also be assessed and the microbiological safe shelf-life can be determined. The model can help to identify microbiological hazards and assess risks in products and processes and it can be used to develop heat process treatments for a range of food formulations (Anon, 1995). Food MicroModel is likely also to be a powerful tool in the HACCP system as a means of identifying hazards (Gould, 1989; Williams *et al.*, 1992).

9.6.1 Types of model in Food MicroModel

There are two groups of models in Food MicroModel: growth or survival models used for determining the safety of stored foods and thermal death models used to determine the lethality of a thermal process. Although a broad range of factors (temperature, water activity (a_w), pH, organic acids, nitrite, modified atmosphere packaging (MAP)) can affect the activity of micro-organisms in foods, temperature, water activity and pH are generally the major factors determining the rate of microbial growth and thermal death (MAFF, 1994). For this reason the majority of models depend upon the measurement of these factors. Some models, however, have additional controlling factors from which a prediction can be made.

9.6.1.1 *Growth or survival models.* These models predict the rate of growth for many of the major food-borne pathogens including *Aeromonas hydrophila*,

Bacillus cereus psychrophilic + mesophilic (vegetative), *B. licheniformis*, *B. subtilis*, *Clostridium botulinum* (non-proteolytic), *C. perfringens*, *Escherichia coli* 0157:H7, *Listeria monocytogenes*, *Salmonella*, *Staphylococcus aureus* and *Yersinia enterocolitica*, and survival only for *Campylobacter jejuni*. Only one food-spoilage micro-organism, *Brochothrix thermosphacta*, is included in the growth model. Growth is predicted with variation in various parameters; for example, a growth model is provided for *Yersinia enterocolitica* in the presence of lactic and acetic acids. These models are usually based on data obtained from a cocktail of strains.

9.6.1.2 *Thermal death models.* These models predict the rate of thermal inactivation of foodborne pathogens including *Clostridium botulinum* (non-proteolytic), *Escherichia coli* 0157:H7, *Listeria monocytogenes*, *Salmonella* and *Yersinia enterocolitica*. The thermal death models are based on decimal reduction (D-) values. A D-value is the time at a given temperature for a decimal reduction in the viable numbers of a spoilage micro-organism. The use of D-value assumes that the logarithmic reduction in organism number with time is linear.

9.6.2 *Use of Food MicroModel in fish transportation*

MAFF (1991) first calculated a time–temperature regime on the growth of *Listeria monocytogenes* determined using the computer-based mathematical models under the Predictive Modelling of Microbial Growth Programme. The results showed that under the time–temperature conditions experienced during distribution by post, growth could occur during the time normally taken for delivery. It is also possible to detect the microbiological safety of smoked salmon. The probable numbers of organisms on these products and time–temperature history can be estimated using Food MicroModel (MAFF, 1994).

Food MicroModel can be used related to fish transportation. For example, given measurements of the times and temperatures to which the fish are exposed during catching, storage, distribution, retailing and in the home prior to consumption, Food MicroModel can be used to make accurate predictions of the possible extent of growth, survival and thermal death of any of the key micro-organisms at any particular stage from catching to consumption. It can also provide information on effective chill temperature to prevent microbial growth. The effect of fluctuating storage and distribution temperatures can also be accommodated within the predictions from the model.

Using Food MicroModel in the fish transportation area will help to provide a safer fish supply for consumers and improved control, shelf-life and stability of fish and their products for processors, retailers and consumers. The fish industry will be well advised to consolidate the work of Food MicroModel with a collaborative initiative into the control of fish spoilage.

9.7 Conclusion

Throughout the world, fresh fish must be transported inside insulated containers (either locally available or imported) with sufficient mass of ice or cooling gels in order to supply a safe and good-quality product to consumers. The use of polystyrene boxes in fish transportation helps to maintain low temperatures, reduces ice meltage and extends the shelf-life of fish dependent on the storage temperatures. Owing to their lower thermal conductivity, polystyrene boxes could be used in many sectors where fish and chilled products are transported to their markets.

TTFI provide data which can be used to inform a processor of the temperature regimes experienced by the product and potential problem areas. If the time–temperature history is known, it is possible to predict the spoilage and shelf-life of product.

Maximum temperature limits and the HACCP concept in fish transportation should exist in every country as food safety standards. The Mailprof computer program is used to predict the time–temperature profile of fish during transportation. It is possible for industry or processors to determine which materials, coolants and conditions to use, and the quantities required to meet a certain time of transportation. When the fish:ice ratio needs to be known at varying ambient temperatures, as may be the case in Summer, this program would be of great importance to developing countries in terms of cost effectiveness. The Mailprof allows processors/manufacturers to assess the likely storage or distribution life of their products. It also facilitates the introduction of HACCP since data on the integrity of the cold chain and identification of the critical control point – control of hazard by elimination or prevention (CCP_1) is made easier. Although it is possible to estimate the ice requirement during transportation with mathematical calculations, this might prove difficult for processors or industry. The Mailprof computer program is easy to use with a PC, provided that the parameters are known.

Alternatives and complements to traditional methods for the predicting and identifying micro-organisms using Food MicroModel will provide fish manufacturers with faster and simpler ways of ensuring fish safety. Food Micro-Model will also encourage a more integrated approach to fish safety and hygiene which will impact on all stages of fish transportation from catching to consumption.

References

Alasalvar, C. (1994), Factors affecting the safety and quality of fish during chill distribution. *PhD Thesis*, University of Humberside, School of Applied Science and Technology, Grimsby, U.K.

Alasalvar, C. and Nesvadba, P. (1994), Measurement of thermal conductivity and density of expanded polystyrene. *Torry Document No. 2612* (Internal document), MAFF, Food Science Laboratory, Torry, Aberdeen, UK.

Alasalvar, C. and Nesvadba, P. (1995), Time/temperature profiles of smoked salmon packaged with cooling gel and shipped at ambient temperature. *J. Food Sci.* **60**(3), 619–621, 626.

Anon. (1995), Food MicroModel. *The World of Ingredients.* **Jan/Feb.**, 57.

ASHRAE (American Society of Heating, Refrigeration and Air-Condition Engineers, Inc.) (1981), *ASHRAE Handbook of 1981 Fundamentals.* ASHRAE, Atlanta, Georgia.

ATP (Agrément de Transport de Périssables) (1987), Consolidated text of the agreement on the international carriage of perishable foodstuffs and on special equipment to be used for such carriage. In *Command No. 25*, November 1987, HMSO, London.

Avdalov, N. and Ripoll, A. (1981), Handling, quality and yield of fresh hake. *Refrig. Sci. Technol.* **49**(4), 71–78.

Bennour, M., Marrakchi, A.EL., Bouchriti, N., Hamama, A. and Quadaa, M.EL. (1991), Chemical and microbiological assessments of mackerel (*Scomber scombrus*) stored in ice. *J. Food Prot.* **54**(10), 784–792.

Bilinski, E., Jonas, R.E.E. and Peters, M.D. (1983), Factors controlling the deterioration of the spiny dogfish (*Squalus acanthias*) during iced storage. *J. Food Sci.* **48**(3), 808–812.

Boeri, R.L., Davidovich, L.A., Giannini, D.H. and Lupin, H.M. (1985), Method to estimate the consumption of ice during fish storage. *Intl. J. Refrig.* **8**(2), 97–101.

Børresen, T. and Strøm, T. (1983), Fish processing. In *Chemistry and World Food Supplies: The New Frontiers*, Shemilt, L.W. (Ed.), Chemrawn II, Pergamon Press, Oxford, pp. 411–419.

Boyd, N.S. and Wilson, N.D.C. (1977), Hypoxanthine concentrations as an indicator of freshness of iced snapper. *New Zealand Journal of Science* **20**(2), 139–143.

Bryan, F.L. (1992), *Hazard Analysis Critical Control Point Evaluations.* World Health organization, Geneva.

BSI (British Standard Institution) 874. (1986), *British Standard Methods for Determining Insulating Properties-Part 2: Test for the Thermal Conductivity and Related Properties, Section 2.1, Guarded Hot-Plate Method, BS 874.* British Standard Institution, London.

Charm, S.E., Learson, R.J., Ronsivalli, L.J. and Schwartz, M. (1972), Organoleptic technique predicts refrigeration shelf life of fish. *J. Food Technol.* **26**(7), 65–68.

Chattopadhyay, P., Raychaudhuri, B.C. and Rose, A.N. (1975), Prediction of temperature of iced fish. *J. Food Sci.* **40**(5), 1080–1084.

Cole, M.B. (1991), Databases in modern food microbiology. *Trends Food Sci. Tech.* **1**, 293–297.

Connell, J.J. (1990), *Control of Fish Quality*, 3rd edn. Fishing News Books, Farnham.

Cox, K. (1982), Capabilities of airfreight containers. In *Proceedings of the First International Conference on Seafood Packaging and Shipping*, Martin, R.E. (Ed.), Science & Technology National Fisheries Institute, Washington, DC, pp. 280–286.

Dept of Health, UK (1990), *Guidelines on the Food Hygiene (Amendment) Regulations 1990.* Department of Health, London.

Dept of Transport (1988), *A Guide to the International Transport of Perishable Foods.* Department of Transport, Publications Sales Unit, Building 1, Victoria Road, South Ruislip, Middlesex, HA4 0NE, U.K.

EC (1989), Council directive amending directive 79/112/EEC on the labelling of foodstuffs, 89/395/EEC. *Official Journal of the European Communities* **32**(L 186), 17–20.

EC (1990) Proposal for a council regulation laying down health conditions for the production and placing on the market of fishery products. *Official Journal of the European Communities* **33** (C 84), 58–70.

Evans, T. (1995), Seafood safety – what exporters must know about HACCP. *Infofish International* **3** (May/Jun.), 48–52.

FAO (1992), Ice in fisheries. In *FAO Fisheries Technical Paper No. 331*, Graham, J., Johston, W.A. and Nicholson, F.J. (Eds), FAO Publications, Rome.

FAO (1994), *FAO/DANIDA Training Project on Fish Technology and Quality Control: FAO Fish Utilisation and Marketing Service No. 15.* FAO, Rome.

Fernandez-Salguero, J. and Mackie, I.M. (1987), Comparative rates of spoilage of fillets and whole fish during storage of haddock (*Melanogrammus aeglefinus*) and herring (*Clupea harengus*) as determined by the formation of non-volatile and volatile amines. *Int. J. Food Sci. Technol.* **22**(4), 385–390.

Gibson, D.M. (1985), Predicting the shelf life of packaged fish from conductance measurements. *J. Appl. Bacteriol.* **58**(5), 465–470.

Gould, G. (1989), Predictive mathematical modelling of microbial growth and survival in foods. *Food Sci. Technol. Today* **3**(2), 89–92.

Hall, P.A. (1994), Scope for rapid microbiological methods in modern food production. In *Rapid Analysis Techniques in Food Microbiology*, Patel, P.D. (Ed.), Blackie, London, pp. 254–267.

Harris, R.W. (1982), Use of gel refrigerants in seafood packaging and shipping. In *Proceedings of the First International Conference on Seafood Packaging and Shipping*, Martin, R.E. (Ed.), Science & Technology National Fisheries Institute, Washington, DC, pp. 139–146.

Heap, R.D. (1992), Refrigeration of chilled foods. In *Chilled Foods: A Comprehensive Guide*, Dennis, C. and Stringer, M. (Eds), Ellis Horwood, Chichester, pp. 59–76.

Huss, H.H. (1992), Development and use of the HACCP concept in fish processing. In *Quality Assurance in the Fish Industry*, Huss, H.H. et al. (Eds), Elsevier Science, Amsterdam, pp. 489–500.

ICMSF (1988), *Microorganisms in Foods 4 – Application of the Hazard Analysis Critical Control Point (HACCP) System to Ensure Microbiological Safety and Quality*. Blackwell Scientific, Oxford.

Jay, J.M. (1992), *Modern Food Microbiology*, 4th edn. Van Nostrand Reinhold, New York.

Jørgensen, B.R., Gibson, D.M. and Huss, H.H. (1988), Microbiological quality and shelf life prediction of chilled fish. *Intl. J. Food Microbiol.* **6**(4), 295–307.

Leu, S.S., Jhaveri, S.N., Karakoltsidis, P.A. and Constantininides, S.M. (1981), Atlantic mackerel (*Scomber scombrus*, L): Seasonal variations in the proximate composition and distribution of chemical nutrients. *J. Food Sci.* **46**(6), 1635–1638.

Lima dos Santos, C.A., Josupeit, H. and Chimisso dos Santos, D. (1994), EEC quality and health requirements and seafood exports from developing countries. *Infofish International* **1** (Jan/Feb.), 21–26.

Lupin, H.M. (1985), Measuring effectiveness of insulated containers. In *FAO Fisheries Report No. 329 Supplement*, Proceedings of the FAO Expert Consultation on Fish Technology in Africa, Lusaka, Zambia, 21–25 January, 1985, FAO Publications, Rome, pp. 36–46.

Lupin, H.M., Giannini, D.H., Soule, C.L., Davidovic, L.A. and Boeri, L. (1980), Storage life of chilled Patagonian hake (*Merluccius hubbsi*). *J. Food Technol.* **15**(3), 285–300.

MAFF (1991), *The Microbiological Status of Some Mail Order Foods*. Food Safety Directorate, MAFF Publications, London.

MAFF (1994), Software marketed by Food Micromodel Ltd., Randalls Road, Leatherland, UK.

Maia, E.L., Rodriguerz-Amaya, D.B. and Moraes, M.A.C. (1983), Sensory and chemical evaluation of the keeping quality of the Brazilian fresh-water fish (*Prochilodus scrofa*) in ice storage. *J. Food Sci.* **48**(4), 1075–1077.

Makene, J. and Mgawe, Y. (1991), Insulation for local insulated containers. In *FAO Fisheries Report No. 467 Supplement*, Proceedings of the FAO Expert Consulation Fish Technology in Africa, Accra, Ghana, 22–25 October, 1991, FAO Publications, Rome, pp. 38–42.

Makene, J., Mgawe, Y. and Mlay, M.L. (1988), Construction and testing of the mbegani fish container. In *FAO Fisheries Report No. 400 Supplement*, Proceedings of the FAO Expert Consulation on Fish Technology in Africa, Abidjan, Côte d'Ivoire, 25–28 April, 1988, FAO Publications, Rome, pp. 1–7.

Matches, J.R. (1982), Microbial changes in packages. In *Proceedings of the First International Conference on Seafood Packaging and Shipping*, Martin, R.E. (Ed.), Science & Technology National Fisheries Institute, Washington, DC, pp. 46–70.

McDonald, I. and Graham, J. (1985), Variations in potential shelf-life resulting from boxing at sea practice. In *Storage Lives of Chilled and Frozen Fish and Fish Products*, I.I.F.-I.I.R. Commissions C_2 and C_3-1985-84, Aberdeen, UK, pp. 275–280.

McDonald, I. and Nesvadba, P. (1988), Measurement of thermal conductance of high density (HD) polyethylene fish box material and the ice meltage in stowage. *Torry Document No. 2251* (Internal document), MAFF, Food Science Laboratory, Torry, Aberdeen, UK.

Mignault, R. (1982), Use of airfright forwarders and freight. In *Proceedings of the First International Conference on Seafood Packaging and Shipping*, Martin, R.E. (Ed.), Science & Technology National Fisheries Institute, Washington, DC, pp. 276–279.

Morita, R.Y. (1973), Psychrophilic bacteria. *Bacteriol. Rev.* **39**(2), 144–167.

Mortimore, S. and Wallace, C. (1994), *HACCP, A Practical Approach*. Chapman & Hall, London.

Myers, M. (1981), Planning and engineering data 1: Fresh fish handling. In *FAO Fisheries Circulate No. 735*, FAO Publications, Rome, p. 64.

National Marine Fisheries Service (1987), *Plan of Operations – Model Seafood Surveillance Project.* Office of Trade and Industry Service, National Seafood Inspection Laboratory, PO Drawer 1207, Pascagoula, MS 39568-1207.

Olley, J. (1978), Current status of the theory of the application of temperature indicators, temperature integrators and temperature function integrators to the food spoilage chain. *Int. J. Refrig.* **1**(2), 81–86.

Olley, J. and Ratkowsky, D.A. (1973), The role of temperature function integration in the monitoring of fish spoilage. *Food Technology in New Zealand* **8**(2), 15–17.

Olley, J. and Quarmby, A.R. (1981), Spoilage of fish from Hong Kong at different storage temperatures; Prediction of storage life at higher temperatures based on storage behaviour at 0°C, and a simple visual technique for comparing taste panel and objective assessment of deterioration. *Tropical Science* **23**(2), 147–153.

Owen, D. and Nesbitt, M. (1984), A versatile time temperature function integrator. *Laboratory Practice* **33**(1), 70–75.

Ratkowsky, D.A., Olley, J., McMeekin, T.A. and Ball, A. (1982), Relationship between temperature and growth rate of bacterial cultures. *J. Bacteriol.* **149**(1), 1–5.

Ratkowsky, D.A., Lowry, R.K., McMeekin, T.A., Stokes, A.N., Chandler, R.E. (1983), Model for bacterial culture growth rate throughout the entire biokinetic temperature range. *J. Bacteriol.* **154**(3), 1222–1226.

Regenstein, J.M. and Regenstein, C.E. (1991), *Introduction to Fish Technology*. Van Nostrand Reinhold, New York.

Roberts, T.A. (1990), Predictive modelling of microbial growth. In *Food Technology International Europe*, Turner, A. (Ed.), Sterling Publications International, London, pp. 231–235.

Rogers, W.F. (1982), Insulated containers for shipping perishables. In *Proceedings of the First International Conference on Seafood Packaging and Shipping*, Martin, R.E. (Ed.), Science & Technology National Fisheries Institute, Washington, DC, pp. 243–248.

Ronsivalli, L.J. and Baker, D.W. (1981), Low temperature preservation of seafoods: A review. *Marine Fisheries Review* **43**(4), 1–15.

Ronsivalli, L.J. and Charm, S.E. (1975), Spoilage and shelf life prediction of refrigerated fish. *Marine Fisheries Review* **37**(4), 32–34.

Siebel, J.E. (1982), Specific heat of various products. *Ice and Refrigeration* **2**, 256–257.

Smith, J.G.R., Hardy, R. and Young, K.W. (1980), A seasonal study of the storage characteristics of mackerel stored at chill and ambient temperatures. In *Advances of Fish Science and Technology*, Connell, J.J. and Torry Research Station (Eds), Fishing News Books, Farnham, pp. 372–378.

Spencer, R. and Baines, C.R. (1964), The effect of temperature on the spoilage of wet white fish. *Food Technology* **18**(5), 769–773.

Storey, R.M. (1985), Time temperature function integration, its relationship and application to chilled fish. In *Storage Lives of Chilled and Frozen Fish and Fish Products*, I.I.F.-I.I.R. Commissions C_2 and C_3-1985-84, Aberdeen, UK, pp. 293–297.

Taylor, M.R. (1993), FDA's plan for food safety and HACCP – institutionalising a philosophy of prevention. Presented at Symposium on Foodborne Microbiological Pathogens, International Life Sciences Institute in Conjunction with the International Association of Milk, Food and Environmental Sanitarians National Meetings, Atlanta.

Turner, A. (1992), Legislation. In *Chilled Foods – A Comprehensive Guide*, Dennis, C. and Stringer, M. (Eds), Ellis Horwood, Chichester, pp. 39–57.

UK (1990), *Food Hygiene (Amendment) Regulations 1990*, (SI 1990 No. 1431). HMSO, London.

US FDA (1992), *Specification for Non-hazardous Materials 29 CFR, 1910-1200*. Food and Drug Administration, USA.

US FDA (1994), *Hazard Analysis Critical Control Point Regulations*, January. Food and Drug Administration, USA.

Walker, S.J. (1992), Chilled foods microbiology. In *Chilled Foods – A Comprehensive Guide*, Dennis, C. and Stringer, M. (Eds), Ellis Horwood, Chichester, pp. 165–195.

Walker, S.J. and Stringer, M.F. (1990), Microbiology of chilled foods. In *Chilled Foods – The State of the Art*, Gormley, T.R. (Ed.), Elsevier Applied Science, Barking, pp. 269–304.

Williams, A.P., Blackburn, C. de W. and Gibbs, P.A. (1992), Advances in the use of predictive techniques to improve the safety and extend the shelf-life of foods. *Food Sci. Technol. Today* **6**(3), 148–151.

Wills, L.A. (1982), Facilitating seafood marketing through modern transportation methods. In *Proceedings of the First International Conference on Seafood Packaging and Shipping*, Martin, R.E. (Ed.), Science & Technology National Fisheries Institute, Washington, DC, pp. 11–22.

Woolfe, M.L. (1992), Temperature monitoring and measurement. In *Chilled Foods – A Comprehensive Guide*, Dennis, C. and Stringer, M. (Eds), Ellis Horwood, Chichester, pp. 77–109.

Index

Page numbers appearing in **bold** refer to figures and page numbers appearing in *italic* refer to tables.

Biochemical dynamics 1–31
   carbohydrates 13–19
      gluconeogensis 14–17
      glycogen 14
      nature of 13
      re-feeding 17–18
      starvation 15
      usage 13–14
   fish condition 3–4
   flavour compounds 23–4
   gonadosomatic index 3
   hepatosomatic index 3
   lipids, role of 4–10
   minerals 24
   pH 19–22
   pigmentation 22–4
   proteins 10–13
   spawning cycle 2–3

Canning 119–59
   aseptic process 133–4
   crustacea 155–6
   exhausting 143–4
   heat transfer in 124
   mackerel 155, **157**
   measuring lethality 124–6
   micro–organisms, thermal resistance of 120–2
   molluscs 155–6
   packaging materials 135–41
   pH effects 119–20
   post-process operations 152–3
   pre-processing operations 139–43
   principles 119–29
   process equipment 146–52
   process factors 134–5
   process prediction 126–9
   quality criteria 129–34
   small pelagics 153
   species designation *140–1*
   species identification 163, 188–9, **189**
   storage aspects 134
   tuna 155, **156**
   see also Species identification

Carbon dioxide, *see* Modified-atmospheric packaging; freezing of fish
CAP, *see* Controlled-atmosphere packaging
Chilling 93–4
   modified-atmosphere packaging 95–7
   spoilage rates 94
   storage life 93–4
   see also Modified-atmosphere packaging
Controlled-atmosphere packaging
   definition 200
   see also Modified-atmosphere packaging
Curing 32–73
   water activity 35–7, *36*, *37*
   water content 32–5
      effect on product quality 40–2

Drying 42–54
   calculations 46–54
      constant rate period 45, 47–8
      falling rate period 45, 49–53
      inaccuracies 49–50
   critical moisture content 49
   processes 42–3
   psychrometrics 43–5
   transport mechanisms 52–545

Electrophoresis
   agarose 169, **193**, **194**
   capillary 171
   isoelectric focusing 171
   of canned fish flesh **189**
   collaborative trial 177, *178*, *180*
   of raw crustacean flesh **182**
   of raw fish flesh **181**, **182**
   polyacrylamide gel 168
   principles 166–72
      of cooked crustacean flesh **187**
      of cooked fish flesh 185, **186**
      fish eggs **195**
      principles 170–1
   zone electrophoresis 169–70

see also Species identification
ELISA, *see* Species identification

Fish eggs 193–6
 identification **195**
Fish mince 88–90
 comparison with surimi 89–90
 products 89
 raw materials 88–9
Fish muscle
 actin 76–9, 175
 connective tissue proteins 176
 effect of heat 78
 effect of salt 78
 myofibrillar proteins 175
 myosin 76–9, 175
 proteins 75–6, 173–6, *175*
 sarcoplasmic proteins 173–5
 structure 172–6, **174**
Flavour compounds 24
Food micromodel
 model types 282
 in transportation 283–4
Freezer burn
 in storage 115
 in transportation 252
Freezing of fish 98–113
 calculations 100–103
  Nagaoka modification 101
  Neumann equation 101–2
  Plank equation 100–1
 freezing systems 103–8
  air-blast 103–5
  carbon dioxide 107
  immersion 106–7
  liquid nitrogen 107–8
  onboard 108–11
  onshore 111–13
  plate/contact 106
  quality aspects 114–16
  spray 107
 see also Freezer burn

Hazard analysis critical control point (HACCP)
 audits *244*
 critical control point 225, 231, **233**, **237**
 decision tree **234**
 defect definition 240–1
 definition 226–8
 description 224

European Union legislation 227
future developments 243–6
GANNT charts 231, **232**
ISO 9000 224, 228
practical aspects 229–38
quality management programme (QMP)
 application 228–9, *238*, 240, **243**
 description 225–6
risk assessment 231, **233**
teams 230
in transportation 260–2
verification 238–43

IEF, *see* Species identification, isoelectric focussing

Lipids 4–10
 human health 6–7
 polyunsaturated fatty acids 6–7
 rancidity in frozen fish 7–10
  in cod 7–8
  in salmon 9–10
 sex differences 5
 spawning and relation to 5–6
Mailprof computer modelling
 introduction 278
 use 278–82, **279**, **280**, *281*
Modified-atmosphere packaging
 Aeromonas hydrophila 206, 207–14, **212**
 carbon dioxide in 201
 chilled fish 93–4
 Clostridium botulinum 205–6
 combination treatments 215
 definition 200
 future developments 214–20
 Listeria monocytogenes 206, 207–14, **211**
 microbial flora 202–4, *203*
 modelling 215–6
 nitrogen in 201
 oxygen in 201
 packaging 210, 217–19
 pathogenic flora 204–6
 present applications 206–7
 quality assurance 219–20
 Salmonella typhimurium 208–14, **211**
 Vibrio parahaemolyticus 206
 Yersinia enterocolitica 206, 207–14, **212**

see also Chilling; Controlled-atmosphere packaging; Vacuum packaging

Overall heat transfer coefficient
 calculation 272–4, **273**
 predicted 269
 in transportation 267

pH 19–22
 gaping 21–2
 texture 20–1
Pigmentation 22–4
 carotenoids 23–4
 in muscle 22
 in skin 22–3

Quality
 biological condition 24–6
 chill storage 113–14
 curing 66–8
 frozen storage 114–16
 quality management programme 225–226
 surimi 86–7
 thermal processing 129–34
 water relations 40–2
 see also Hazard analysis critical control point

Salting 54–62
 fish wood 62
 maturing process 59–61
 salt-boiled fish 61–2
 salting processes 55–6
 salt types 56–8
  quality effects 57–8
 spoilage 60–1
  factors affecting 58–9
Smoking 62–72
 nutrition 66–8
 process equipment 68–9
 purposes 62–4
 quality 66–8
 safety 66–8
 smoke components 64–6, 66
 smoke production 63–4
Species identification 160–99
 adulteration 161, 182, 185
 canned fish 163
 capillary chromatography 171, 190
 cooked fish 184–8, **186**, **187**

crustacea 182–3, **186**, **187**
deoxyribonucleic acid 165,191, **193**, **194**, 196
electrophoresis 164, 166–72
frozen storage 183
heat treated fish 188–9, **189**
high pressure liquid chromatography 191
iced storage effect 183
immunoassay 164, 190, 196
isoelectric focussing 171, *178*, *180*, **181**, **182**, **184**, **189**
legislation 160
method requirement 164–6
raw fish 176–84, **181**, **182**
sardine 163
substitution 162, 182
surimi 161, 187–8
tuna 163
see also Electrophoresis
Surimi 74–92, 161, 187–8
 cryoprotectants 77–8, 84
 definitions 74
 frozen storage 77–9
 microbial aspects 87–8
 process stages 80–5
 quality aspects 86–7
 specifies appropriate for 85–6
 species identification 161, 187–8

Temperature relationships in transportation
 by air 251–2
 containers for 252
 cooling gels 249, 251, 255–7, **256**
 in developed countries 255
 in developing countries 253–5
 food micromodel 282–3
 hazard analysis critical control point 260–2, **261**
 heat transfer relationships 265–75
 ice consumption 253
  calculation 269
  prediction 270–2, *271*
 ice melting rate 272
  calculation 268, *273*
 importance 249
 latent heat of ice 274
 legislation 259–60
 microbial growth 257–9
 modelling 250, 275–8

Temperature relationships in
  transportation *contd*
    temperature control 259–60
    thermal conductivity 268, 275
    by road 250–1
    by sea 252–3
    shelf life 262–3
    specific heat 269
    thawing 116–7
Time–temperature function indicators
  (TTFI) 263–5

rate of spoilage 264

Vacuum packaging 200

Water activity 33–5, *36*
  adsorption isotherms 37–41
  shelf-life 54–5
  spoilage 35, *37*
  water relations 36–7, *37*